Enantioselective Multicatalysed Tandem Reactions

RSC Catalysis Series

Editor-in-Chief:
Professor James J Spivey, *Louisiana State University, Baton Rouge, USA*

Series Editors:
Professor Chris Hardacre, *Queen's University Belfast, Northern Ireland*
Professor Zinfer Ismagilov, *Boreskov Institute of Catalysis, Novosibirsk, Russia*
Professor Umit Ozkan, *Ohio State University, USA*

Titles in the Series:

How to obtain future titles on publication:
A standing order plan is available for this series. A standing order will bring
delivery of each new volume immediately on publication.

For further information please contact:
Book Sales Department, Royal Society of Chemistry, Thomas Graham House,
Science Park, Milton Road, Cambridge, CB4 0WF, UK
Telephone: +44 (0)1223 420066, Fax: +44 (0)1223 420247
Email: booksales@rsc.org

Visit our website at www.rsc.org/books

Enantioselective Multicatalysed Tandem Reactions

Hélène Pellissier
CNRS and Aix Marseille Université, Centrale Marseille, Marseille, France
Email: h.pellissier@univ-amu.fr

THE QUEEN'S AWARDS
FOR ENTERPRISE:
INTERNATIONAL TRADE
2013

RSC Catalysis Series No. 20

Print ISBN: 978-1-84973-988-7
PDF eISBN: 978-1-78262-135-5
ISSN: 1757-6725

A catalogue record for this book is available from the British Library

Published by The Royal Society of Chemistry,
Thomas Graham House, Science Park, Milton Road,
Cambridge CB4 0WF, UK

Registered Charity Number 207890

For further information see our web site at www.rsc.org

Printed in the United Kingdom by CPI Group (UK) Ltd, Croydon, CR0 4YY, UK

Preface

Although asymmetric synthesis is sometimes viewed as a subdiscipline of organic chemistry, actually this topical field transcends any narrow classification and pervades essentially all chemistry.[1] Indeed, the preparation of chiral compounds is an important and challenging area of contemporary synthetic organic chemistry, mainly in connection with the fact that most natural products are chiral and their physiological or pharmacological properties depend upon their recognition by chiral receptors.[2] The broad utility of synthetic chiral molecules in medicines and materials has made asymmetric catalysis a prominent area of investigation. Indeed, of the methods available for preparing chiral products, catalytic asymmetric synthesis has attracted most attention. In particular, asymmetric transition metal catalysis has emerged as a powerful tool to perform reactions in a highly enantioselective fashion over the past few decades in spite of the common drawbacks of metals, such as moisture sensitivity, recoverability, and toxicity, particularly for heavy metals.[3] Indeed, it is traditionally assumed that catalysis has been the realm of metals. While the end of the last century has been dominated by the use of metal[1,4] and also biocatalysis,[5] a change in perception occurred during the last decade when several reports confirmed that relatively simple organic molecules, such as proline, could be highly effective and remarkably enantioselective catalysts of a variety of fundamentally important transformations.[6] This rediscovery has initiated an explosive growth of research activities in organocatalysis, both in industry and in academia. Organocatalysts have several important advantages, since they are usually robust, inexpensive, readily available, and non-toxic. Enantioselective organocatalytic processes have reached maturity in recent years with an impressive and steadily increasing number of works regarding the applications of this type of reaction, which paint a comprehensive picture of their real possibilities in organic synthesis.[7] Hence, the application of

RSC Catalysis Series No. 20
Enantioselective Multicatalysed Tandem Reactions
By Hélène Pellissier
© Hélène Pellissier 2014
Published by the Royal Society of Chemistry, www.rsc.org

chiral organocatalysts has permitted the preparation of a number of very valuable chiral products with the exclusion of any trace of hazardous metals and with several advantages from an economical and environmental point of view.[8] The ability of organocatalysts to promote a wide range of reactions by different activation modes makes organocatalysis ideal for its application in tandem reactions, which proceed in a one-pot procedure to build complex frameworks from simple starting compounds. These organocatalysed one-pot reactions are often highly efficient and follow, in some way, different biomimetic pathways, with the same principles that are found in bio-synthesis in nature. Chemists have devoted more and more effort to the development of new and powerful strategies in tandem reactions that avoid the use of costly and time-consuming protection–deprotection processes, as well as purification procedures of intermediates. The efficiency of these processes can be evaluated by the bonds formed, the stereoselectivity achieved and the complexity of the newly formed molecules. Tandem cata-lysed reactions refer to the synthetic strategies of modular combination of catalytic reactions into one synthetic operation, occurring one after the other and working in conjunction with each other with minimum workup or change in conditions. Compared with stepwise synthesis, tandem reactions avoid time, energy, and yield losses associated with the isolation and puri-fication of intermediates. Furthermore, they help lower the risk in the storage, transportation, and handling of toxic, unstable, or explosive inter-mediates. Tandem reactions include one-, two-, and multicomponent domino reactions,[9] which have been defined by Tietze and co-workers as reactions involving two or more bond-forming transformations, taking place under the same reaction conditions, without adding additional reagents and catalysts, and in which the subsequent reactions result as a consequence of the functionality formed by bond formation or fragmentation in the previous step.[10] It must be recognised that a relatively narrow distinction exists be-tween domino and consecutive cascade or tandem reactions. From the point of view of an operator, the only difference between the two lies in the point along the sequence at which one or more catalysts (or reagents) had to be added to effect either the initiation of a sequence (that is, domino reaction) or propagation to the next step (that is, consecutive or tandem reaction). Tandem catalysis refers to the synthetic strategies of modular combination of catalytic reactions into one synthetic operation with minimum workup or change in conditions. The use of tandem reactions in organic synthesis is increasing constantly, since they allow the synthesis of a wide range of complex molecules, including natural products and biologically active compounds in an economically favourable way.[11] Indeed, decreasing the number of laboratory operations required and the quantities of chemicals and solvents used have made tandem reactions unavoidable processes. The proliferation of these reactions is evidenced by the number of recent reviews covering the literature since 1986.[9,12] In the last few years, an explosive number of multiple-catalyst systems for various organic transformations have been developed.[13] The driving force in combining catalysts is to discover more efficient approaches for complex molecule synthesis with good chemo- and stereoselectivity inaccessible through the use of single specific catalytic systems. With increasing attention being given to this new area, momentous chemical transformations are being discovered. It must be noted that this novel methodology is particularly adapted to enantioselective tandem reactions, allowing a rapid and economic construction of highly functionalised chiral molecules from simple and readily available starting materials in one pot. A reaction catalysed by multiple catalysts (two or three organocatalysts, two metals, two (or more) enzymes, a metal and an organocatalyst, a metal and enzyme(s), or an organocatalyst and enzyme(s) at the same time can allow a reactivity and a selectivity to be achieved otherwise not possible by using a single catalyst alone. The main problem, however, lies in finding the proper catalyst which should not only be compatible with the other catalysts but also tolerate reagents, solvent and intermediates generated during the course of the reaction. Unlike in biological processes in which nature takes advantage of enzyme architecture to facilitate a multiple reaction manifold, it is very difficult to exploit such a process in one pot. For example, chiral metal catalysts are generally sensitive to species with co-ordinating ability in the reaction media, while organocatalysts are relatively robust and have high compatibility and, moreover, some organocatalysts can promote several types of reactions through different activating modes. The event of coordination between the two catalysts can occasionally be avoided, for example by their sequential addition to the reaction media (sequential catalysis). In general, the use of multiple catalyst systems enlarges the substrate and reaction scope for the reaction design, improves the reactivity, and benefits the control of the selectivity. The goal of this book is to provide researchers and professionals in academic and industrial laboratories with a broad overview of all the developments in multicatalysed enantioselective tandem reactions. It is divided into two sections, dealing with asymmetric tandem reactions catalysed by multiple catalysts from the same discipline in the first one, and asymmetric tandem reactions catalysed by multiple catalysts from different disciplines in the second one. The first section is subdivided into five chapters, dealing successively after an introduction (Chapter 1) with reactions catalysed by multiple organocatalysts (Chapter 2), reactions catalysed by two metals (Chapter 3), and multienzyme-catalysed reactions (Chapter 4), followed by conclusions (Chapter 5). The second section of the book, which concerns tandem reactions catalysed by multiple catalysts from different disciplines, is also divided into five chapters, dealing successively after an introduction (Chapter 6) with reactions catalysed by a combination of metals and organocatalysts (Chapter 7), reactions catalysed by a combination of metals and enzymes (Chapter 8), and finally reactions catalysed by a combination of organocatalysts and enzymes (Chapter 9), followed by conclusions (Chapter 10). The two catalysts can interact in a cooperative, relay or sequential manner; these three types of catalysis will be treated successively in most of the sections of the book. In cooperative catalysis, both the two catalysts are present at the onset of the reaction, and share the same catalytic cycle, activating two different functional groups cooperatively to achieve the bond-formation steps. On the other hand, in relay or sequential catalysis, the substrate first reacts with one catalyst to give an intermediate through a first catalytic cycle. Then, this former intermediate reacts with the second catalyst to provide, through a second catalytic cycle, the final product or an intermediate for subsequent transformations. The difference between relay and sequential catalysis consists of the presence or not of the two catalysts at the onset of the reaction. Thus, relay as well as sequential catalysis involves a set of reactions independently catalysed by two catalysts in a consecutive manner but, while in relay catalysis the two compatible catalysts are both present from onset, in sequential catalysis the addition of the second catalyst during the course of the reaction is necessary to avoid compatibility issues.

Hélène Pellissier

References

1. *Comprehensive Asymmetric Catalysis*, ed. E. N. Jacobsen, A. Pfaltz and H. Yamamoto, Springer, Berlin, 1999.
2. M. Nogradi, in *Stereoselective Synthesis*, VCH, Weinheim, 1995.
3. (a) M. Beller and C. Bolm, *Transition Metals for Organic Synthesis*, Wiley-VCH, Weinheim, 1998, vol. 1 and vol. 2; (b) G. Poli, G. Giambastiani and A. Heumann, *Tetrahedron*, 2000, **56**, 5959–5989; (c) E. Negishi, *Handbook of Organopalladium Chemistry for Organic Synthesis*, John Wiley & Sons, Inc., Hoboken, NJ, 2002, 2, 1689–1705; (d) A. de Meijere, P. von Zezschwitz, H. Nüske and B. Stulgies, *J. Organomet. Chem.*, 2002, **653**, 129–140; (e) L. F. Tietze, I. Hiriyakkanavar and H. P. Bell, *Chem. Rev.*, 2004, **104**, 3453–3516.
4. (a) R. Noyori, in *Asymmetric Catalysts in Organic Synthesis*, Wiley, New York, 1994; (b) *Catalytic Asymmetric Synthesis*, ed. I. Ojima, Wiley-VCH, New York, 2nd edn, 2000; (c) *Transition Metals for Organic Synthesis*, ed. M. Beller and C. Bolm, Wiley-VCH, Weinheim, 2nd edn, 2004; (d) D. J. Ramon and M. Yus, *Chem. Rev.*, 2006, **106**, 2126–2208.
5. (a) *Biocatalysts for Fine Chemicals Synthesis*, ed. S. M. Roberts, Wiley-VCH, New York, 1999; (b) *Enzyme Catalysis in Organic Synthesis*, ed. K. Drauz and H. Waldmann, Wiley-VCH, Weinheim, 2nd edn, 2002; (c) A. S. Bommarius and B. R. Riebel, *Biocatalysis*, Wiley-VCH, Weinheim, 2004.
6. (a) E. Knoevenagel, *Chem. Ber.*, 1896, **29**, 172–174; (b) U. Eder, G. Sauer and R. Wiechert, *Angew. Chem., Int. Ed. Engl.*, 1971, **10**, 496–497; (c) Z. G. Hajos and D. R. Parrish, *J. Org. Chem.*, 1974, **39**, 1615–1621; (d) K. A. Ahrendt, C. J. Borths and D. W. C. MacMillan, *J. Am. Chem. Soc.*, 2000, **122**, 4243–4244; (e) B. List, R. A. Lerner and C. F. Barbas, *J. Am. Chem. Soc.*, 2000, **122**, 2395–2396.

7. (a) P. I. Dalko and L. Moisan, *Angew. Chem., Int. Ed.*, 2001, **40**, 3726–3748; (b) B. List, *Tetrahedron*, 2002, **58**, 5573–5590; (c) E. R. Jarvo and L. Miller, *Angew. Chem., Int. Ed.*, 2002, **58**, 2481–2495; (d) B. Westermann, *Angew. Chem., Int. Ed.*, 2003, **42**, 151–153; (e) P. I. Dalko and L. Moisan, *Angew. Chem., Int. Ed.*, 2004, **43**, 5138–5175; (f) A. Berkessel and H. Gröger, in *Asymmetric Organocatalysis—From Biomimetic Concepts to Powerful Methods for Asymmetric Synthesis*, Wiley-VCH, Weinheim, 2005; (g) J. Seayad and B. List, *Org. Biomol. Chem.*, 2005, **3**, 719–724; (h) M. S. Taylor and E. N. Jacobsen, *Angew. Chem., Int. Ed.*, 2006, **45**, 1520–1543; (i) B. List, *Chem. Commun.*, 2006, 819–824; (j) G. Lelais and D. W. C. MacMillan, *Aldrichimica Acta*, 2006, **39**, 79–87; (k) D. Enders, O. Niemeier and A. Henseler, *Chem. Rev.*, 2007, **107**, 5606–5655; (l) P. I. Dalko, in *Enantioselective Organocatalysis*, Wiley-VCH, Weinheim, 2007; (m) P. I. Dalko, *Chimia*, 2007, **61**, 213–218; (n) H. Pellissier, *Tetrahedron*, 2007, **63**, 9267–9331; (o) A. G. Doyle and E. N. Jacobsen, *Chem. Rev.*, 2007, **107**, 5713–5743; (p) M. G. Gaunt, C. C. C. Johansson, A. McNally and N. C. Vo, *Drug Discovery Today*, 2007, **2**, 8–27; (q) *Chem. Rev.*, 2007, **107**(12), 5413–5883, Special Issue on Organocatalysis, ed. B. List; (r) T. Akiyama, *Chem. Rev.*, 2007, **107**, 5744–5758; (s) S. Mukherjee, J. W. Yang, S. Hoffmann and B. List, *Chem. Rev.*, 2007, **107**, 5471–5569; (t) R. P. Wurz, *Chem. Rev.*, 2007, **107**, 5570–5595; (u) M. J. Gaunt and C. C. C. Johansson, *Chem. Rev.*, 2007, **107**, 5596–5605; (v) E. A. Colby Davie, S. M. Mennen, Y. Xu and S. J. Miller, *Chem. Rev.*, 2007, **107**, 5759–5812; (w) D. W. C. MacMillan, *Nature*, 2008, **455**, 304–308; (x) X. Yu and W. Wang, *Chem. – Asian. J.*, 2008, **3**, 516–532; (y) A. Dondoni and A. Massi, *Angew. Chem., Int. Ed.*, 2008, **47**, 4638–4660; (z) P. Melchiorre, M. Marigo, A. Carlone and G. Bartoli, *Angew. Chem., Int. Ed.*, 2008, **47**, 6138–6171; (aa) F. Peng and Z. Shao, *J. Mol. Catal. A: Chem.*, 2008, **285**, 1–13; (ab) C. F. Barbas, *Angew. Chem., Int. Ed.*, 2008, **47**, 42–47; (ac) C. Palomo, M. Oiarbide and R. Lopez, *Chem. Soc. Rev.*, 2009, **38**, 632–653; (ad) S. Bertelsen and K. A. Jorgensen, *Chem. Soc. Rev.*, 2009, **38**, 2178–2189; (ae) L.-W. Xu, J. Luo and Y. Lu, *Chem. Commun.*, 2009, 1807–1821; (af) M. Bella and T. Gasperi, *Synthesis*, 2009, 1583–1614; (ag) L. Gong, in Special Topic: Asymmetric Organocatalysis, , in *Chin. Sci. Bull.*, 2010, **55**(17); (ah) B. List, in Asymmetric Organocatalysis, in *Top. Curr. Chem.*, 2010, **291**, 1–456; (ai) H. Pellissier, in *Recent Developments in Asymmetric Organocatalysis*, Royal Society of Chemistry, Cambridge, 2010; (aj) *Enantioselective Organocatalysed Reactions, Vols. I and II*, ed. R. Mahrwald, Springer, Berlin, 2011.
8. (a) R. M. de Figueiredo and M. Christmann, *Eur. J. Org. Chem.*, 2007, 2575–2600; (b) E. Marquès-Lopez, R. P. Herrera and M. Christmann, *Nat. Prod. Rep.*, 2010, **27**, 1138–1167.
9. (a) L. F. Tietze and A. Modi, *Med. Res. Rev.*, 2000, **20**, 304–322; (b) *Multicomponent Reactions*, ed. J. Zhu and H. Bienaymé, Wiley-VCH, Weinheim, 2005; (c) D. J. Ramon and M. Yus, *Angew. Chem., Int. Ed.*, 2005, **44**, 1602–1634; (d) G. Guillena, D. J. Ramon and M. Yus,

Tetrahedron: Asymmetry, 2007, **18**, 693–700; (e) *Synthesis of Heterocycles via Multicomponent Reactions*, ed. R. V. A. Orru and E. Ruijter, Topics in Heterocyclic Chemistry, vol. I and II, Springer, Berlin, 2010.

10. (a) L. F. Tietze and U. Beifuss, *Angew. Chem., Int. Ed. Engl.*, 1993, **32**, 131–163; (b) L. F. Tietze, *Chem. Rev.*, 1996, **96**, 115–136; (c) L. F. Tietze, G. Brasche and K. Gericke, *Domino Reactions in Organic Synthesis*, Wiley-VCH, Weinheim, 2006.

11. (a) C. Hulme and V. Gore, *Curr. Med. Chem.*, 2003, **10**, 51–80; (b) A. Padwa and S. K. Bur, *Tetrahedron*, 2007, **63**, 5341–5378; (c) M. Colombo and I. Peretto, *Drug Discovery Today*, 2008, **13**, 677–684; (d) K. C. Nicolaou and J. S. Chen, *Chem. Soc. Rev.*, 2009, **38**, 2993–3009; (e) B. B. Touré and D. G. Hall, *Chem. Rev.*, 2009, **109**, 4439–4486; (f) C. Vaxelaire, P. Winter and M. Christmann, *Angew. Chem., Int. Ed.*, 2011, **50**, 3605–3607; (g) Z. Zhang and J. C. Antilla, *Angew. Chem., Int. Ed.*, 2012, **51**, 11778–11782.

12. (a) G. H. Posner, *Chem. Rev.*, 1986, **86**, 831–844; (b) T.-L. Ho, in *Tandem Organic Reactions*, Wiley, New York, 1992; (c) H. Waldmann, *Nachr. Chem., Tech. Lab.*, 1992, **40**, 1133–1140; (d) K. Fukumoto, *Synth. Org. Chem. Jpn.*, 1994, **52**, 2–18; (e) R. A. Bunce, *Tetrahedron*, 1995, **51**, 13103–13159; (f) P. J. Parsons, C. S. Penkett and A. J. Shell, *Chem. Rev.*, 1996, **96**, 195–206; (g) A. Padwa and M. D. Weingarten, *Chem. Rev.*, 1996, **96**, 223–269; (h) S. E. Denmark and A. Thorarensen, *Chem. Rev.*, 1996, **96**, 137–165; (i) P. C. F. Balaure and P. I. A. Filip, *Rev. Roum. Chim.*, 2002, **46**(8), 809–833; (j) E. Capdevila, J. Rayo, F. Carrion, I. Jové, J. I. Borrell and J. Teixido, *Afinidad*, 2003, **506**, 317–337; (k) K. C. Nicolaou, T. Montagnon and S. A. Snyder, *Chem. Commun.*, 2003, 551–564; (l) L. F. Tietze and N. Rackelmann, *Pure Appl. Chem.*, 2004, **76**, 1967–1983; (m) D. E. Fogg and E. N. dos Santos, *Coord. Chem. Rev.*, 2004, **248**, 2365–2379; (n) J.-C. Wasilke, S. J. Obrey, R. T. Baker and G. C. Bazan, *Chem. Rev.*, 2005, **105**, 1001–1020; (o) K. C. Nicolaou, D. J. Edmonds and P. G. Bulger, *Angew. Chem., Int. Ed.*, 2006, **45**, 7134–7186; (p) H. Pellissier, *Tetrahedron*, 2006, **62**, 2143–2173; (q) H. Pellissier, *Tetrahedron*, 2006, **62**, 1619–1665; (r) D. Enders, C. Grondal and M. R. M. Hüttl, *Angew. Chem., Int. Ed.*, 2007, **46**, 1570–1581; (s) C. J. Chapman and C. G. Frost, *Synthesis*, 2007, 1–21; (t) D. M. D'Souza and T. J. J. Müller, *Chem. Soc. Rev.*, 2007, **36**, 1095–1108; (u) A.-N. Alba, X. Companyo, M. Viciano and R. Rios, *Curr. Org. Chem.*, 2009, **13**, 1432–1474; (v) *Chem. Soc. Rev.*, 38(11), 2009, Special Issue on Rapid formation of molecular complexity in organic synthesis; (w) T. Boddaert, M. Presset, D. Mailhol and J. Rodriguez, in *Ideas in Chemistry and Molecular Sciences*, Advances in Synthetic Chemistry, ed. B. Pignataro, Wiley-VCH, Weinheim, 2010, vol. 1, ch. 9, p. 187; (x) J. E. Biggs-Houck, A. Younai and J. T. Shaw, *Curr. Opin. Chem. Biol.*, 2010, **14**, 371–382; (y) M. Ruiz, P. Lopez-Alvarado, G. Giorgi and J. C. Menéndez, *Chem. Soc. Rev.*, 2011, **40**, 3445–3454; (z) L. Albrecht, H. Jiang and K. A. Jorgensen, *Angew. Chem., Int. Ed.*, 2011, **50**, 8492–8509; (aa) H. Pellissier, *Adv. Synth. Catal.*, 2012, **354**, 237–294;

(ab) C. De Graaff, E. Ruijter and R. V. A. Orru, *Chem. Soc. Rev.*, 2012, **41**, 3969–4009; (ac) H. Clavier and H. Pellissier, *Adv. Synth. Catal.*, 2012, **354**, 3347–3403; (ad) H. Pellissier, *Chem. Rev.*, 2013, **113**, 442–524.

13. (a) J.-A. Ma and D. Cahard, *Angew. Chem., Int. Ed.*, 2004, **43**, 4566–1508; (b) M. Kanai, N. Kato, E. Ichikawa and M. Shibasaki, *Synlett*, 2005, 1491–1508; (c) D. H. Paull, C. J. Abraham, M. T. Scerba, E. Alden-Danforth and T. Leckta, *Acc. Chem. Res.*, 2008, **41**, 655–663; (d) Z. Shao and H. Zhang, *Chem. Soc. Rev.*, 2009, **38**, 2745–2755; (e) C. Zhong and X. Shi, *Eur. J. Org. Chem.*, 2010, 2999–3025; (f) M. Rueping, R. M. Koenigs and I. Atodiresei, *Chem. – Eur. J*, 2010, **16**, 9350–9365; (g) J. Zhou, *Chem. – Asian J.*, 2010, **5**, 422–434; (h) L. M. Ambrosini and T. H. Lambert, *ChemCatChem*, 2010, **2**, 1373–1380; (i) S. Piovesana, D. M. Scarpino Schietroma and M. Bella, *Angew. Chem., Int. Ed.*, 2011, **50**, 6216–6232; (j) N. T. Patil, V. S. Shinde and B. Gajula, *Org. Biomol. Chem.*, 2012, **10**, 211–224; (k) A. E. Allen and D. W. C. MacMillan, *Chem. Sci.*, 2012, **3**, 633–658; (l) Z. Du and Z. Shao, *Chem. Soc. Rev.*, 2013, **42**, 1337–1378.

Contents

RSC Catalysis Series No. 20
Enantioselective Multicatalysed Tandem Reactions
By Hélène Pellissier
© Hélène Pellissier 2014
Published by the Royal Society of Chemistry, www.rsc.org

Abbreviations

AADH	amino acid dehydrogenase
AAO	amino acid oxidase
Acac	acetylacetone
ADCA	7-aminodesacetoxycephalosporanic acid
ADH	alcohol dehydrogenase
ADP	adenosine diphosphate
Aib	α-aminoisobutyric acid
AIBN	2,2′-azobisisobutyronitrile
Ala	alanine
AlaDH	alanine dehydrogenase
AP	acid phosphatase
aq	aqueous
AQN	anthraquinone
Ar	aryl
Asp	aspartate
AspAT	aspartate transaminase
ATP	adenosine triphosphate
BCAAT	branched amino acid transaminase
BDHP	1,1′-binaphth-2,2′-diyl hydrogen phosphate
BEMP	*tert*-butylimino-2-diethylamino-1,3-dimethylperhydro-1,2,3-diazaphosphorine
BINAP	2,2′-bis(diphenylphosphino)-1,1′-binaphthyl
BINIM	binapthyldiimine
BINOL	1,1′-bi-2-naphthol
Biphep	2,2′-bis(diphenylphosphino)-1,1′-biphenyl
BMIm	1-butyl-3-methylimidazolium
Bn	benzyl
Boc	*tert*-butoxycarbonyl

RSC Catalysis Series No. 20
Enantioselective Multicatalysed Tandem Reactions
By Hélène Pellissier
© Hélène Pellissier 2014
Published by the Royal Society of Chemistry, www.rsc.org

Bpy	2,2′-bipyridyl
Bz	benzoyl
CAL	*Candida Antarctica* lipase
Cbz	benzyloxycarbonyl
CDMO	cyclododecanone monooxygenase
CLEA	cross-linked enzyme aggregate
CLEC	cross-linked crystal
CMP	cytidine 5′-monophosphate
cod	cyclooctadiene
Cp	cyclopentadienyl
CSA	camphorsulfonic acid
CTP	cytidine 5′-triphosphate
Cy	cyclohexyl
DABCO	1,4-diazabicyclo[2.2.2]octane
DadA	D-amino acid dehydrogenase
DBU	1,8-diazabicyclo[5.4.0]undec-7-ene
de	diastereomeric excess
DERA	2-deoxy-D-ribose 5-phosphate aldolase
DHAP	dihydroxyacetone phosphate
DHQD	dihydroquinidine
DIPEA	diisopropylethylamine
DKR	dynamic kinetic resolution
DMAP	4-(dimethylamino)pyridine
DMF	dimethylformamide
DMSO	dimethylsulfoxide
DPP	diphenylphosphate
Dppf	1,1′-bis(diphenylphosphanyl)ferrocene
DYKAT	dynamic kinetic asymmetric transformation
E	electrophile
ee	enantiomeric excess
EMIm	1-ethyl-3-methylimidazolium
ESI	electrospray ionisation
EWG	electron-withdrawing
FAD	flavin adenine dinucleotide
FDH	formate dehydrogenase
Flu	fluorenyl
Fu	furyl
Galk	galactokinase
GDH	glucose dehydrogenase
Goase	galactose oxidase
HEPES	4-(2-hydroxyethyl)-1-piperazineethanesulfonic acid
Hex	hexyl
Hhe	halohydrin dehalogenase
KRED	ketoreductase
L	ligand
LDH	lactate dehydrogenase

Leu	leucine
LeuDH	leucine dehydrogenase
MDH	mandelate dehydrogenase
Mes	mesyl
MOM	methoxymethyl
MS	mass spectroscopy or molecular sieves
MTBE	methyl *tert*-butyl ether
NAD	nicotinamide adenine dinucleotide
NADH	nicotinamide adenine dinucleotide hydrogenase
NADP	nicotinamide adenine dinucleotide phosphate
Naph	naphthyl
NeuAc	*N*-acetyl neuraminic acid
NFSI	*N*-fluorobenzenesulfonimide
NHC	*N*-heterocyclic carbine
NMM	*N*-methylmorpholine
NMO	*N*-methylmorpholine-*N*-oxide
NMP	*N*-methylpyrrolidinone
NMR	nuclear magnetic resonance
Ns	nosyl
NTTL	1,8-naphthanoyl-*tert*-leucine
Nu	nucleophile
PAL	phenylalanine ammonia lyase
PAM	phenylalanine aminomutase
PBG	phosphobilinogen
PEP	*p*-ethoxyaniline
Ph	phenyl
PHAL	1,4-phthalazinediyl
PHOX	phosphinooxazoline
Pin	pinacolato
PMP	*para*-methoxyphenyl
PPL	porcine pancreatic lipase
Pro	proline
PS-BEMP	2-*tert*-butylimino-2-diethylamino-1,3-dimethylperhydro-1,3,2-diazaphosphorine polymer supported
PTSA	*p*-toluenesulfonic acid
Py	pyridine
QN	8-quinoline
RAMA	1,6-diphosphate aldolase
RhaD	rhamnulose-1-phosphate aldolase
rt	room temperature
SEM	2-(trimethylsilyl)ethoxymethyl
TA	transaminase
TACA	tumor-associated carbohydrate antigen
TBD	1,5,7-triazabicyclo[4.4.0]dec-5-ene
TBDPS	*tert*-butyldiphenylsilyl
TBHP	*tert*-butylhydroperoxide

TBS	*tert*-butyldimethylsilyl
TEA	triethylamine
TEMPO	2,2,6,6-tetramethylpiperidinyloxy
TES	triethylsilyl
Tf	trifluoromethanesulfonyl
THF	tetrahydrofuran
Thr	threonine
TMBA	tetramethylenebisacetamide
TMS	trimethylsilyl
Tol	tolyl
TPA	tris(2-pyridylmethyl)amine
TPAP	tetrapropylammonium perruthenate
TRIP	triisopropylphenyl
Trp	tryptophan
Ts	4-toluenesulfonyl (tosyl)
Tyr	tyrosine
UDP	uridine diphosphate
XPhos	2-(dicyclohexylphosphino)-2′,4′,6′-triisopropylbiphenyl)

SECTION I
Asymmetric Tandem Reactions Catalysed by Multiple Catalysts from the Same Discipline

CHAPTER 1

Introduction

The first section of the book illustrates how much asymmetric multicatalysis based on the use of catalysts belonging to the same discipline has contributed to the development of various types of powerful enantioselective tandem reactions. It collects all the major progress in the field of enantioselective tandem reactions promoted by multiple (two or three) organocatalysts, two metal catalysts, or two or more biocatalysts. It demonstrates the power of these remarkable one-pot processes of two or more bond-forming reactions, occurring with minimum workup or change in conditions in which the subsequent transformation takes place at the functionalities obtained in the former transformation, following the same principles that are found in biosynthesis in nature. The first section of the book is subdivided into five chapters, dealing successively after this introduction (Chapter 1) with reactions catalysed by multiple organocatalysts (Chapter 2), reactions catalysed by two metals (Chapter 3), and multienzyme-catalysed reactions (Chapter 4) followed by conclusions (Chapter 5). The two catalysts can interact in a cooperative, relay or sequential manner; these three types of catalysis will be treated successively in chapters 2 and 3. In cooperative catalysis, both the two catalysts are present at the onset of the reaction, and share the same catalytic cycle, activating two different functional groups cooperatively to achieve the bond-formation steps. On the other hand, in relay or sequential catalysis, the substrate first reacts with one catalyst to give an intermediate through a first catalytic cycle. Then, this former intermediate reacts with the second catalyst to provide, through a second catalytic cycle, the final product or an intermediate for subsequent transformations. The difference between relay and sequential catalysis consists of the presence or not of the two catalysts at the onset of the reaction. Thus, relay as well as sequential catalysis involves a set of reactions independently catalysed by two catalysts in a consecutive manner but, while in relay catalysis the two compatible catalysts are both present from onset,

RSC Catalysis Series No. 20
Enantioselective Multicatalysed Tandem Reactions
By Hélène Pellissier
© Hélène Pellissier 2014
Published by the Royal Society of Chemistry, www.rsc.org

in sequential catalysis the addition of the second catalyst during the course of the reaction is necessary to avoid compatibility issues. Chapter 4 dealing with multienzyme-catalysed reactions is divided into three sections concerning multienzymatic synthesis of chiral alcohols, multienzymatic synthesis of chiral amines and amino acids, and other multienzymatic reactions.

CHAPTER 2

Reactions Catalysed by Multiple Organocatalysts

2.1 Introduction

Organocatalysts are known to be the most robust catalysts, well tolerating impurities and traces of water. Additionally, they are usually readily available, easy to handle, and present a high compatibility which is a significant advantage. Furthermore, a wide variety of organocatalysts is able to induce different types of transformations through different activation models. These advantages allow the combination of different organocatalysts to be achieved to design novel enantioselective tandem reactions.[1] The two (or three) organocatalysts can interact in a cooperative, relay or sequential manner; these three types of catalysis will be treated successively in the text. In cooperative catalysis, both the two catalysts are present at the onset of the reaction, and share the same catalytic cycle, activating two different functional groups cooperatively to achieve the bond-formation steps. On the other hand, in relay or sequential catalysis, the substrate first reacts with one catalyst to give an intermediate through a first catalytic cycle. Then, this former intermediate reacts with the second catalyst to provide, through a second catalytic cycle, the final product or an intermediate for subsequent transformations. The difference between relay and sequential catalysis consists of the presence or not of the two catalysts at the onset of the reaction. Thus, relay as well as sequential catalysis involves a set of reactions independently catalysed by two catalysts in a consecutive manner but, while in relay catalysis the two compatible catalysts are both present from onset, in sequential catalysis the addition of the second catalyst during the course of the reaction is necessary to avoid compatibility issues. Various types of organocatalysts, such as phosphoric acids, L-proline and its derivatives, cinchona alkaloids, ureas, thioureas, amino acids, *N*-heterocyclic carbenes,

RSC Catalysis Series No. 20
Enantioselective Multicatalysed Tandem Reactions
By Hélène Pellissier
© Hélène Pellissier 2014
Published by the Royal Society of Chemistry, www.rsc.org

pyrrolidines, and various (di)amines, have already been combined to induce enantioselective domino and multicomponent domino reactions evolving through cooperative as well as relay catalysis, but also used sequentially to achieve enantioselective tandem reactions. In a number of examples, both the two organocatalysts employed are chiral which is possible when one does not interfere with the activity of the other. Their simultaneous use is particularly useful when a synergistic effect is present as in the case of cooperative catalysis which is the most developed.

2.2 Cooperative Catalysis

In 2007, a chiral tertiary amine, such as (−)-spartein, was used by Hong *et al.* to enhance the nucleophilic character of an enamine intermediate by deprotonation and effectively shield one of the enantiotopic faces of this intermediate, thus improving the stereoselectivity of a reaction.[2] This enamine was the intermediate of an enantioselective Robinson condensation occurring between two α,β-unsaturated aldehydes, providing the corresponding chiral cyclohexadienes in good yields and enantioselectivities of up to 93% ee, as shown in Scheme 2.1.

In the same year, Zhou and List reported a novel one-pot tandem reaction which, for the first time, combined chiral Brønsted acid catalysis with enamine and iminium catalysis.[3] Later, on the basis of control experiments and ESI-MS/MS analysis,[4] a reasonable mechanism was proposed (Scheme 2.2). The initial step of this tandem reaction was mediated by achiral *p*-ethoxyaniline (PEP-NH$_2$) and chiral phosphoric acid (*R*)-TRIP; either reagent alone was inefficient in promoting this aldol condensation to afford the first iminium intermediate. The following step was a conjugate reduction which was also Brønsted acid and amine co-catalysed, and no further conversion took place in the absence of either catalyst. The final step was an acid-catalysed reductive amination. This novel sequence allowed the highly enantioselective synthesis of pharmaceutically active chiral *cis*-3-substituted cyclohexyl or heterocyclohexyl amines in high diastereo- and

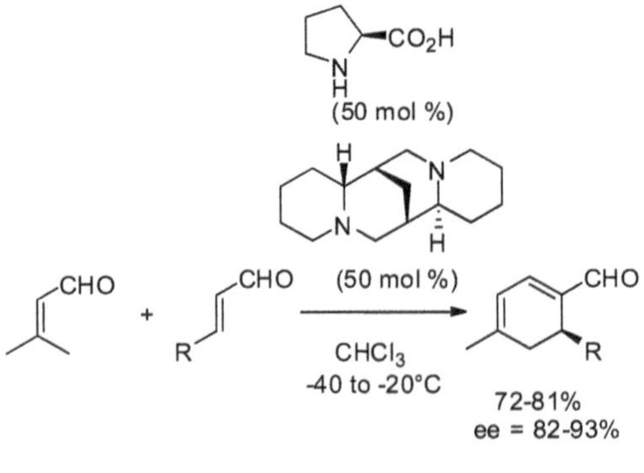

Scheme 2.1   Robinson condensation reaction catalysed by a combination of L-proline and (−)-spartein.

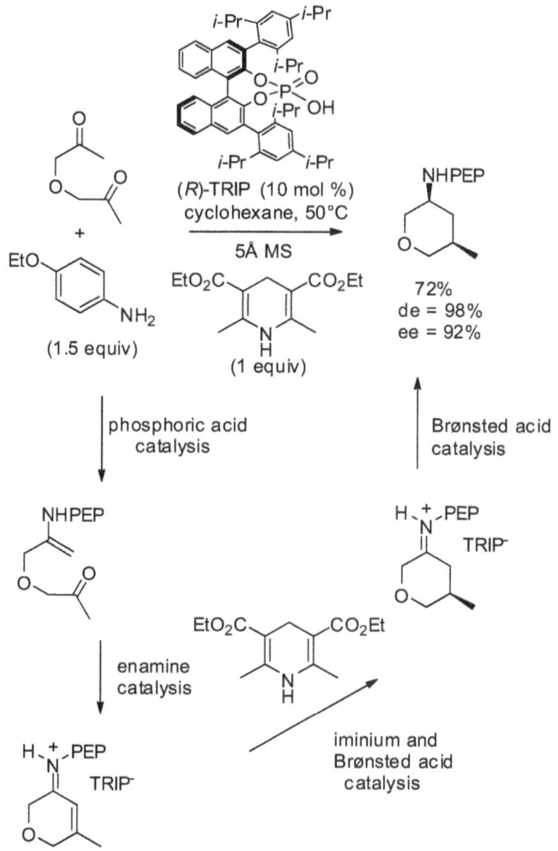

Scheme 2.2 Domino aldol–conjugate reduction–reductive amination reaction catalysed by a combination of a Brønsted acid and a chiral phosphoric acid.

enantioselectivities of up to 98% de and 92% ee, as shown in Scheme 2.2. It should be noted that the stereocontrol of the conjugate reduction and reductive amination step was accomplished by the chiral phosphoric acid TRIP, as demonstrated by the control experiments.

In 2009, Xie *et al.* developed an enantioselective synthesis of highly functionalised chiral 2-amino-2-chromene derivatives based on a domino reaction occurring between α,β-unsaturated enones, such as 2-hydroxybenzalacetones, and malononitrile.[5] This novel process was catalysed by a cinchona alkaloid-derived primary amine, such as 9-amino-9-deoxyepiquinine, in combination with (*R*)-1,1′-binaphth-2,2′-diyl hydrogen phosphate ((*R*)-BDHP). As shown in Scheme 2.3, excellent enantioselectivities of up to 96% ee were obtained in combination with high yields of up to 84% for a range of β-substituted 2-hydroxybenzalacetones. The scope of the reaction could be extended to other 2-hydroxychalcones, providing the corresponding 2-amino-2-chromene derivatives in good yields and

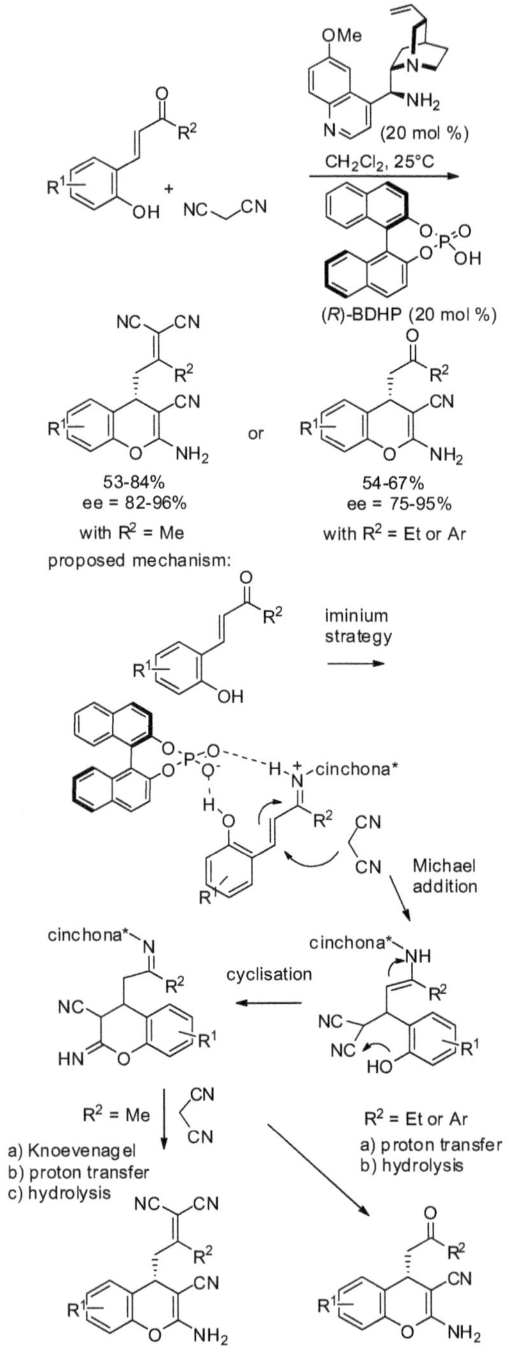

Scheme 2.3 Domino Michael–cyclisation reactions catalysed by a combination of a chiral cinchona alkaloid-derived primary amine and a chiral phosphoric acid.

enantioselectivities (75–95% ee), as shown in Scheme 2.3. A hypothetical mechanism is depicted in Scheme 2.3 in which the chiral primary amine was an effective catalyst for the formation of the iminium from the enone. This iminium could be stabilised through hydrogen-bonding with the chiral phosphoric acid. Then, the higher reactivity of this iminium was used to facilitate the Michael reaction between the enone and malononitrile to produce the corresponding enamine intermediate, and the final products were obtained from the following Knoevenagel condensation ($R^2 = Me$) and the nucleophilic addition of the phenolic OH group on the cyano moiety and proton transfer. On the other hand, when the iminium intermediate bore a bulkier imine moiety ($R^2 = Et$, Ar), ketones were obtained instead of the alkylidene malononitriles arising from Knoevenagel condensation under the same reaction conditions.

In 2009, Bella *et al.* reported a formal [4 + 2] cycloaddition of substituted arylacetaldehydes and 2-cyclohexen-1-one which was promoted by a chiral thiazolidine catalyst and chiral quinine *via* enamine formation and spontaneous intramolecular aldol reaction (Scheme 2.4).[6] The stereoselection depended upon the secondary amine catalyst, whereas the secondary catalyst was involved in the enhancement of the nucleophilicity of the derived enamine, probably through deprotonation of the carboxylic group. There

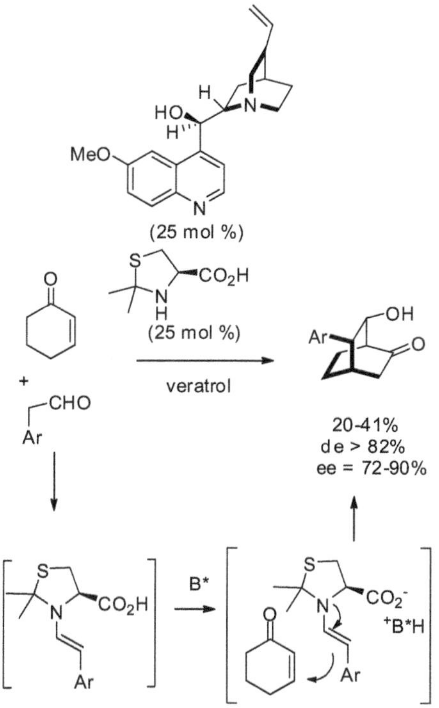

Scheme 2.4 Domino Michael–aldol reaction catalysed by a combination of a chiral thiazolidine catalyst and chiral quinine.

was a synergistic effect in the contemporary use of the two catalysts because neither of them was able to efficiently promote the reaction alone. As shown in Scheme 2.4, the domino products were reached in low to moderate yields and good to high enantioselectivities of up to 90% ee.

In 2010, Kotsuki *et al.* used another combination of organocatalysts to achieve a new powerful strategy for the asymmetric construction of a quaternary carbon stereogenic centre at the 4-position of cyclohexenone derivatives.[7] Indeed, the simultaneous employment of (1S,2S)-1,2-cyclohexanediamine and (1S,2S)-1,2-cyclohexanedicarboxylic acid as organocatalysts in the Robinson-type annulation of methyl vinyl ketone or ethyl vinyl ketone with α-aryl-substituted aldehydes allowed the corresponding chiral cyclohexenones bearing a quaternary carbon centre to be achieved in moderate yields and good to excellent enantioselectivities of up to 97% ee, as shown in Scheme 2.5. The authors have proposed the mechanism depicted in Scheme 2.5 to explain the results. First, condensation of (1S,2S)-1,2-cyclohexanediamine catalyst with both α-aryl-substituted aldehyde and enone in the presence of (1S,2S)-1,2-cyclohexanedicarboxylic acid proceeded through the formation of an enamine-iminium double activation intermediate **1**, which then underwent an intramolecular Michael addition to afford the corresponding cyclic enamine-iminium ion intermediate **2**. This intermediate collapsed spontaneously *via* hydrolysis to give the corresponding keto-aldehyde precursor **3** and regenerated the catalytic system. In this sequence, the vicinal *trans*-diamine arrangement in (1S,2S)-1,2-cyclohexanediamine was essential not only to activate both the Michael donor and acceptor components but also to bring them together in close proximity to achieve carbon–carbon bond formation with the observed enantiocontrol. The synthetic utility of this novel method was demonstrated by its application to a short synthesis of (+)-sporochnol A which is a natural chemical fish deterrent.

In the same year, Xu *et al.* developed an efficient example of asymmetric cooperative catalysis applied to a domino oxa-Michael–Mannich reaction of salicylaldehydes with cyclohexenones.[8] The process was catalysed by a combination of two chiral catalysts, such as a chiral pyrrolidine and amino acid D-*tert*-leucine. The authors assumed that there was protonation of the aromatic nitrogen atom of the pyrrolidine catalyst by D-*tert*-leucine, which spontaneously led to the corresponding ion-pair assembly (Scheme 2.6). This self-assembled catalyst possessed dual activation centres, enabling the catalysis of the electrophilic and nucleophilic substrates simultaneously. The domino oxa-Michael–Mannich reaction provided a range of versatile chiral tetrahydroxanthenones in high yields and high to excellent enantioselectivities of up to 98% ee, as shown in Scheme 2.6.

In 2011, Wang *et al.* disclosed a highly enantioselective domino double Michael reaction of dienones with 3-nonsubstituted oxindoles to access chiral spirocyclic oxindoles in high yields of up to 98% and excellent diastereo- and enantioselectivities of up to >90% de and 98% ee, respectively.[9] This novel reaction was performed in the presence of a cinchona-based

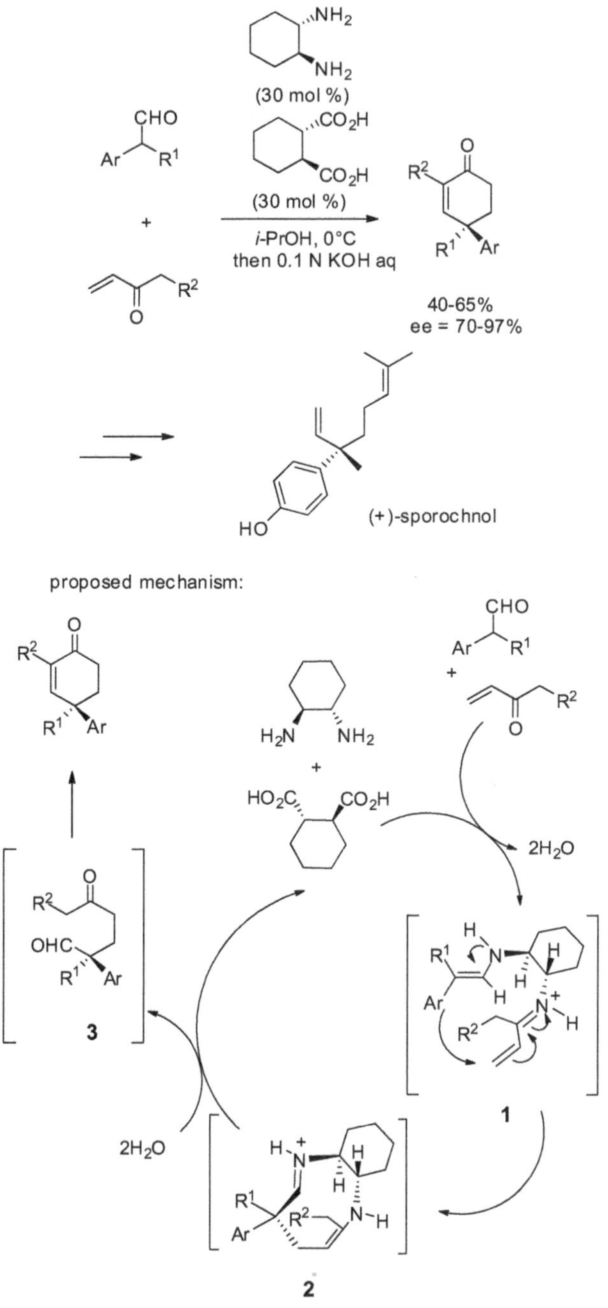

Scheme 2.5 Domino Michael–aldol–dehydration reaction catalysed by a combination of a chiral diamine and a chiral dicarboxylic acid.

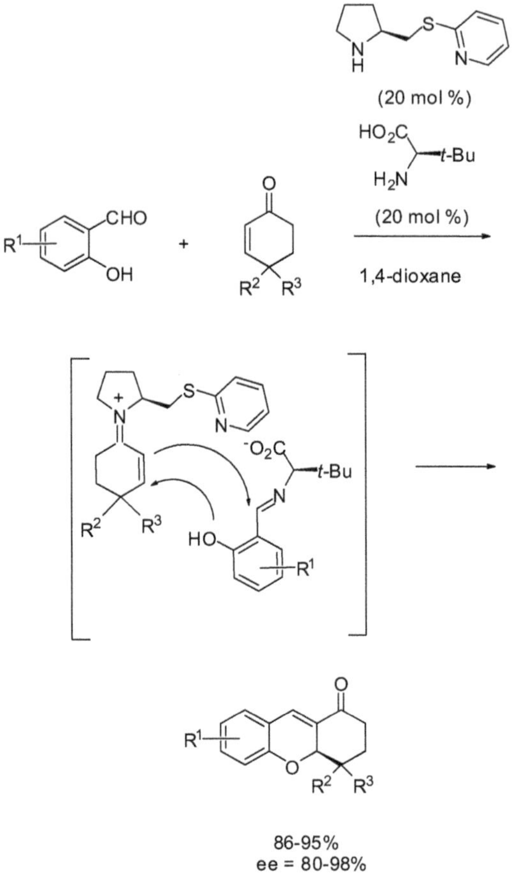

Scheme 2.6 Domino oxa-Michael–Mannich reaction catalysed by a combination of a chiral pyrrolidine and D-*tert*-leucine.

primary amine in combination with a chiral phosphoric acid, as shown in Scheme 2.7. This reactivity pattern was also applied to other pronucleophiles such as pyrazolones.[10]

The Povarov reaction, an inverse electron-demand aza-Diels–Alder reaction of 2-azadienes with electron-rich olefins, allows a rapid construction of polysubstituted tetrahydroquinolines.[11] It must be noted that enantioselective versions of the Povarov reaction remain rare. Actually, the first highly enantioselective example of this type of reaction was developed by Zhu *et al.*, in 2009.[12] Later, Jacobsen *et al.* reported another enantioselective Povarov reaction catalysed by a combination of a strong Brønsted acid, such as *o*-nitrobenzene sulfonic acid, with a chiral urea.[13] As shown in Scheme 2.8, the reaction of electron-rich alkenes with imines provided the corresponding tricyclic products in good yields, moderate diastereoselectivities of up to 62% de, and generally high enantioselectivities ranging from 90 to 98% ee.

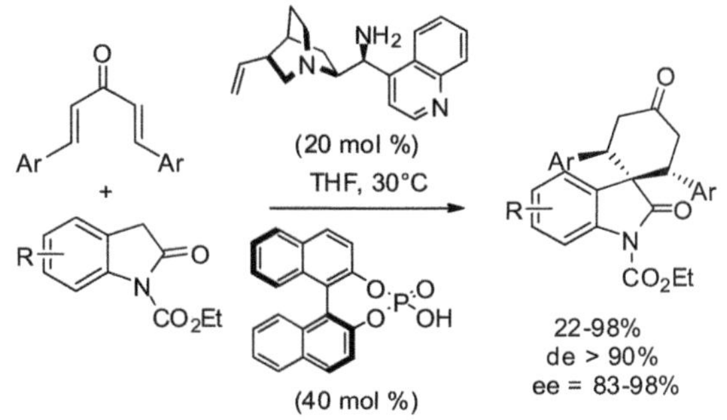

Scheme 2.7 Domino Michael–Michael reaction catalysed by a combination of a chiral amine catalyst and a chiral phosphoric acid.

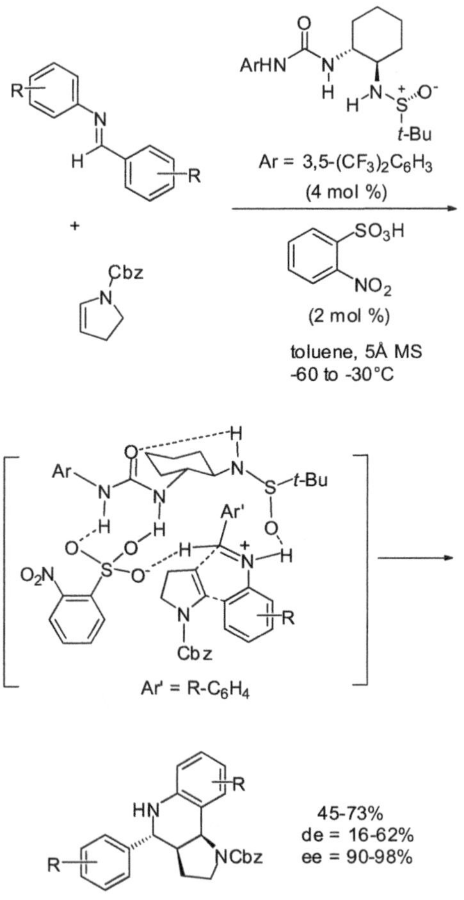

Scheme 2.8 Povarov reaction catalysed by a combination of a chiral urea and *o*-nitrobenzene sulfonic acid.

In 2010, Melchiorre *et al.* demonstrated the compatibility of a chiral quinidine derivative with a chiral BINOL-derived phosphoric acid as catalysts (Scheme 2.9) to induce the reaction of α,β-unsaturated aldehydes with an aromatic secondary alcohol.[14] The reaction evolved through the formation of an enamine from the corresponding enal and the chiral quinidine catalyst, which further added to the carbocation arising from the aromatic alcohol to give the corresponding γ-alkylated α,β-unsaturated aldehyde. As shown in Scheme 2.9, a range of functionalised chiral α,β-unsaturated

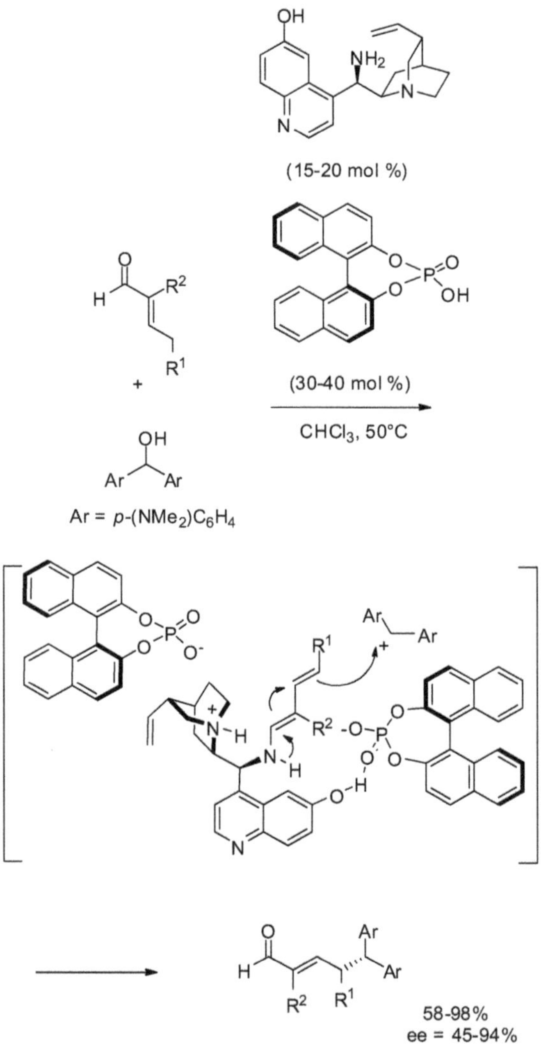

Scheme 2.9   Reaction of α,β-unsaturated aldehydes with aromatic secondary alcohols catalysed by a combination of a chiral quinidine and a chiral phosphoric acid.

aldehydes could be achieved in good to excellent yields and moderate to high enantioselectivities of up to 94% ee.

The utility of *N*-heterocyclic carbenes as organocatalysts in domino reactions has received growing attention in the past few years.[15] In this context, Rovis *et al.* developed in 2011 a novel efficient synthesis of chiral *trans*-γ-lactams by using a combination of a chiral catalyst of this type with a Brønsted acid, such as *o*-chlorobenzoic acid.[16] Under this cooperative catalysis, strongly electrophilic ethyl *trans*-4-oxo-2-butenoate reacted with unactivated imines to provide the corresponding chiral *trans*-γ-lactams in good to high yields, high diastereoselectivities of up to >90% de, combined with good to high enantioselectivities of up to 93% ee, as shown in Scheme 2.10. A plausible mechanism for this process could involve the generation of a Breslow intermediate arising from the reaction of ethyl *trans*-4-oxo-2-butenoate with the carbene. This intermediate attacked the acid-activated imine *via* hydrogen-bonding intermediate **4**. Steric hindrance led to an *anti*-orientation of the CO_2Et group and the alkenyl group. Proton transfer then

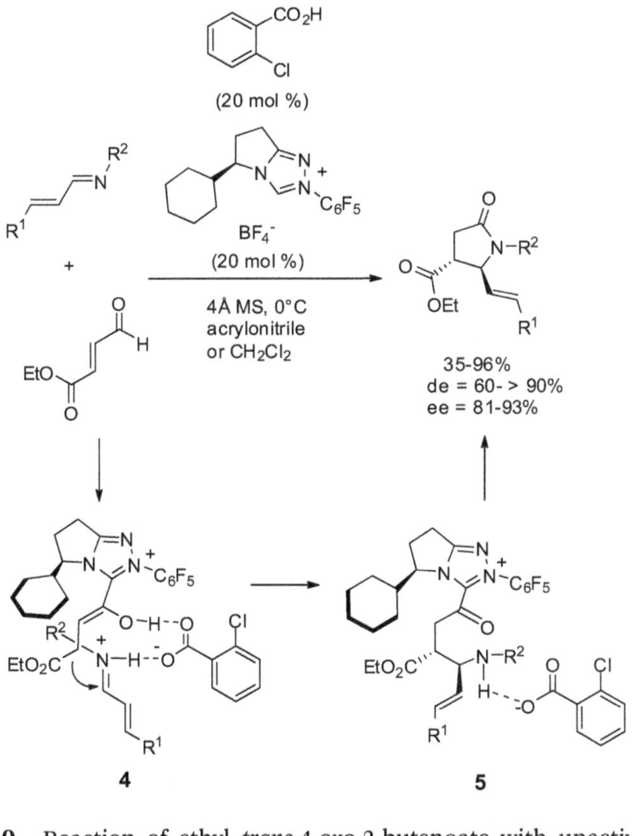

Scheme 2.10 Reaction of ethyl *trans*-4-oxo-2-butenoate with unactivated imines catalysed by a combination of a chiral *N*-heterocyclic carbene and *o*-chlorobenzoic acid.

resulted in the formation of acyl carboxylate **5**. The nitrogen species replaced the carbene to afford the final product and the free carbene (Scheme 2.10). The scope of the reaction could be extended to a variety of imines and enals other than ethyl *trans*-4-oxo-2-butenoate. Indeed, several less nucleophilic *p*-nitrocinnamaldehydes provided the corresponding lactams in yields ranging from 48 to 99%, diastereoselectivities of 86 to 90% de, and high enantioselectivities of 90 to 93% ee. On the other hand, using a ketone as a substituent on the enal (Et instead of OEt) gave the corresponding lactam in lower enantioselectivity (66% ee).

Even though the history of multicomponent reactions dates back to the second half of the 19th century with the reactions of Strecker, Hantzsch, and Biginelli, it was only in the last few decades with the work of Ugi *et al.* that the concept of the multicomponent reaction emerged as a powerful tool in synthetic chemistry.[17] In this context, Feng *et al.* investigated in 2008 the Biginelli reaction of aromatic aldehydes, urea, and β-ketoesters catalysed by a combination of a chiral *trans*-4-hydroxyproline-derived secondary amine and a Bronsted acid, such as 2-chloro-4-nitrobenzoic acid, as catalyst system.[18] The dual-catalysed process was performed at 25 °C with 5 mol% of catalyst loading of each of the two catalysts in 1,4-dioxane and in the presence of an additive such as *t*-BuNH$_2$·TFA. Under these conditions, the Biginelli products were obtained in moderate to good yields (34–73%) and good to excellent enantioselectivities (70–98% ee). Later, another example of cooperative catalysis was developed by Cordova *et al.* in a dynamic one-pot three-component asymmetric transformation between aldehydes, protected α-cyanoglycine esters, and α,β-unsaturated aldehydes.[19] The domino reaction afforded the corresponding chiral cyano-, formyl-, and ester-functionalised α-quaternary proline derivatives bearing four contiguous stereocentres. When it was induced by a combination of a chiral amine catalyst, such as chiral diphenylprolinol triethylsilyl ether, and an hydrogen-bond-donating catalyst, such as the oxime derived from methyl cyanoacetate, the products were obtained in good to high yields, good to high *endo/exo* ratios of up to >19 : 1, good diastereoselectivities of up to >90% de, and generally excellent enantioselectivities of 93 to 98% ee, as shown in Scheme 2.11. The authors have demonstrated that the hydrogen-bond-donating catalyst was crucial for the formation of the imine and thus pushed the equilibrium of the reaction towards the imine formation. Most likely this occurred by activation of the carbonyl group of the aldehyde. Next, the intramolecular hydrogen-bonding network, which can be created by the oxime catalyst, activated the imine and locked its conformation to **6** (Scheme 2.11). The imine then underwent proton shifts to form intermediates **7–9**, which were stabilised by the hydrogen-bonding network. In parallel, the chiral amine catalyst formed the iminium intermediate **10**, which was efficiently shielded at the *Si* face by the bulky aryl groups. Next, the activated species **7–9**, for which the conformation was locked by hydrogen-bonding with the oxime catalyst, approached the opposite face, and stereoselective C–C bond formation occurred from the *Re* face of intermediate

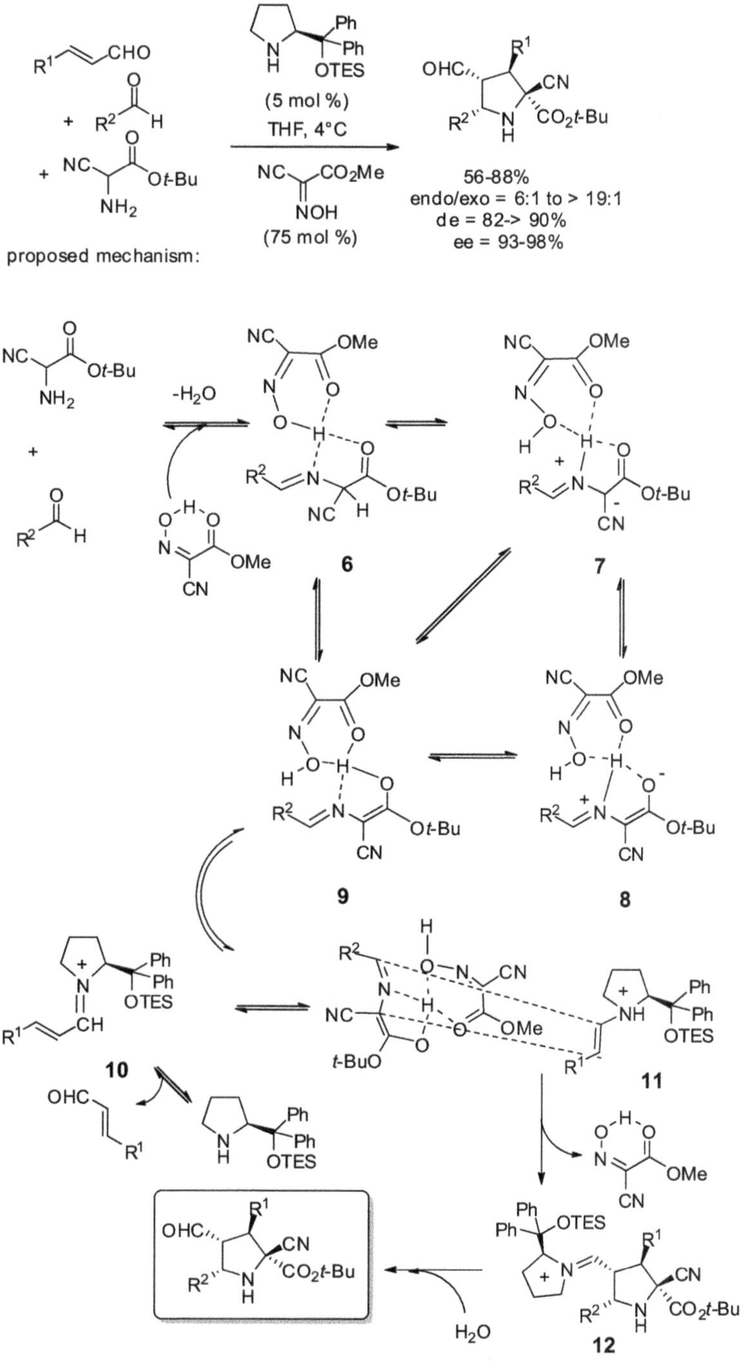

Scheme 2.11　Three-component reaction of aldehydes, protected α-cyanoglycine esters and α,β-unsaturated aldehydes catalysed by a combination of chiral diphenylprolinol triethylsilyl ether and an oxime.

11 either by a concerted *endo*-selective mechanism (cycloaddition) or a stepwise (Michael–Mannich) mechanism to give iminium intermediate **12**. Subsequent hydrolysis regenerated the amine catalyst and gave the poly-substituted proline product.

In 2012, the same authors reported the cooperative catalysis of a pseudo three-component domino Michael–aldol reaction of two equivalents of γ-nitro ketones with α,β-unsaturated aldehydes, leading to the corresponding chiral bicyclic products in high yields and moderate to excellent diastereo- and enantioselectivities of up to >90% de, and >99% ee, respectively.[20] The catalytic system consisted of the combination of chiral diphenylprolinol triethyl- or trimethysilyl ether with an achiral thiourea, as shown in Scheme 2.12. To account for the stereoselectivity of the reaction, the authors have proposed the mechanism depicted in Scheme 2.12. The catalytic cas-cade cycle started with the formation of hydrogen-bond-donating thiourea-accelerated iminium intermediate **13** from the chiral amine catalyst and the enal. The bulky chiral group efficiently shielded the *Re*-face of this inter-mediate **13**. Next, of all the possible stereomers of the starting γ-nitro ketone, the fast reacting (*S,R*)-enantiomer performed a nucleophilic *Si*-facial conju-gate attack on **13** to form an enamine intermediate **14**, which involved hydrogen-bond activation by the thiourea catalyst. The subsequent cascade intramolecular *exo–trig* aldol reaction at the *Re*-face of the keto moiety of **14** led to intermediate **15**. Subsequent hydrolysis regenerated the chiral amine catalyst and delivered the final chiral product. In parallel to this catalytic proposed cycle, the achiral hydrogen-bond-donating thiourea catalyst toge-ther with the chiral amine catalyst catalysed the dynamic conversion of the (*R,S*)-enantiomer of the γ-nitro ketone to its (*S,R*)-enantiomer so that all of the starting racemic mixture was converted to the final product. The pres-ence of intermediates **13**, **14** and **15** was experimentally confirmed by HR-MS analysis.

In 2012, a chiral cinchona alkaloid-derived primary amine was associated by Wang *et al.* to a (*R*)-BINOL-derived phosphoric acid to induce a three-component domino Knoevenagel–Michael reaction between isatins, malononitrile, and acetone, providing the corresponding domino products in generally excellent yields and enantioselectivities, as shown in Scheme 2.13.[21] A hypothetic cooperative catalysis can be envisaged to explain these excellent results.

Earlier in 2008, Fréchet *et al.* designed non-interpenetrating star polymer catalysts to combine iminium, enamine and hydrogen-bond catalysts in one pot for an enantioselective three-component domino double Michael re-action.[22] The process occurred between *N*-methyl indole, 2-hexenal and methyl vinyl ketone, providing the corresponding domino product in 89% yield with a good diastereoselectivity of 85% de and a total enantioselectivity (ee >99%), as shown in Scheme 2.14. These results were reached by using two chiral amine catalysts encapsulated in the core of star polymers in combination with an achiral hydrogen-bond donor, which were normally incompatible catalysts.

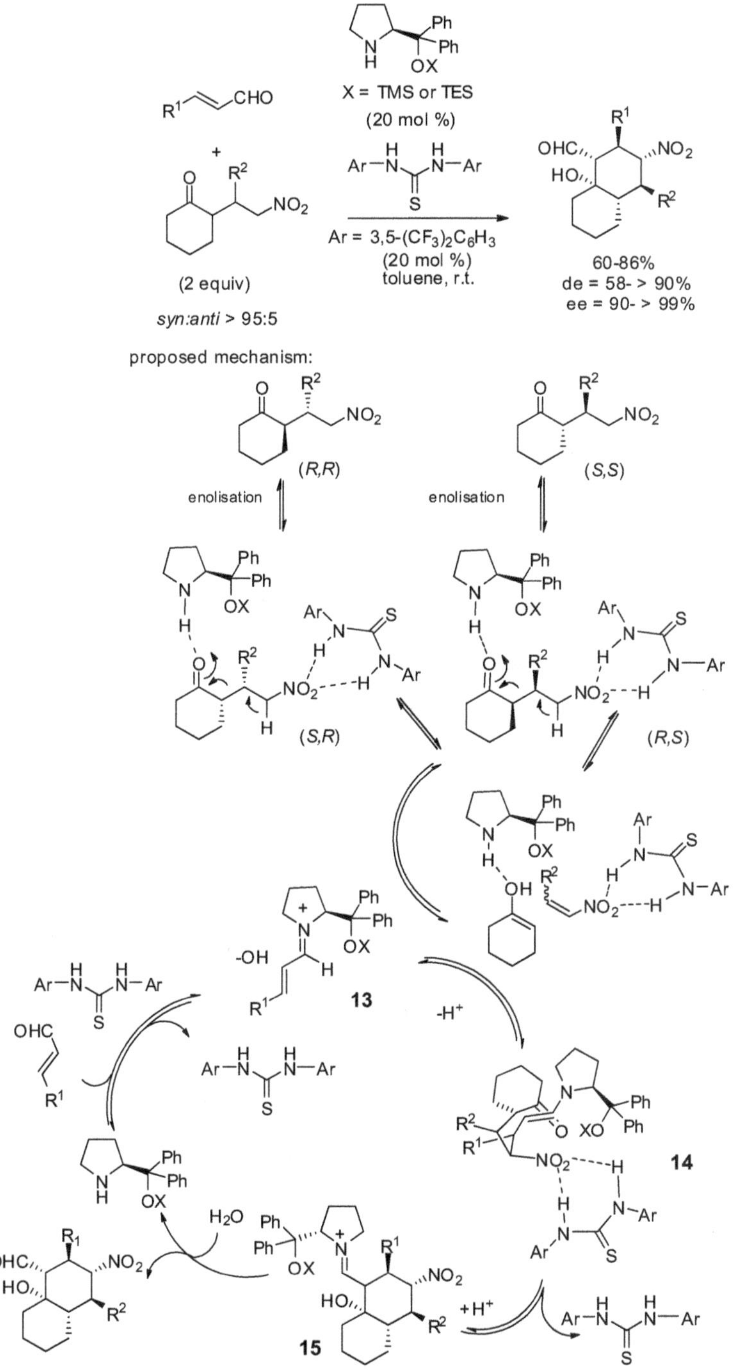

Scheme 2.12 Pseudo three-component domino Michael–aldol reaction catalysed by a combination of chiral diphenylprolinol trimethyl- or triethylsilyl ether and a thiourea.

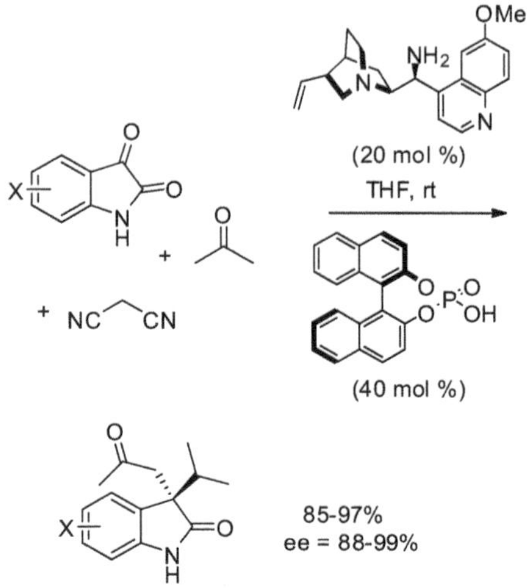

Scheme 2.13   Three-component domino Knoevenagel–Michael reaction catalysed by a combination of a chiral cinchona alkaloid-derived primary amine and a chiral phosphoric acid.

2.3   Relay Catalysis

In 2009, Rovis and Lathrop developed a multicatalytic asymmetric domino reaction for the preparation of chiral α-hydroxycyclopentanones containing three contiguous stereocentres.[23] This process was based on the reaction of 1,3-diketones with α,β-unsaturated aldehydes, which generated through Michael addition induced by a chiral diarylprolinol trimethylsilyl ether, followed by hydrolysis the corresponding aldehydes bearing a tethered ketone (Scheme 2.15). These intermediates then underwent an intramolecular crossed benzoin reaction in the presence of the second carbene catalyst depicted in Scheme 2.15 to afford final cyclopentanones *via* a formal [3 + 2] process. In spite of moderate diastereoselectivities, these highly functionalised products were obtained in high enantioselectivities of up to 97% ee, as shown in Scheme 2.15.

Later, Rovis *et al.* reported an enantioselective domino oxa-Michael–Stetter reaction occurring between salicylaldehydes and electron-deficient alkynes to provide the corresponding chiral benzofuranones in moderate to good yields and good to high enantioselectivities of up to 94% ee.[24] The first step of the process was catalysed by a tertiary amine, such as DABCO or quinuclidine, to give the corresponding Michael adduct, which was further submitted to a Stetter reaction induced by the second catalyst, being the chiral *N*-heterocyclic carbene depicted in Scheme 2.16. It must be noted that the best enantioselectivities were reached in the case of symmetrical activated alkynes.

In 2011, these authors also developed an enantioselective synthesis of complex cyclopentanones on the basis of a domino Michael–benzoin

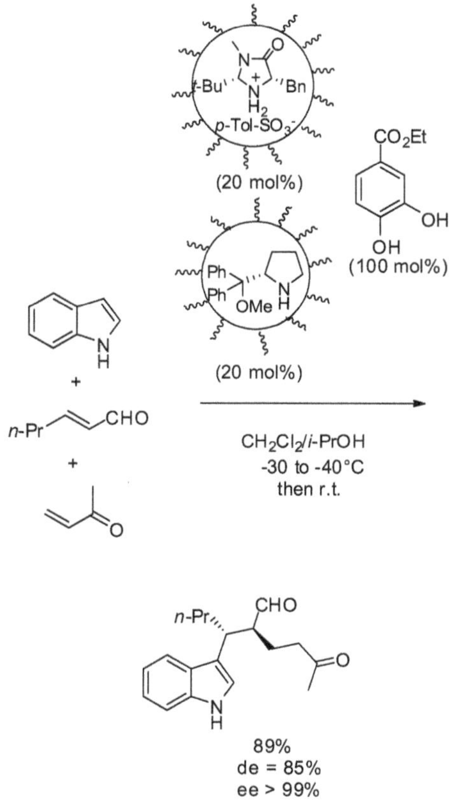

Scheme 2.14 Three-component domino double Michael reaction catalysed by a combination of two soluble star chiral polymers and a hydrogen-bond donor.

reaction of aliphatic aldehydes with activated enones.[25] With the combination of the same chiral *N*-heterocyclic carbene catalyst with chiral diphenylprolinol trimethylsilyl ether, moderate to high diastereoselectivities of up to 92% de and moderate to excellent enantioselectivities of up to 98% ee could be achieved, as shown in Scheme 2.17.

In 2012, the strong and available Schwesinger base, 2-*tert*-butylimino-2-diethylamino-1,3-dimethylperhydro-1,3,2-diazaphosphorine, polymer supported (PS-BEMP), and a chiral bulky BINOL-derived phosphoric acid were used by Dixon *et al.* as combined catalysts in an enantioselective domino Michael–*N*-acyliminium cyclisation reaction, allowing the preparation of structurally complex β-carbolines with moderate to good enantiocontrol.[26] Indeed, the reaction of substituted indole-derived malonamides with methyl or ethyl vinyl ketone began with a base-catalysed Michael addition followed by an acid-catalysed *N*-acyliminium cyclisation cascade, which provided a range of highly functionalised chiral tetracyclic products in good yields and enantioselectivities of up to 82% ee, as shown in Scheme 2.18. This process

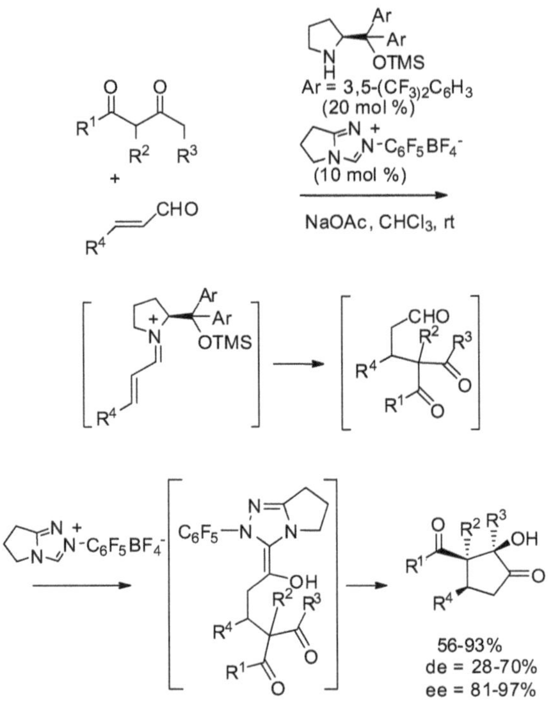

Scheme 2.15   Domino Michael–crossed benzoin reaction catalysed by a combination of a chiral diarylprolinol trimethylsilyl ether and a triazolium
salt.

took advantage of a novel size exclusion phenomenon between PS-BEMP and
a sterically bulky BINOL-derived phosphoric acid. Thus, the latter was suf-
ficiently large so as to prevent full penetration of the microporous lattice,
thus leaving unquenched both the internal basic residues and the BINOL-
derived phosphoric acid in solution. The smaller substrates penetrated
through to the internal catalytically active basic sites; this size exclusion
phenomenon allowed both strong base and strong chiral acid catalysts to co-
exist and function simultaneously in the same vessel.

In another context, Cao and Qu showed that an enantioselective acylation
catalysed by a chiral thioamide modified 1-methylhistidine methyl ester
(Scheme 2.19) in combination with a DABCO-mediated racemisation of the
substrate led to the efficient dynamic kinetic resolution (DKR) of *meso*-1,2-
diol monodichloroacetates.[27] As shown in Scheme 2.19, both cyclic and
acyclic *meso*-1,2-diol monodichloroacetates could be transformed to the
corresponding enantiomerically enriched (1*S*,2*R*)-heterosubstituted diol
esters in good yields and moderate enantioselectivities of up to 74% ee.

In addition to simple two-component domino reactions based on relay
enantioselective multiorganocatalysis, a number of multicomponent reactions have recently been successfully developed. As an example, in 2008

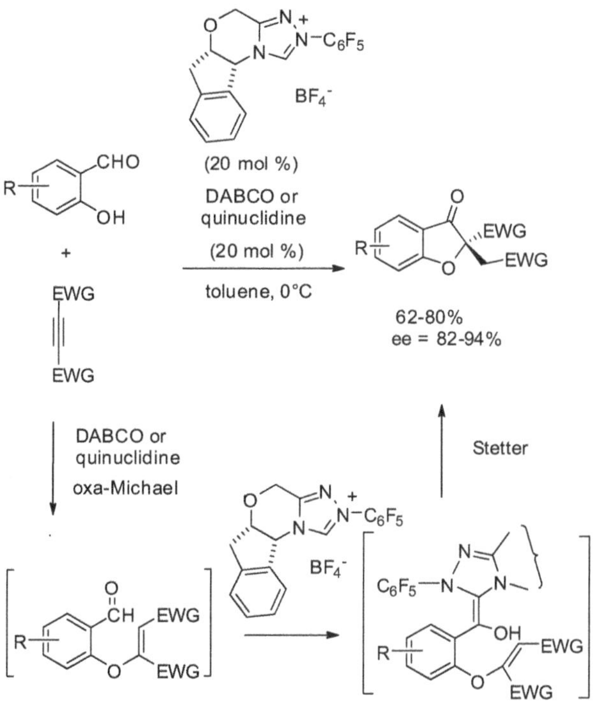

Scheme 2.16 Domino oxa-Michael–Stetter reaction catalysed by a combination of a chiral *N*-heterocyclic carbene and a tertiary amine.

Ramachary and Sakthidevi developed a novel strategy to reach chiral highly functionalised bicyclo[4.4.0]decane-1,6-diones based on an enantioselective multicomponent domino Michael–Robinson annulation–hydrogenation reaction of a variety of Wieland–Miescher ketones, Hajos–Parrish ketones and their analogs with organic hydrides such as Hantzsch esters, as the hydrogen source.[28] This astonishingly simple process was catalysed by four catalysts used in one-pot, such as L-proline, a chiral diamine, TEA, and HClO$_4$. As shown in Scheme 2.20, the reaction of 2-methylcyclohexane-1,3-dione with methyl vinyl ketone and Hantzsch ester provided the corresponding bicyclic dione in 45% yield as a single diastereomer and with a good enantioselectivity of 75% ee. Probably, the tandem Michael–Robinson annulation reaction was catalysed by L-proline and TEA while the final hydrogenation was catalysed by the chiral diamine in combination with HClO$_4$.

When nitrostyrenes and dimethylmalonate were reacted with α,β-unsaturated aldehydes in the presence of a combination of chiral diphenylprolinol triethylsilyl ether and a chiral cinchona alkaloid catalyst through a three-component reaction, Xu *et al.* showed that they afforded, according to a Michael–Michael–aldol domino reaction, the corresponding chiral functionalised cyclohexanes in moderate to good yields, good to excellent enantioselectivities of up to >99% ee, albeit moderate diastereoselectivities

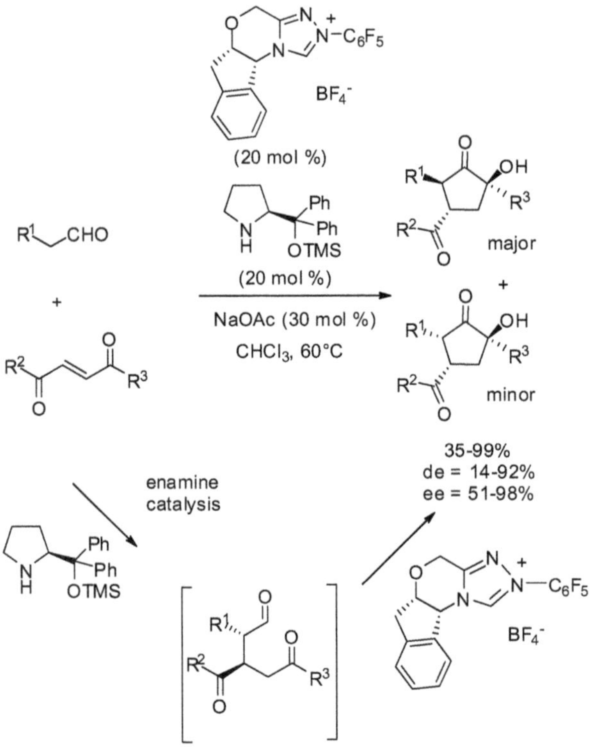

Scheme 2.17 Domino Michael–benzoin reaction catalysed by a combination of chiral diphenylprolinol trimethylsilyl ether and a chiral *N*-heterocyclic carbene.

($\leq$54% de).[29] The first Michael addition occurred between dimethylmalonate and nitroalkenes affording the corresponding nitroalkanes, which were subsequently added onto α,β-unsaturated aldehydes to give the corresponding aldehydes. These aldehyde intermediates then cyclised into the final cyclohexanes through an intramolecular aldol condensation. Importantly, the authors have demonstrated that it was possible to achieve different stereoisomers by changing the combination of catalysts used in the cascade reaction. Some of the best results are collected in Scheme 2.21.

In 2010, Xu *et al.* developed a chemo-, diastereo- and enantioselective three-component organo-relay cascade reaction catalysed by a combination of chiral diphenylprolinol triethylsilyl ether and a chiral thiourea.[30] It involved an aldehyde, a nitroalkene, and an imine as substrates, providing the corresponding chiral fully substituted piperidines in good yields, and generally almost complete enantioselectivities (>99% ee) as mixtures of α- and β-diastereomers arising from the hemiaminal stereocentre, as shown in Scheme 2.22. The process evolved through three steps, beginning with the initial activation of the aldehyde by the proline catalyst (enamine activation), which reacted through a Michael addition with the nitroalkene to give the

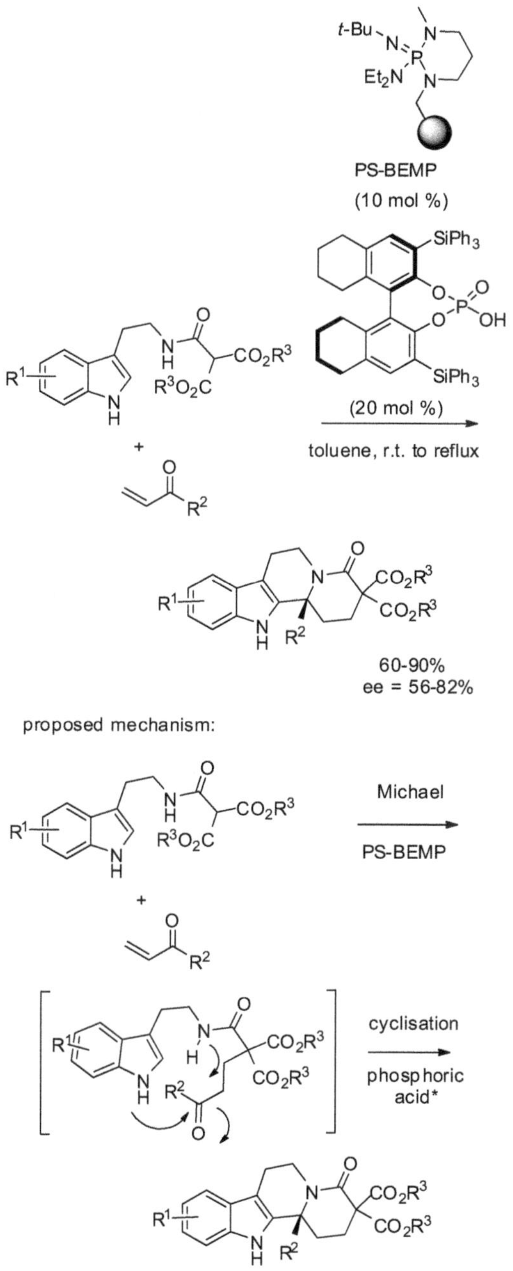

Scheme 2.18 Domino Michael–*N*-acyliminium cyclisation reaction catalysed by a combination of a chiral phosphoric acid and a polymer supported Schwesinger base.

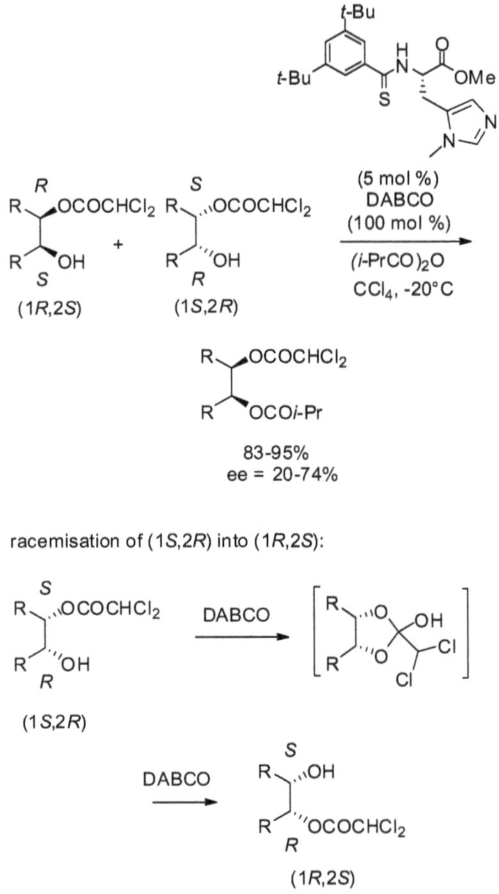

Scheme 2.19 DKR of *meso*-1,2-diol monodichloroacetates catalysed by a chiral thioamide and DABCO.

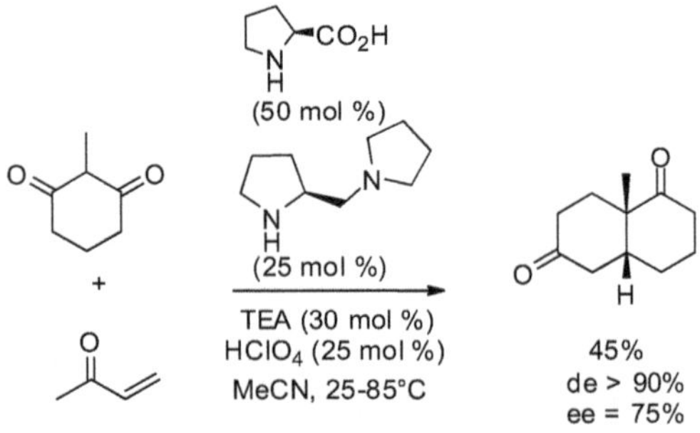

Scheme 2.20 Three-component domino Michael–Robinson annulation–hydrogenation reaction catalysed by a combination of L-proline, a chiral diamine, TEA and HClO₄.

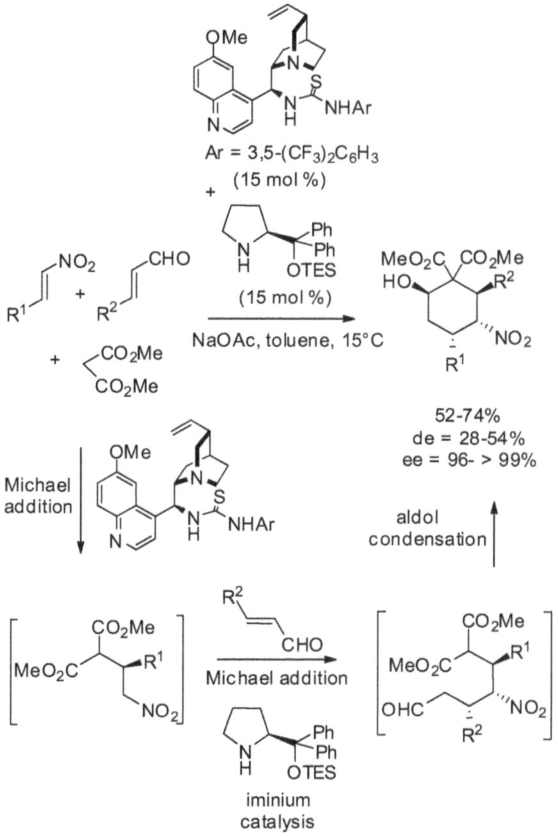

Scheme 2.21 Three-component domino Michael–Michael–aldol reaction catalysed by a combination of chiral diphenylprolinol triethylsilyl ether
and a chiral cinchona thiourea.

corresponding intermediate nitroalkane, which could then participate in the
relay catalysis cycle. The thiourea catalyst promoted the subsequent nitro-
Mannich reaction of the nitroalkane with the imine to generate the corresponding persubstituted *N*-Tos-protected aminoaldehyde, which could then
undergo cyclisation to give the final hemiaminal (Scheme 2.22). The merit of
this process was highlighted by its high efficiency in producing three new
bonds and five new stereogenic centres in one operation.

In 2012, Wang *et al.* reported a pseudo-three-component highly
enantioselective domino Michael–Michael–Henry reaction catalysed by a
combination of chiral diphenylprolinol trimethylsilyl ether and a chiral
quinine-derived thiourea.[31] The reaction occurred between aliphatic
aldehydes and two equivalents of nitroalkenes, providing the corresponding
hexasubstituted chiral cyclohexanols in moderate to good yields and diastereoselectivities of up to 60% de, combined with generally excellent
enantioselectivities ranging from 96 to >99% ee (Scheme 2.23). The authors

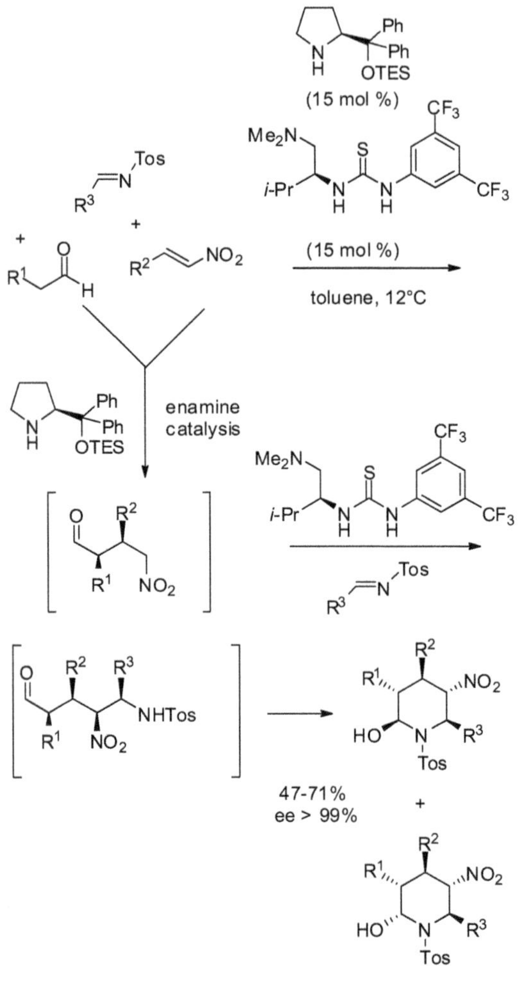

Scheme 2.22 Three-component domino Michael–Mannich–cyclisation reaction catalysed by a combination of chiral diphenylprolinol triethylsilyl ether and a chiral thiourea.

assumed that the first Michael reaction was induced by the proline-derived catalyst while the second Michael reaction was catalysed by the chiral thiourea catalyst.

The same year, Lin *et al.* employed a combination of L-proline with chiral diphenylprolinol trimethylsilyl ether to catalyse a remarkable highly diastereo- and enantioselective domino α-aminoxylation–aza-Michael–aldol reaction of aliphatic aldehydes, α,β-unsaturated aldehydes and nitroso compounds. The reaction afforded the corresponding almost enantiopure 1,2-oxazine derivatives in moderate to good yields, as shown in Scheme 2.24.[32] A plausible mechanism of the reaction is depicted in Scheme 2.24, beginning with a α-oxyamination of the aldehyde with the

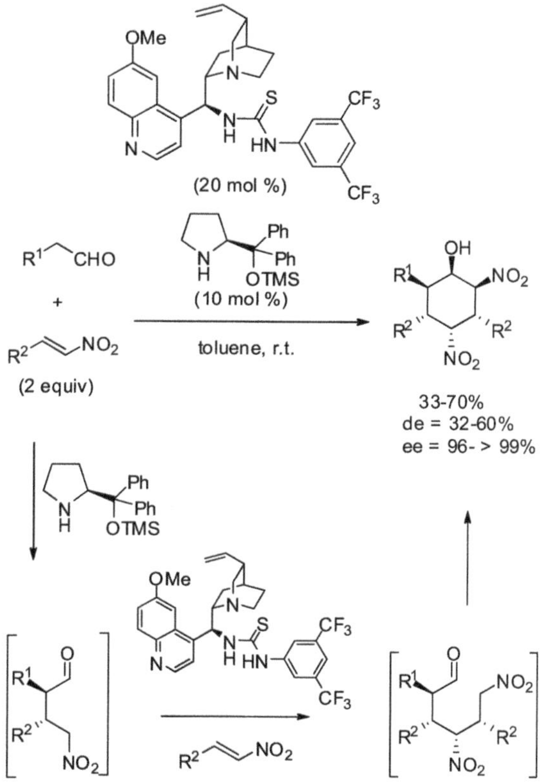

Scheme 2.23 Pseudo three-component domino Michael–Michael–Henry reaction catalysed by a combination of chiral diphenylprolinol trimethylsilyl ether and a chiral thiourea.

nitroso compound catalysed by a proline-based catalyst *via* an enamine process, leading to intermediate **16**. This intermediate underwent a aza-Michael addition *via* the same catalyst through an iminium process to provide the corresponding intermediate **17**, which then was subjected to an aldol condensation again *via* enamine catalysis induced by one of the two catalysts to finally afford the corresponding trisubstituted functionalised dihydro-1,2-oxazine bearing two newly formed stereogenic centres. Although proline is known to be an excellent enamine catalyst, this amino acid is generally ineffective as an iminium catalyst in the case of enals. Since the proposed cascade reaction involved a succession of enamine and iminium catalysis, the combination of L-proline, dedicated to enamine catalysis, with chiral diphenylprolinol trimethylsilyl ether, more dedicated to iminium catalysis than proline, was highly effective.

In another context, Fréchet *et al.* have described a polarity-directed three-component domino Henry–Michael reaction of two aliphatic aldehydes with different polarities and nitromethane catalysed by a combination of chiral diphenylprolinol trimethylsilyl ether and L-proline.[33] In this original process,

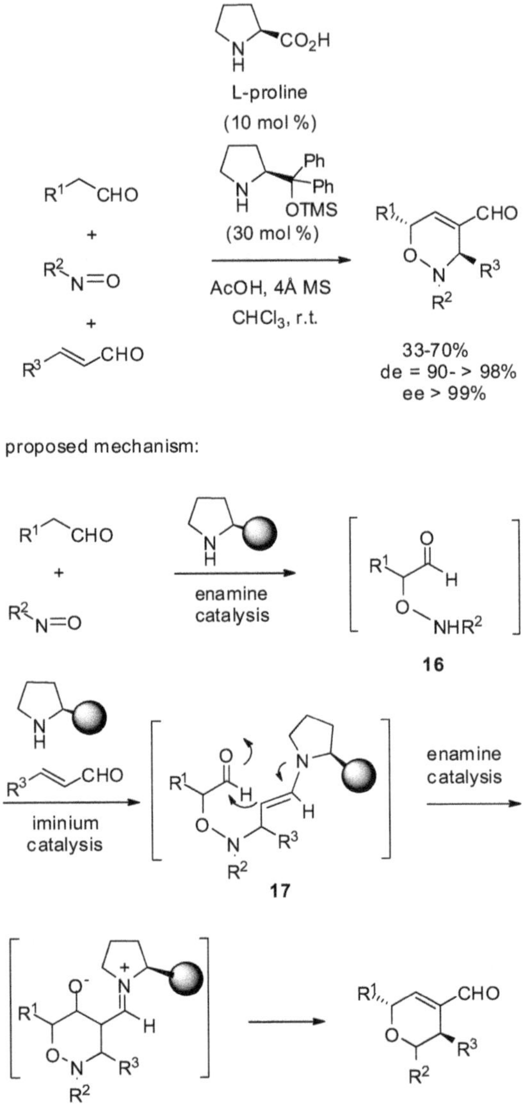

Scheme 2.24   Three-component domino α-aminoxylation–aza-Michael–aldol reaction catalysed by a combination of chiral diphenylprolinol trimethylsilyl ether and ʟ-proline.

each catalyst mediated an individual reaction in either the aqueous or organic phase, such as oily droplets. Indeed, the first step (Henry reaction) was catalysed by ʟ-proline in an aqueous medium to give the Henry product, arising from the reaction of nitromethane with the aldehyde having the smallest size and consequently a significantly greater miscibility with the aqueous phase. This nitroalkene intermediate formed was then converted through Michael reaction with the second aldehyde catalysed by chiral

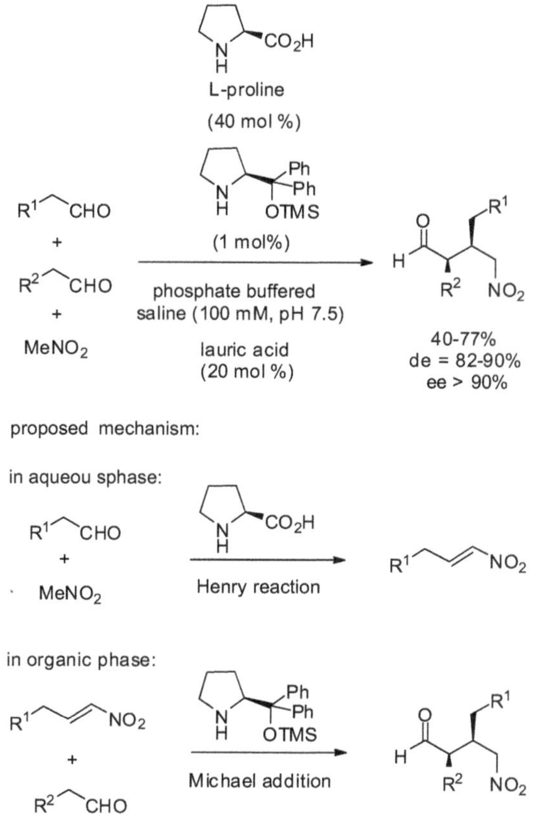

Scheme 2.25 Three-component domino Henry–Michael reaction catalysed by a combination of chiral diphenylprolinol trimethylsilyl ether and L-proline.

diphenylprolinol trimethylsilyl ether in the organic substrate phase to provide the final product. As shown in Scheme 2.25, a range of chiral cross-products could be achieved in good yields and high diastereo- and enantioselectivities of up to 90% de, and > 90% ee, respectively. This interesting system highlighted an often-ignored approach to developing chemoselective reactions by using properties other than chemical reactivity, such as polarity, inherent to the substrates or catalysts.

In 2013, Li *et al.* reported a novel synthesis of chiral six-membered spirocyclic oxindoles bearing five consecutive stereocentres on the basis of an enantioselective three-component domino Michael–Michael–aldol reaction catalysed by a combination of a chiral diphenylprolinol silyl ether and a chiral bifunctional quinine thiourea.[34] As shown in Scheme 2.26, the reaction between *N*-substituted oxindoles, nitrostyrenes, and enals led to the corresponding highly substituted spirocyclic oxindoles in high yields, moderate diastereoselectivities of up to 44% de, and enantioselectivities of up to >99% ee. A large variety of substrates could be used well in this

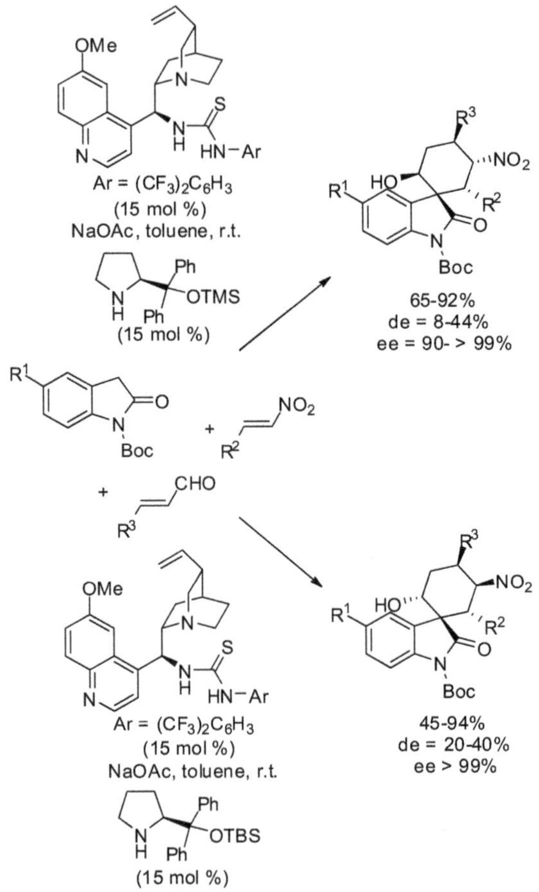

Scheme 2.26 Three-component domino Michael–Michael–aldol reaction catalysed by a combination of chiral diphenylprolinol silyl ethers and
a chiral bifunctional quinine thiourea.

reaction through relay catalysis. Importantly, the authors found that through
judicious choice of the diphenylprolinol silyl ether employed, the reaction
could be readily adapted to predominantly afford an alternative major diastereomer of the product. As shown in Scheme 2.26, using the same chiral
quinine thiourea catalyst in combination with (*S*)-diphenylprolinol *tert*-
butyldimethylsilyl ether instead of (*S*)-diphenylprolinol trimethylsilyl ether
resulted in the formation of another diastereomer of the same chiral six-
membered spirocyclic oxindoles in generally perfect enantioselectivities of
>99% ee in all cases of substrates studied.

2.4   Sequential Catalysis

In 2005, MacMillan *et al.* disclosed the concept of cycle-specific catalysis,
wherein each cycle in a tandem reaction is moderated by a different chiral

amine catalyst, which allows selective synthesis of any product enantiomer or diastereomer by choosing an appropriate catalyst.[35] These authors demonstrated the versatility of this concept by controlling the diastereoselectivity in the asymmetric addition of HF across the 3-phenyl-but-2-enal by using two different chiral amine catalysts. A chiral secondary amine acted as iminium catalyst for the asymmetric transfer hydrogenation of the enal, and the other chiral amine catalyst served as enamine catalyst for the α-fluorination of the aldehyde intermediate. To ensure high yield, the enamine catalyst and the electrophile were added after the consumption of the Hantsch ester. It should be noted that the structure of the enamine catalyst significantly influenced the diastereoselectivity of the reaction. Using the same chiral secondary amine as both the iminium and enamine catalyst only afforded a moderate ratio of *syn* to *anti* (3 : 1). The best *anti*-diastereo- and enantioselectivities of 88% de and 99% ee were reached by using the two different catalysts depicted in Scheme 2.27.

Later, Ramachary and Sakthidevi reported for the first time the organocatalytic cascade approach to the asymmetric synthesis of functionalised chromans *via* Barbas–List aldol–acetalisation reaction, as depicted in Scheme 2.28.[36] The reaction of acetone with 2-hydroxybenzaldehyde under *trans*-4-OH-L-proline-catalysis in NMP as solvent furnished the corresponding aldol/lactol intermediate which upon treatment with *p*-TSA in methanol in one-pot furnished the selectively *trans*-2-methoxy-2-methyl-chroman-4-ol in 55% yield and 77% ee, as shown in Scheme 2.28.

In 2009, Jørgensen and co-workers reported a novel one-pot efficient synthesis of chiral 4,5-disubstituted isoxazoline-*N*-oxides from simple commercially available starting materials using the combination of an iminium

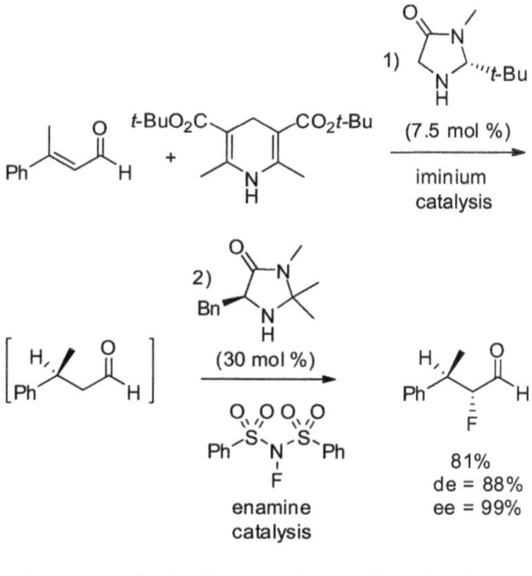

Scheme 2.27   Tandem transfer hydrogenation–α-fluorination reaction catalysed by two chiral amines.

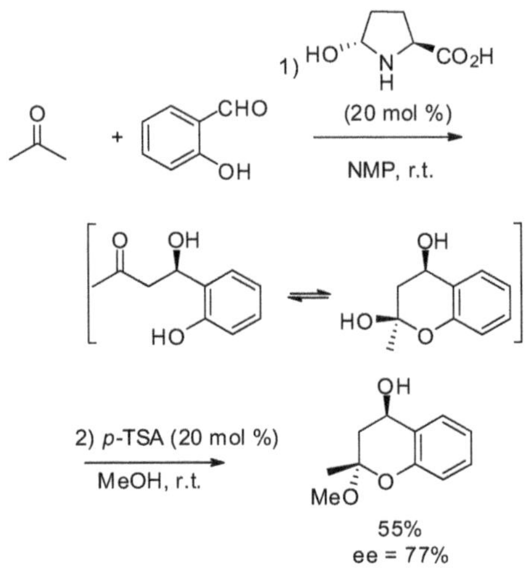

Scheme 2.28　Tandem aldol–acetalisation reaction catalysed by chiral 4-hydroxy-proline and *p*-TSA.

catalyst, such as a chiral diarylprolinol trimethylsilyl ether, and a chiral phase transfer catalyst.[37] This cascade was initiated by the asymmetric epoxidation of a α,β-unsaturated aldehyde, such as cinnamaldehyde, followed by a Henry reaction induced by the chiral ammonium salt catalyst under phase-transfer-catalysis conditions using CsOH as base, and then cyclisation through *O*-alkylation. It must be noted that the use of the chiral ammonium salt was essential for achieving excellent enantioselectivity of 99% ee, good yield (65%), and diastereoselectivity of 56% de, as shown in Scheme 2.29. The chiral product could be further easily transformed into the corresponding β,γ,δ-trihydroxylated α-amino acid derivatives.

In 2011, Diez *et al.* reported a novel synthesis of chiral 2-alkylidene cyclohexenones based on an enantioselective tandem Michael–Morita–Baylis–Hillman;–Knoevenagel; reaction between a Nazarov reagent and α,β-unsaturated aldehydes sequentially catalysed by a chiral diarylprolinol trimethylsilyl ether and L-proline.[38] The authors postulated that the chiral diarylprolinol trimethylsilyl ether firstly induced the Michael addition of the Nazarov reagent to the α,β-unsaturated aldehyde, providing the corresponding aldehyde intermediate. This aldehyde then underwent a Morita–Baylis–Hillman reaction followed by a Knoevenagel condensation catalysed by L-proline, which afforded the final cyclised product as a 2 : 1 mixture of *E*/*Z* diastereomers obtained in moderate to good yields (41–77%) and good to high enantioselectivities of 82 to 96% ee for both diastereomers (Scheme 2.30).

In 2011, Enders *et al.* reported a sequential organocatalytic cascade reaction between α,β-unsaturated aldehydes and β-oxosulfones, using a combination of a chiral diarylprolinol trimethylsilyl ether and an achiral *N*-heterocyclic carbene as a catalytic system.[39] This sequential

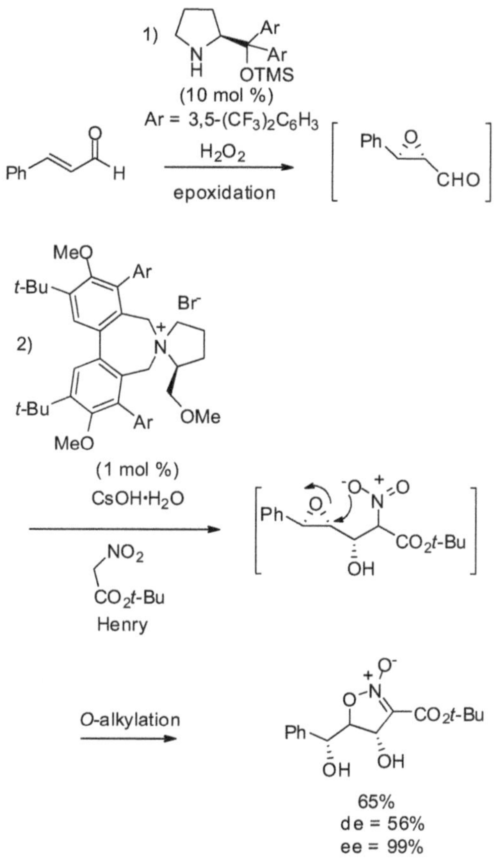

Scheme 2.29 Tandem epoxidation–Henry–*O*-alkylation reaction catalysed by a chiral diarylprolinol trimethylsilyl ether and a chiral ammonium salt.

Michael–cross-benzoin cyclisation reaction afforded the corresponding chiral polyfunctionalised cyclopentanones bearing three contiguous stereocentres in good yields, moderate to excellent diastereoselectivities of up to 98% de, and good to high enantioselectivities of up to 96% ee, as shown in Scheme 2.31.

In 2012, Melchiorre *et al.* reported a novel stereoselective access to chiral *trans*-fused tetracyclic indole-based products having four stereogenic centres on the basis of a multicatalytic tandem Diels–Alder–benzoin reaction involving *N*-Boc protected 3-(2-methyl-indol-3-yl)acrylaldehyde derivative and *trans*-1,2-dibenzoylethylene derivative as substrates.[40] As shown in Scheme 2.32, the process was successively induced by chiral diphenylprolinol trimethylsilyl ether in the presence of bulky 2,4,6-trimethylbenzoic acid (TMBA) as co-catalyst for the Diels–Alder reaction (trienamine catalysis), and an *N*-heterocyclic carbene for the following cross-benzoin condensation

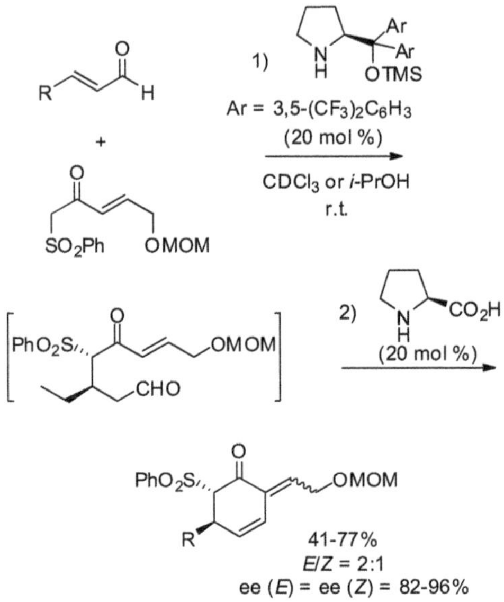

Scheme 2.30 Tandem Michael–Morita–Baylis–Hillman;–Knoevenagel; reaction catalysed by a chiral diarylprolinol trimethylsilyl ether and ʟ-proline.

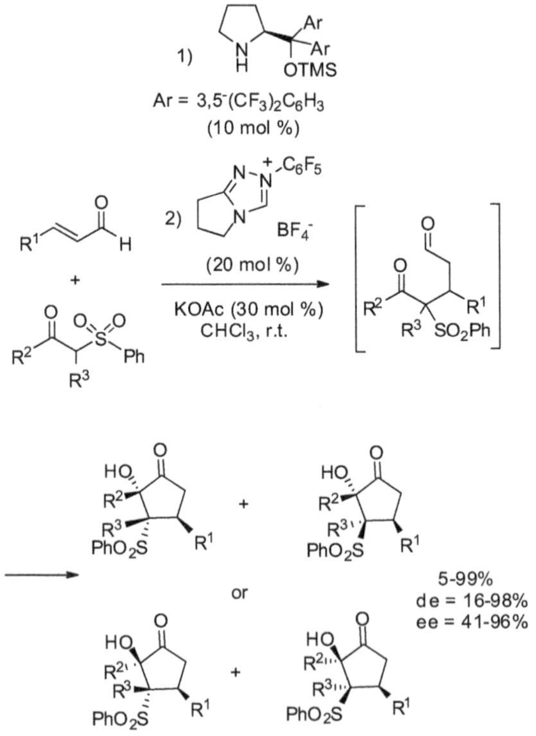

Scheme 2.31 Tandem Michael–cross-benzoin cyclisation reaction catalysed by a chiral diarylprolinol trimethylsilyl ether and an *N*-heterocyclic carbene.

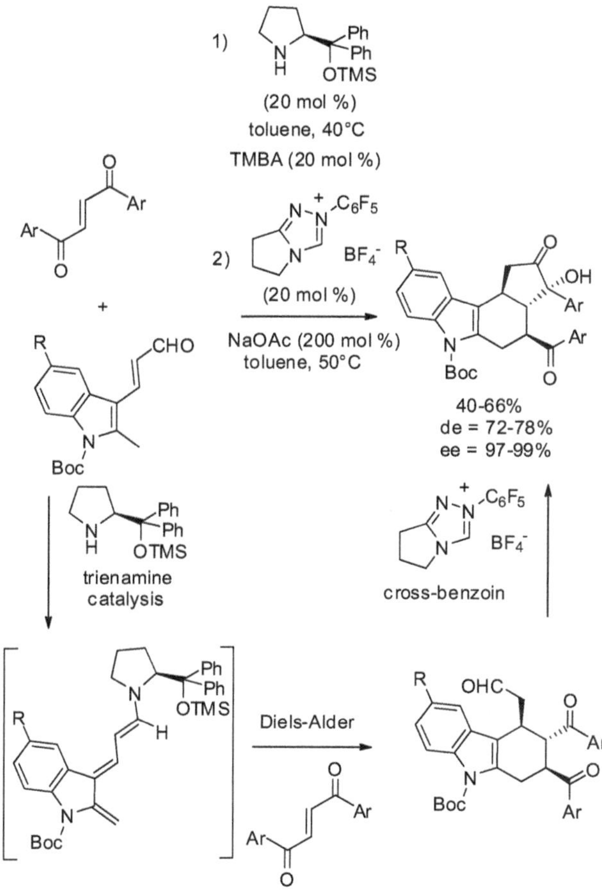

Scheme 2.32 Tandem Diels–Alder–benzoin reaction catalysed by chiral diphenyl-prolinol trimethylsilyl ether and an *N*-heterocyclic carbene.

(carbene catalysis). The corresponding complex tetrahydrocarbazole derivatives were achieved in moderate to good yields and diastereoselectivities combined with generally excellent enantioselectivities of 97 to 99% ee.

In 2012, a comparable combination of catalysts was employed by Jørgensen *et al.* to induce a Michael-initiated cascade reaction of aryloxyacetaldehydes to give the corresponding chiral 3,4-dihydrocoumarins.[41] As shown in Scheme 2.33, these important products were produced in good yields and moderate to high enantioselectivities of up to 96% ee.

In 2013, Zhao *et al.* described a highly diastereo- and enantioselective synthesis of trisubstituted cyclohexanols based on a one-pot sequential tandem Henry–Michael reaction catalysed by a combination of a chiral cinchona alkaloid thiourea and 1,1,3,3-tetramethyl guanidine.[42] This process occurred between nitromethane and 7-oxo-hept-5-en-1-als, providing the corresponding tandem chiral products as almost single diastereomers

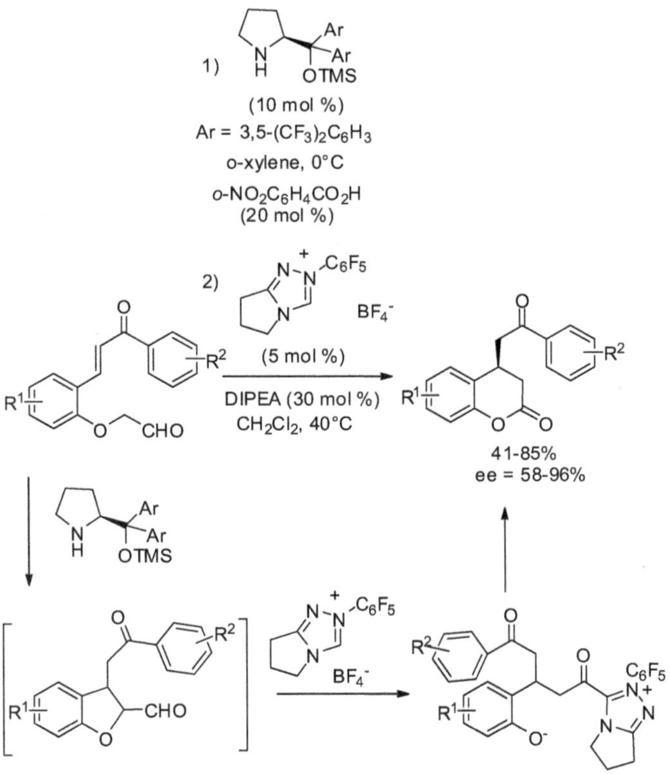

Scheme 2.33 Tandem Michael-initiated reaction catalysed by a chiral diarylprolinol trimethylsilyl ether and an *N*-heterocyclic carbene.

(de >98%) in excellent yields (90–99%) as well as high enantioselectivities of
up to 96% ee, as shown in Scheme 2.34. The scope of the reaction was extended to a wide number of enals bearing a substituted phenyl group
(R = Ph). It was shown that the electronic nature of the substituent of this
phenyl group has no effect on both the diastereo- and enantioselectivity of
the reaction. Nonetheless, when the phenyl group on the enal was replaced
with an alkyl group, such as methyl or *tert*-butyl, the corresponding cyclohexanes were obtained in much lower enantioselectivities of 70 and 67% ee,
respectively, although the diastereoselectivity values remained high (de
>98%).

In addition to simple enantioselective two-component tandem multi-
organocatalysed reactions, several groups have recently reported multicomponent reactions based on the enantioselective sequential
multiorganocatalysis concept. For example, Cordova *et al.* disclosed in
2011 a highly enantioselective one-pot cascade sequence based on the
combination of asymmetric amine catalysis and *N*-heterocyclic carbene
catalysis, which allowed the synthesis of a range of chiral *N-tert*-butox-
ycarbonyl and *N*-carbobenzyloxy-protected β-amino acid esters to be

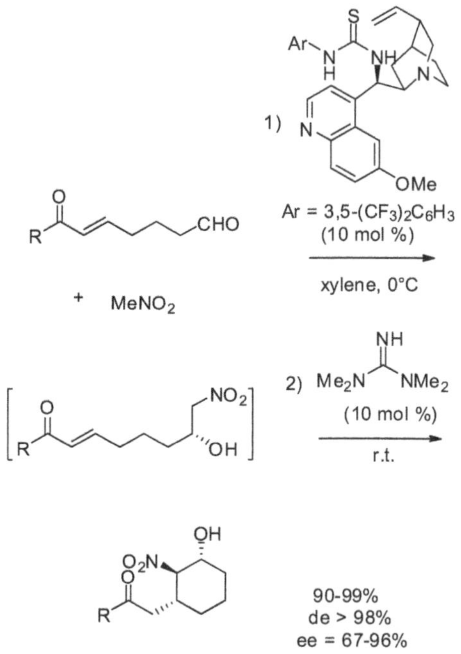

Scheme 2.34 Tandem Henry–Michael–retro-Henry reaction catalysed by a chiral cinchona alkaloid thiourea and 1,1,3,3-tetramethyl guanidine.

achieved in enantioselectivities ranging from 92 to 99% ee.[43] This remarkable three-component tandem process, occurring between an enal, a hydroxylamine derivative and an alcohol, began with a domino aza-Michael–intramolecular nucleophilic substitution reaction catalysed by chiral diphenylprolinol trimethylsilyl ether to give the corresponding aziridine intermediate. This aziridine was then treated by the second *N*-heterocyclic carbene catalyst and provided, by reaction with the alcohol as the third component, the corresponding final chiral *N-tert*-butoxycarbonyl and *N*-carbobenzyloxy-protected β-amino acid ester in moderate to good yields and generally high enantioselectivities, as shown in Scheme 2.35. The mechanistic proposal for this second step of the cascade reaction involved the formation of zwitterionic species **18**. Then, C–N bond cleavage/ring opening occurred and intermediate **19** was generated. Activated intermediate **21** was formed by keto–enol tautomerisation *via* intermediates **19** and **20**. Final transesterification by the alcohol component gave the corresponding final β-amino acid ester and released the carbene catalyst (Scheme 2.35).

In another context, Guo and co-workers have used a combination of a chiral thiourea, an achiral *N*-Boc-protected amino acid, and a chiral phosphoric acid to promote a remarkable enantioselective three-component tandem reaction occurring between an aldehyde, a 3-indolylmethanol compound, and an *N*-benzyl-protected indole.[44] The process began with the α-alkylation of the aldehyde with the 3-indolylmethanol catalysed by the

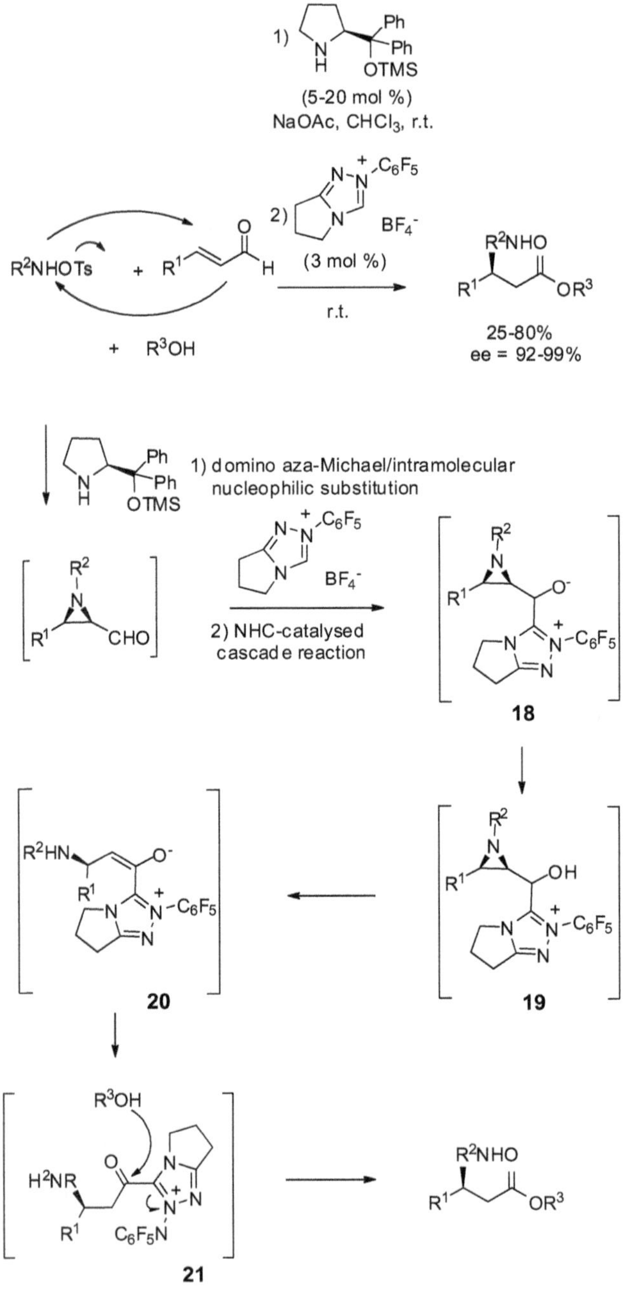

Scheme 2.35 Three-component tandem reaction catalysed by chiral diphenylpro-linol trimethylsilyl ether and an *N*-heterocyclic carbene.

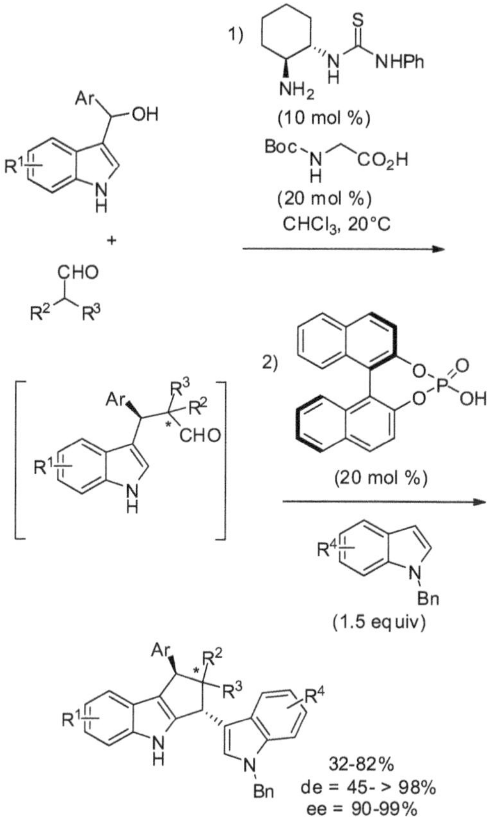

Scheme 2.36 Three-component tandem α-alkylation–double Friedel–Crafts reaction catalysed by a chiral thiourea, an *N*-Boc-protected amino acid, and a chiral phosphoric acid.

chiral thiourea, followed by two consecutive Friedel–Crafts reactions catalysed by the amino acid and the chiral phosphoric acid, respectively, to give the corresponding chiral polysubstituted cyclopenta[*b*]indoles in high yields and enantioselectivities of up to 99% ee, as shown in Scheme 2.36. The substrate scope was large since indoles bearing electron-withdrawing as well as electron-donating groups were all suitable reaction partners. Moreover, symmetric as well as unsymmetric α,α-disubstituted aldehydes and α-unsubstituted aldehydes gave excellent results.

Finally, a remarkable four-component tandem Michael–aza-Henry–hemiaminalisation–dehydration tandem reaction was recently developed by Lin and co-workers on the basis of a dual organocatalysis involving a chiral diaryl prolinol trimethylsilyl ether and a chiral cinchona alkaloid.[45] As shown in Scheme 2.37, the reaction began with the Michael addition of an aldehyde to a nitroalkene catalysed by the L-proline-derived catalyst, giving the corresponding intermediate aldehyde. The latter intermediate

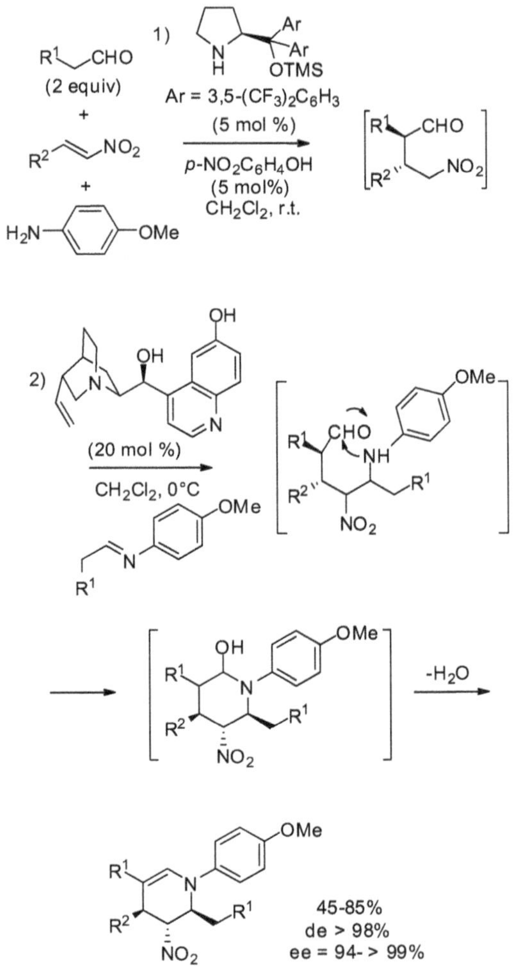

Scheme 2.37 Four-component tandem Michael–aza-Henry–hemiaminalisation–dehydration reaction catalysed by a chiral diarylprolinol trimethyl-silyl ether and a chiral cinchona alkaloid catalyst.

subsequently reacted through aza-Henry reaction with an imine, generated *in situ* from a second equivalent of aldehyde and *para*-methoxyaniline, to provide the corresponding intermediate amine, which further cyclised through hemiaminalisation to give the corresponding hemiaminol. This intermediate finally dehydrated to provide the functionalised chiral tetra-hydropyridine, obtained as a single diastereomer in all cases of the sub-strates studied and in good yields (45–85%) and very high enantioselectivities ranging from 94 to >99% ee. In addition, the authors have shown that the scope of this methodology could be extended to the use of two different aldehydes, providing excellent chemoselectivities and diastereoselectivities greater than 20 : 1 and 99 : 1, respectively, in combination with high enantioselectivities of up to 94% ee.

References

1. (a) M. L. Clarke and J. A. Fuentes, *Angew. Chem., Int. Ed.*, 2007, **46**, 930–933; (b) R. C. Wende and P. R. Schreiner, *Green Chem.*, 2012, **14**, 1821–1849.
2. B.-C. Hong, M.-F. Wu, H.-C. Tseng, G.-F. Huang, C.-F. Su and J.-H. Liao, *J. Org. Chem.*, 2007, **72**, 8459–8471.
3. (a) J. Zhou and B. List, *J. Am. Chem. Soc.*, 2007, **129**, 7498–7499; (b) J. Zhou and B. List, *Synlett*, 2007, 2037–2040.
4. W. Schrader, P. P. Handayani, J. Zhou and B. List, *Angew. Chem., Int. Ed.*, 2009, **48**, 1463–1466.
5. J.-W. Xie, X. Huang, L.-P. Fan, D.-C. Xu, X.-S. Li, H. Su and Y.-H. Wen, *Adv. Synth. Catal.*, 2009, **351**, 3077–3082.
6. M. Bella, D. M. Scarpino Schietroma, P. P. Cusella, T. Gasperi and V. Visca, *Chem. Commun.*, 2009, 597–599.
7. Y. Inokoishi, N. Sasakura, K. Nakano, Y. Ichikawa and H. Kotsuki, *Org. Lett.*, 2010, **12**, 1616–1619.
8. A.-B. Xia, D.-Q. Xu, S.-P. Luo, J.-R. Jiang, J. Tang, Y.-F. Wang and Z.-Y. Xu, *Chem. – Eur. J.*, 2010, **16**, 801–804.
9. L.-L. Wang, L. Peng, J.-F. Bai, L.-N. Jia, X.-Y. Luo, Q.-C. Huang, X.-Y. Xu and L. X. Wang, *Chem. Commun.*, 2011, **47**, 5593–5595.
10. B. Wu, J. Chen, M.-Q. Li, J.-X. Zhang, X.-P. Xu, S.-J. Ji and X.-W. Wang, *Eur. J. Org. Chem.*, 2012, 1318–1327.
11. L. S. Povarov, *Russ. Chem. Rev.*, 1967, **36**, 656–670.
12. H. Liu, G. Dagousset, G. Masson, P. Retailleau and J. Zhu, *J. Am. Chem. Soc.*, 2009, **131**, 4598–4599.
13. H. Xu, S. J. Zuend, M. G. Woll, Y. Tao and E. N. Jacobsen, *Science*, 2010, **327**, 986–990.
14. G. Bergonzini, S. Vera and P. Melchiorre, *Angew. Chem., Int. Ed.*, 2010, **49**, 9685–9688.
15. (a) K. Zeitler, *Angew. Chem., Int. Ed.*, 2005, **44**, 7506–7510; (b) N. Marion, S. Diez-Gonzalez and S. P. Nolan, *Angew. Chem., Int. Ed.*, 2007, **46**, 2988–3000; (c) D. Enders, O. Niemeier and A. Henseler, *Chem. Rev.*, 2007, **107**, 5606–5655; (d) S. E. Denmark and G. L. Beutner, *Angew. Chem., Int. Ed.*, 2008, **47**, 1560–1638; (e) E. M. Phillips, A. Chan and K. A. Scheidt, *Aldrichimica Acta*, 2009, **42**, 55–66; (f) C. Grondal, M. Jeanty and D. Enders, *Nat. Chem.*, 2010, **2**, 167–178; (g) A. Grossmann and D. Enders, *Angew. Chem., Int. Ed.*, 2011, **50**, 2–14; (h) A. Grossman and D. Enders, *Angew. Chem., Int. Ed.*, 2012, **51**, 314–325.
16. X. Zhao, D. A. DiRocco and T. Rovis, *J. Am. Chem. Soc.*, 2011, **133**, 12466–12469.
17. (a) I. Ugi, R. Meyr, U. Fetzer and C. Steinbrückner, *Angew. Chem.*, 1959, **71**, 386–388; (b) R. V. A. Orru and M. de Greef, *Synthesis*, 2003, 1471–1499.

18. J. Xin, L. Chan, Z. Hou, D. Shang, X. Liu and X. Feng, *Chem. – Eur. J.*, 2008, **14**, 3177–3181.
19. S. Lin, L. Deiana, G.-L. Zhao, J. Sun and A. Cordova, *Angew. Chem., Int. Ed.*, 2011, **50**, 7624–7630.
20. G. Ma, S. Lin, I. Ibrahem, G. Kubik, L. Liu, J. Sun and A. Cordova, *Adv. Synth. Catal.*, 2012, **354**, 2865–2872.
21. L. Liu, D. Wu, X. Li, S. Wang, H. Li, J. Li and W. Wang, *Chem. Commun.*, 2012, **48**, 1692–1694.
22. Y. Chi, T. Scroggins and J. M. J. Fréchet, *J. Am. Chem. Soc.*, 2008, **130**, 6322–6323.
23. S. P. Lathrop and T. Rovis, *J. Am. Chem. Soc.*, 2009, **131**, 13628–13630.
24. C. M. Filloux, S. P. Lathrop and T. Rovis, *Proc. Natl. Acad. Sci. U. S. A.*, 2010, **107**, 20666–20671.
25. K. E. Ozboya and T. Rovis, *Chem. Sci.*, 2011, **2**, 1835–1838.
26. M. E. Muratore, L. Shi, A. W. Pilling, R. I. Storer and D. J. Dixon, *Chem. Commun.*, 2012, **48**, 6351–6353.
27. J.-L. Cao and J. Qu, *J. Org. Chem.*, 2010, **75**, 3663–3670.
28. D. B. Ramachary and R. Sakthidevi, *Org. Biomol. Chem.*, 2008, **6**, 2488–2492.
29. Y. Wang, R.-G. Han, Y.-L. Zhao, S. Yang, P.-F. Xu and D. J. Dixon, *Angew. Chem., Int. Ed.*, 2009, **48**, 9834–9838.
30. Y. Wang, D.-F. Yu, Y.-Z. Liu, H. Wei, Y.-C. Luo, D. J. Dixon and P.-F. Xu, *Chem. – Eur. J.*, 2010, **16**, 3922–3925.
31. Z. Mao, Y. Jia, Z. Xu and R. Wang, *Adv. Synth. Catal.*, 2012, **354**, 1401–1406.
32. H. Lin, Y. Tan, X.-W. Sun and G.-Q. Lin, *Org. Lett.*, 2012, **14**, 3818–3821.
33. S. T. Scroggins, Y. Chi and J. M. J. Fréchet, *Angew. Chem., Int. Ed.*, 2010, **49**, 2393–2396.
34. B. Zhou, Y. Yang, J. Shi, Z. Luo and Y. Li, *J. Org. Chem.*, 2013, **78**, 2897–2907.
35. Y. Huang, A. M. Walji, C. H. Larsen and D. W. C. MacMillan, *J. Am. Chem. Soc.*, 2005, **127**, 15051–15052.
36. (a) D. B. Ramachary and R. Sakthidevi, *Chem. – Eur. J.*, 2009, **15**, 4516–4522; (b) D. B. Ramachary and R. Sakthidevi, *Org. Biomol. Chem.*, 2010, **8**, 4259–4265.
37. H. Jiang, E. Elsner, K. L. Jensen, A. Falcicchio, V. Marcos and K. A. Jørgensen, *Angew. Chem., Int. Ed.*, 2009, **48**, 6844–6848.
38. J. Pena, A. B. Anton, R. F. Moro, I. S. Marcos, N. M. Garrido and D. Diez, *Tetrahedron*, 2011, **67**, 8331–8337.
39. D. Enders, A. Griossmann, H. Huang and G. Raabe, *Eur. J. Org. Chem.*, 2011, 4298–4301.
40. Y. Liu, M. Nappi, E. C. Escudero-Adan and P. Melchiorre, *Org. Lett.*, 2012, **14**, 1310–1313.
41. C. B. Jacobsen, L. Albrecht, J. Udmark and K. A. Jørgensen, *Org. Lett.*, 2012, **14**, 5526–5529.
42. Q. Dai, H. Arman and J. C.-G. Zhao, *Chem. – Eur. J.*, 2013, **19**, 1666–1671.

43. L. Deiana, P. Dziedzic, G.-L. Zhao, J. Vesely, I. Ibrahem, R. Rios, J. Sun and A. Cordova, *Chem. – Eur. J.*, 2011, **17**, 7904–7917.
44. B. Xu, Z.-L. Guo, W.-Y. Jin, Z.-P. Wang, Y.-G. Peng and Q.-X. Guo, *Angew. Chem., Int. Ed.*, 2012, **51**, 1059–1062.
45. H. Lin, Y. Tan, W.-J. Liu, Z.-C. Zhang, X.-W. Sun and G.-Q. Lin, *Chem. Commun.*, 2013, **49**, 4024–4026.

CHAPTER 3

Reactions Catalysed by Two Metals

3.1 Introduction

Although non-metallic chiral organocatalysts have received much attention lately, metal-based catalysts still remain a main player for various important asymmetric reactions owing to their efficiency and generally broad substrate scope. The use of two different metal catalysts is possible when the relative stability of each complex prevents cross-exchange between their ligands. With respect to mixing two purely organic catalysts, this strategy is less common and so far only limited examples are present in the literature. Conventional metal-based catalysts consist of a single metal centre equipped with proper chiral ligands. As a result, the single activation of one reactant is generally attributed to the observed catalytic activity. While remarkable advances have been made with this approach over the years, there are still a number of important asymmetric transformations that lack efficient catalytic methods. The development of multimetallic catalytic systems and their application to asymmetric catalysis is an emerging area in modern organic synthesis.[1] The use of a multimetallic entity in catalyst design is a viable approach to the construction of a multifunctional catalyst, which can activate multiple substrates simultaneously. The bimetallic cooperative catalysts use various different types of metals such as alkali metals, transition metals, and lanthanides. The proper arrangement of those metals in close proximity is probably the key to success for efficient catalysis. From a mechanistic point of view, one metal generally plays a role as a Lewis acid for activating electrophiles, while the other metal ion serves as the counterion of nucleophiles. In the last few years, cooperative bimetallic catalysts have emerged as a new strategy in asymmetric catalysis to achieve high efficiency and selectivity. Moreover, the dual activation of reactants by multiple metal centres

RSC Catalysis Series No. 20
Enantioselective Multicatalysed Tandem Reactions
By Hélène Pellissier
© Hélène Pellissier 2014
Published by the Royal Society of Chemistry, www.rsc.org

can render reaction conditions milder. Although there are challenges to rationally designing efficient bimetallic catalysts, a number of outstanding chiral bimetallic catalysts have been developed to showcase the unique power of such a bio-inspired approach. Cooperative work of multiple catalytically active sites in the asymmetric multimetallic catalysts allows for enhanced catalytic activity and stereoselectivity over monometallic catalysts.

3.2 Cooperative Catalysis

In 2007, Shibasaki *et al.* reported a three-component domino reaction induced by cooperative catalysis between copper and zinc.[2] This novel domino reaction occurred between acetophenone, diethylzinc and an allenic ester, providing the corresponding chiral lactone in high yield and excellent enantioselectivity of 96% ee, as shown in Scheme 3.1. The process began with the addition of diethylzinc to the allenic ester, which was followed by an aldolisation with acetophenone, and a lactonisation reaction gave the final product. Initially, the Cu(II) was reduced to Cu(I) in the presence of

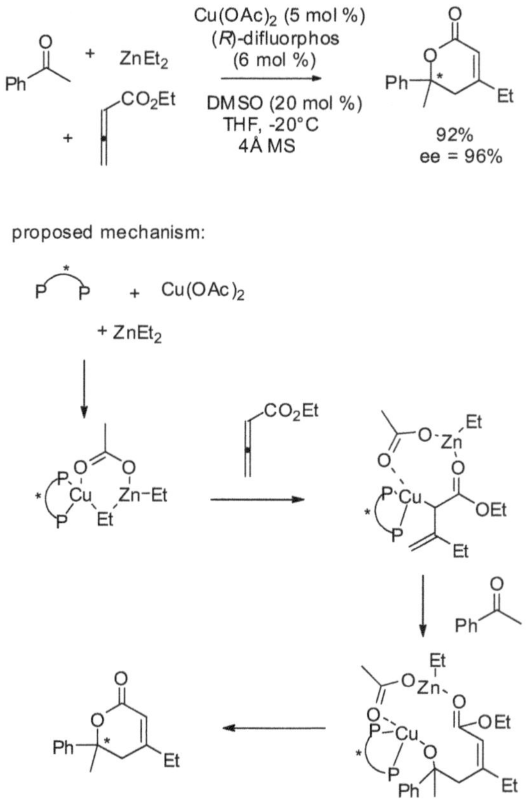

Scheme 3.1 Three-component reaction catalysed by a combination of chiral copper catalysis and zinc catalysis.

diethylzinc to produce an alkylcopper-phosphine complex. Because the use of Cu(OAc)$_2$ afforded significantly higher enantioselectivity and yield than CuBr and CuCl, the authors proposed that the acetate might act as a bridging ligand between Zn(II) and Cu(I), and thus Zn(II) might serve as Lewis acid to activate carbonyl groups, stabilising the oxygen anion intermediate, and directing the substrate (Scheme 3.1). As such, copper and zinc cooperated efficiently to afford the product in 96% ee.

In 2010, Hu *et al.* developed an enantioselective three-component coupling reaction of α,β-unsaturated 2-acyl imidazoles, water and α-diazoesters based on a cooperative catalytic system consisting of 2 mol% of Rh$_2$(OAc)$_4$ and 30 mol% of chiral (*S*)-*t*-Bu-BOX-Zn(OTf)$_2$ in the presence of 40 mol% of TsOH.[3] The process evolved through the *in situ* generation of oxonium ylides from the aryl diazoacetates and water, which were subsequently trapped by the α,β-unsaturated 2-acyl imidazoles through a Michael addition to give the corresponding chiral γ-hydroxyketones in good yields, high diastereoselectivities of up to 98% de, and excellent enantioselectivities of up to 99% ee, as shown in Scheme 3.2.

In 2013, Riant *et al.* developed a novel enantioselective domino conjugate reduction–allylic alkylation reaction based on a cooperative dual catalysis involving a chiral copper catalyst and an achiral palladium catalyst.[4] The reaction occurred between cyclic α-substituted α,β-unsaturated ketones and allyl methylcarbonate, providing the corresponding chiral cyclic α-allylic

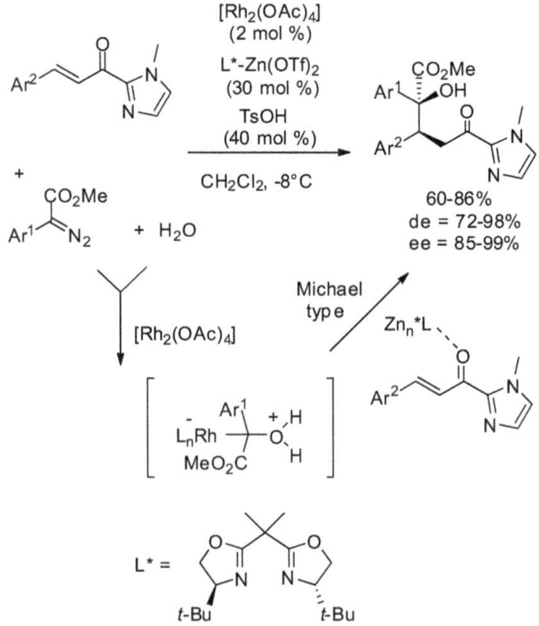

Scheme 3.2   Three-component reaction catalysed by a combination of rhodium catalysis and chiral zinc catalysis.

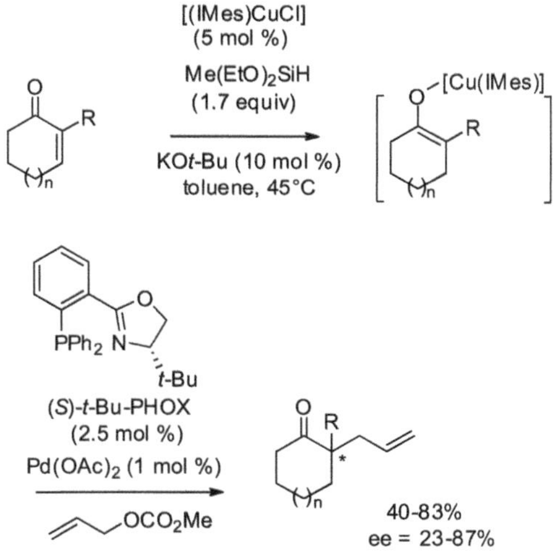

Scheme 3.3 Domino conjugate reduction–allylic alkylation reaction catalysed by a combination of copper catalysis and chiral palladium catalysis.

ketones in moderate to good yields and enantioselectivities of up to 87% ee, as shown in Scheme 3.3. The reaction began with the conjugate reduction of the enone by an *in situ* generated copper(I) hydride to give the corresponding copper enolate, which subsequently reacted with a π-allyl–palladium complex to form the final product. The optimal conditions using (S)-*t*-Bu-PHOX as chiral ligand were applied to several cyclic enones, and the reaction was shown to be tolerant toward alkyl as well as benzyl groups on the enones.

3.3 Relay Catalysis

The combination of different metal catalysts for catalytic asymmetric reactions is rather limited, owing to the fact that the presence of multiple metal catalysts will lead to the competitive coordination of metals to the chiral ligand used, which makes the chiral environment unpredictable and unsuitable for a given reaction. Nevertheless, there are some examples to elucidate the power of this concept. In 1998, Takahashi and co-workers reported a double asymmetric hydrogenation performed in the presence of both rhodium(I) and ruthenium(II) chiral phosphine complexes.[5] As shown in Scheme 3.4, the domino asymmetric hydrogenation reaction of γ-(acylamino)-γ,δ-unsaturated-β-ketoesters provided quantitatively the corresponding statin analogues as almost single *trans*-diastereomers (de >98%) in the presence of both Rh(I) and Ru(II) chiral catalysts of (S)-BINAP. Remarkably, the major *trans*-diastereomers were reached in excellent enantioselectivities of >95% ee in almost all cases of the substrates studied.

In 2010, Suga *et al.* reported high enantioselectivities of up to 97% ee for the inverse electron-demand 1,3-dipolar cycloaddition between cyclohexyl or butyl vinyl ethers and carbonyl ylides *in situ* generated *via* rhodium-catalysed

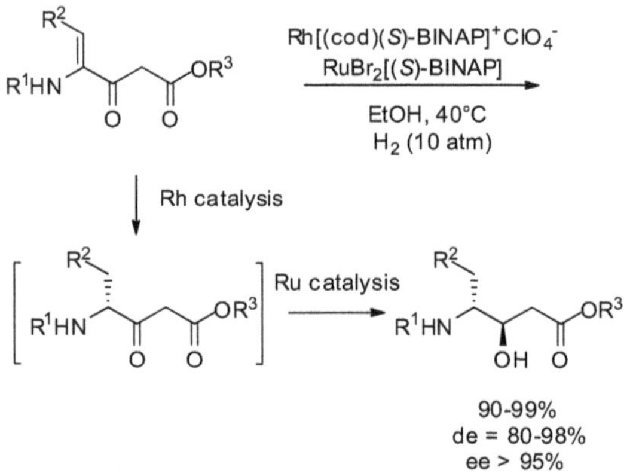

Scheme 3.4   Domino double hydrogenation reaction catalysed by a combination of chiral rhodium catalysis and chiral ruthenium catalysis.

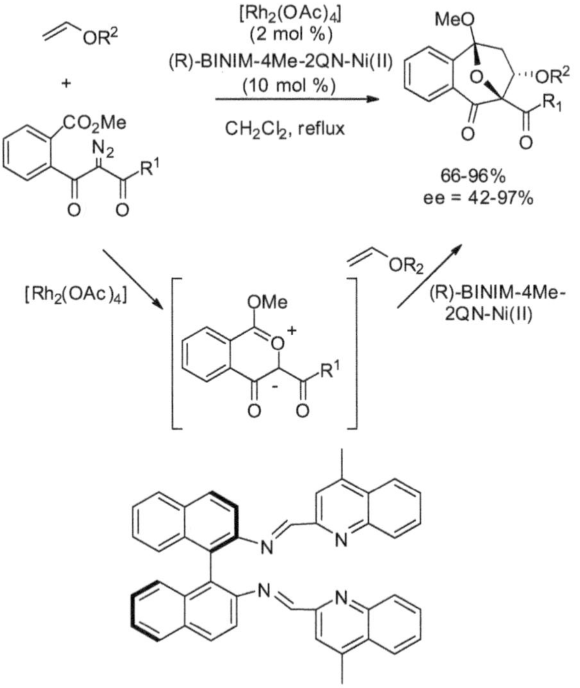

(R)-BINIM-4Me-2QN

Scheme 3.5   Domino carbonyl ylide formation–1,3-dipolar cycloaddition reaction catalysed by a combination of rhodium catalysis and chiral nickel catalysis.

decomposition of *o*-methoxycarbonyl-α,α′-dicarbonyl diazo compounds.[6] As shown in Scheme 3.5, these high levels of asymmetric induction were achieved using BINIM–Ni(II) complexes, previously prepared from nickel triflate and (R)-BINIM-4Me-2QN ligand, as the chiral Lewis acid catalysts in combination with $Rh_2(OAc)_4$. It must be noted that a combination of $Rh_2(OAc)_4$ with chiral Pybox-lanthanoid metals(III) were also shown to be effective to induce this type of tandem reaction, providing enantioselectivities ranging from 60 to 97% ee.

In 2013, Sigman *et al.* reported a highly enantioselective domino Heck–oxidation reaction of acyclic alkenyl alcohols and arylboronic acids induced by a combination of a palladium complex of a chiral pyridine oxazoline with copper(II) triflate under O_2 atmosphere.[7] As depicted in Scheme 3.6, the reaction delivered remotely functionalised arylated carbonyl products with high enantioselectivities of up to 98% ee and generally high $\gamma:\beta$ regioselectivity ratios of up to 98 : 2 combined with moderate to good yields (16–85%). The authors found that the enantioselectivity of the process was essentially independent of the nature of both reaction partners. In contrast, the regioselectivity of the reaction was controlled both by the nature of the arylboronic acid and the substitution and chain length of the alkenyl alcohol. Therefore, a clear trend emerged in terms of site selectivity for alkene insertion: as the alcohol was further removed from the alkene the selectivity was diminished.

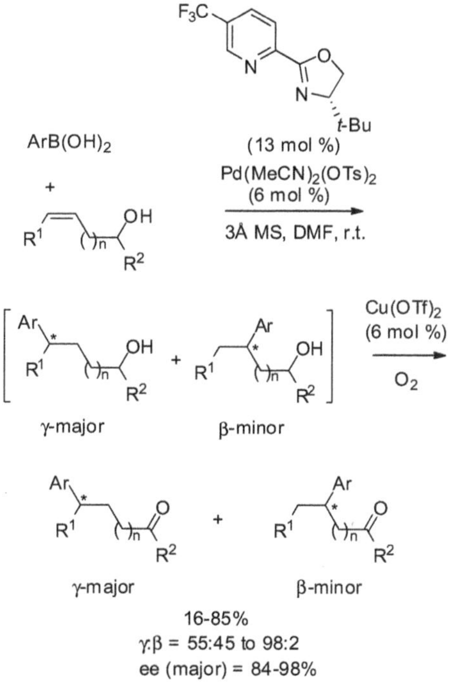

Scheme 3.6 Domino Heck–oxidation reaction catalysed by a combination of chiral palladium catalysis and copper catalysis.

3.4   Sequential Catalysis

In 2002, Hayashi *et al.* described an elegant strategy for the asymmetric synthesis of chiral 1,2-diols from terminal alkynes by use of successive double hydrosilylations with platinum and palladium catalysts in one-pot followed by Tamao oxidation of the crude bis(silyl)ethanes (Scheme 3.7).[8] Because the regioselectivity in the palladium-catalysed hydrosilylation of terminal alkynes was low, platinum was employed as the first hydrosilylation catalyst, generating (*E*)-vinylsilanes which were directly subjected to the subsequent palladium-catalysed asymmetric hydrosilylation. With the platinum catalyst, the second hydrosilylation of vinyl silane intermediates did not occur under the reaction conditions. The asymmetric second hydrosilylation of alkenyl silanes was performed by using a palladium catalyst with (*R*)-2-bis[3,5-bis(trifluoromethyl)phenyl]phosphine-1,1-binaphthyl (L*) with a complete regioselectivity and high enantioselectivity of 95% ee. It must be noted that when the tandem reaction was initially tried with only a single palladium catalyst in the presence of the same chiral ligand, the desired diol product was obtained in low yield (30%) and the same enantioselectivity, demonstrating the beneficial effects from the use of multicatalytic systems.

The same year, Shibasaki *et al.* showed that asymmetric epoxidation of α,β-unsaturated amides by chiral lanthanide catalysts could be efficiently combined with a subsequent palladium-catalysed epoxide opening process in one-pot to give the corresponding β-aryl α-hydroxy amides in excellent

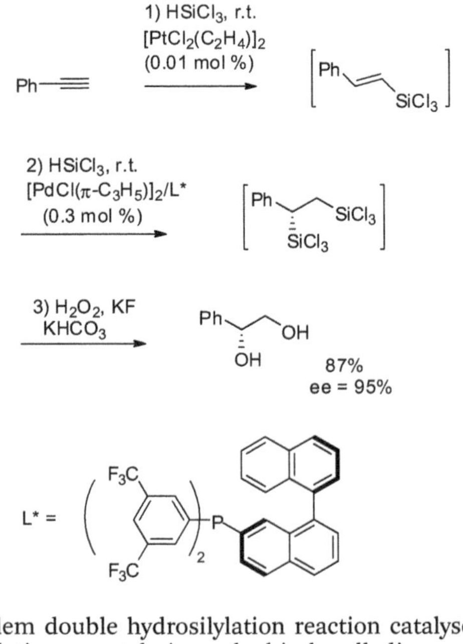

Scheme 3.7   Tandem double hydrosilylation reaction catalysed by a combination of platinum catalysis and chiral palladium catalysis followed by oxidation.

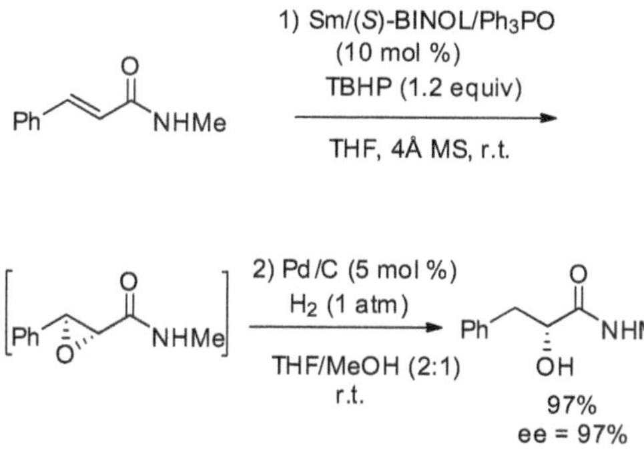

1) Sm/(S)-BINOL/Ph₃PO
(10 mol %)
TBHP (1.2 equiv)

$$\text{Ph} \diagup\diagdown \overset{O}{\underset{}{\bigg\Vert}} \text{NHMe} \xrightarrow[\text{THF, 4Å MS, r.t.}]{}$$

$$\left[\text{Ph} \overset{\cdots}{\underset{O}{\diagup\diagdown}} \overset{O}{\underset{}{\bigg\Vert}} \text{NHMe} \right] \xrightarrow[\substack{\text{THF/MeOH (2:1)} \\ \text{r.t.}}]{\substack{\text{2) Pd/C (5 mol %)} \\ \text{H}_2 \text{ (1 atm)}}} \text{Ph} \overset{O}{\underset{\overset{|}{OH}}{\diagup\diagdown}} \overset{O}{\underset{}{\bigg\Vert}} \text{NHMe}$$

97%
ee = 97%

Scheme 3.8 Tandem epoxidation–ring-opening reaction catalysed by a combination of chiral samarium catalysis and palladium catalysis.

yields and enantioselectivity of up to 97% ee, as shown in Scheme 3.8.[9] The epoxidation of unsaturated amides was readily performed with chiral lanthanide catalysts, such as Sm-(S)-BINOL-Ph₃PO, in the presence of TBHP to afford the corresponding α,β-epoxy amides. Upon completion of the epoxidation reaction, the epoxide opening process was efficiently achieved by using Pd/C catalyst in THF and methanol as co-solvent. It was interesting that selectivity in the epoxide opening was dependent on the components of the first step. When Pd/C was used alone, two side products, dehydroxylated adduct and α-oxoamide moiety were also generated in about 8% from the reaction. In contrast, formation of the undesired molecules was almost completely suppressed when the reaction was carried out in the presence of all of the reagents for the first epoxidation. The authors suggested that beneficial modifications of the palladium catalyst were achieved by the samarium catalyst system of the first epoxidation reaction, producing a more suitable palladium catalytic system for the subsequent epoxide opening step.

In 2004, Feringa *et al.* reported a short, catalytic asymmetric synthesis of (−)-pumiliotoxin C, the key step of which was a tandem asymmetric Michael addition–allylic substitution reaction catalysed by a combination of a copper catalyst of chiral aminophosphine with achiral palladium catalyst Pd(PPh₃)₄ (Scheme 3.9).[10] To prevent possible catalyst interference, the palladium catalyst was added after the completion of the catalytic asymmetric Michael addition of ZnMe₂ to cyclohexenone. The tandem product was achieved in high yield and good diastereoselectivity of 78% de combined with very high enantioselectivity of 96% ee. The advantage of this tandem reaction was demonstrated by the fact that allylic substitution exclusively took place at the 2-position of the 3-methylcyclohexenone, while it is difficult to control the regioselectivity if using enantiopure 3-methylcyclohexanone to prepare the enolate.

In 2006, Trost and co-workers reported a one-pot synthesis of enantiopure *N*- and *O*-heterocyclic compounds using the combination of an achiral ruthenium catalyst and a chiral palladium complex.[11] After the completion of the ruthenium-catalysed alkene–alkyne cross-coupling reaction between the

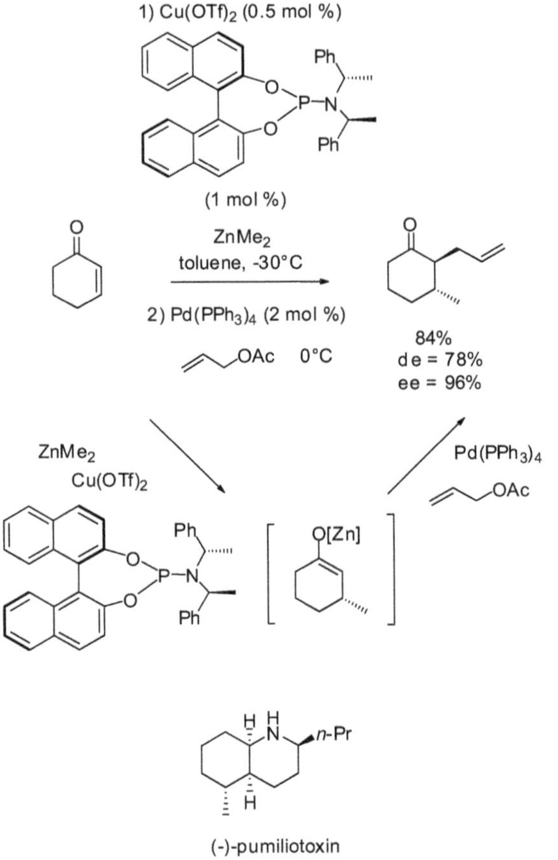

Scheme 3.9 Tandem Michael–allylic substitution reaction catalysed by a combination of chiral copper catalysis and palladium catalysis.

two substrates, the chiral ligand and the palladium catalyst were added to promote the enantioselective intramolecular heterocyclisation reaction, which provided the corresponding chiral functionalised pyrrolidine in high yield and enantioselectivity of 91% ee, as shown in Scheme 3.10. This sequential tandem reaction was employed for the concise synthesis of the B ring of bryostatin, a potent antitumor agent.

In 2006, Nishibayashi *et al.* demonstrated that iridium catalysts can be compatible with ruthenium catalysts in the same medium to promote a tandem asymmetric α-alkylative reduction of ketones, such as acetophenone, with alcohols such as *n*-butanol.[12] This tandem reaction probably began with the catalytic dehydrogenation of *n*-butanol by the achiral iridium complex to form the corresponding aldehyde, which underwent a base-catalysed aldol condensation to afford the corresponding α,β-unsaturated ketone. The following iridium-catalysed hydrogenation gave the corresponding α-alkylated ketone, which was converted into the final chiral

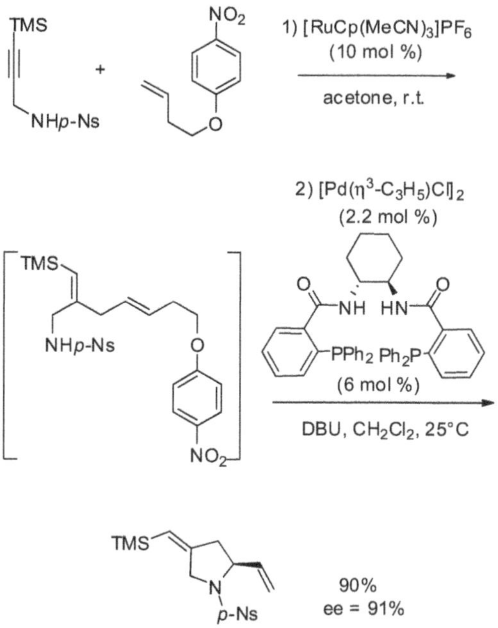

Scheme 3.10 Tandem alkene–alkyne cross-coupling reaction–intramolecular heterocyclisation reaction catalysed by a combination of ruthenium catalysis and chiral palladium catalysis.

alcohol in 75% yield and high enantioselectivity of 94% ee by an enantioselective transfer hydrogenation performed with a chiral ruthenium catalyst (Scheme 3.11). This catalyst and reagents were added after the completion of the α-alkylation of the ketone to avoid the hydrogenation of the substrate ketone.

In 2007, the same authors reported the deracemisation of secondary benzylic alcohols on the basis of a two-step process with the combination of two different chiral ruthenium catalysts (Scheme 3.12).[13] The initial step of this sequential process was a kinetic resolution of the starting secondary alcohol by the selective oxidation of the *S*-alcohol to the corresponding ketone catalysed by a first chiral ruthenium complex. This ketone intermediate was then selectively reduced to the *R*-alcohol by the second chiral ruthenium catalyst. As compared with kinetic resolution, this two chiral ruthenium system provided a convenient and efficient approach for the synthesis of chiral alcohols in high yields and excellent enantioselectivities of up to 92% ee, as shown in Scheme 3.12.

Later, Hajra and Sinha reported an enantioselective tandem aziridoarylation of aryl cinnamyl ethers to give the corresponding *N*-sulfonyl-protected *trans*-3-amino-4-arylchromans using a combination of two copper catalysts.[14] The first step of the sequence involved the asymmetric aziridination of the aryl cinnamyl ether catalysed by a copper catalyst of a chiral

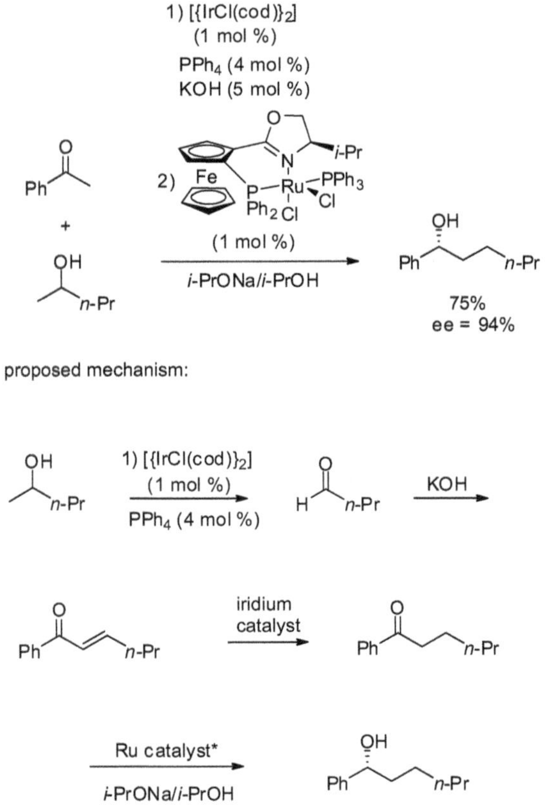

Scheme 3.11   Tandem dehydrogenation–aldol–dehydration–hydrogenation;–dehydrogenation; reaction catalysed by a combination of iridium catalysis
and chiral ruthenium catalysis.

indanolamine-derived Box ligand (Scheme 3.13). The *in situ* generated aziridine was subsequently submitted to an intramolecular arylation by addition
of Cu(OTf)$_2$, which provided the corresponding chiral aminochroman with
high regio-, diastereo-, and enantioselectivities, as shown in Scheme 3.13.
These products constituted key intermediates for the synthesis of biologically active chromenoisoquinoline compounds such as doxanthrine, a potent
and selective full antagonist for the dopamine-D$_1$ receptor.

   It must be noted that a number of enantioselective reactions using two
metal catalysts in which the role of one catalyst consists of activating the
other one are not included in this review.[15]

   In 2013, Fox and co-workers reported an enantioselective synthesis of
cyclobutanes based on a sequential rhodium-catalysed bicyclobutanation–
copper-catalysed homoconjugate addition–enolate trapping reaction.[16] Indeed, a range of enantiomerically enriched functionalised cyclobutanes were
constructed through this one-pot three-component process in which *t*-butyl

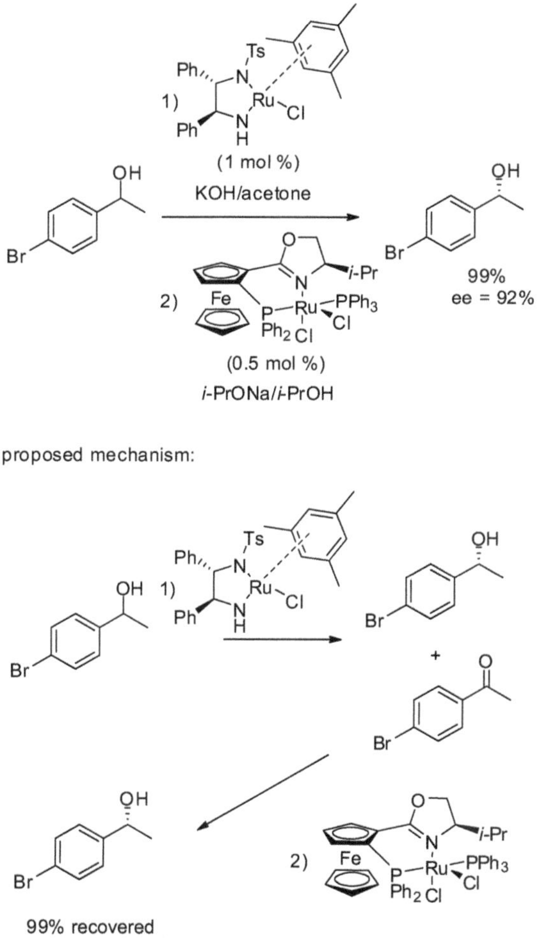

Scheme 3.12 Tandem kinetic resolution–reduction reaction catalysed by two chiral ruthenium catalysts.

(*E*)-2-diazo-5-arylpent-4-enoates were treated with Rh₂(*S*-NTTL)₄ to provide the corresponding chiral bicyclobutanes, which were subsequently submitted to a copper-catalysed Grignard reagent homoconjugate addition–enolate trapping sequence to give the corresponding densely functionalised chiral cyclobutanes with moderate to high diastereoselectivities of up to 94% de (Scheme 3.14). This one-pot reaction was applied to various Grignard reagents, such as phenylmagnesium chloride, phenylmagnesium bromide, methylmagnesium chloride, ethylmagnesium chloride, benzylmagnesium chloride, *p*-fluorophenylmagnesium bromide and *p*-methoxyphenylmagnesium bromide. Moreover, the intermediate enolates could be trapped by water as well as various electrophiles, such as allyliodide, ethyliodide, benzylbromide, diphenyldisulfide, 4-bromobenzoyl chloride.

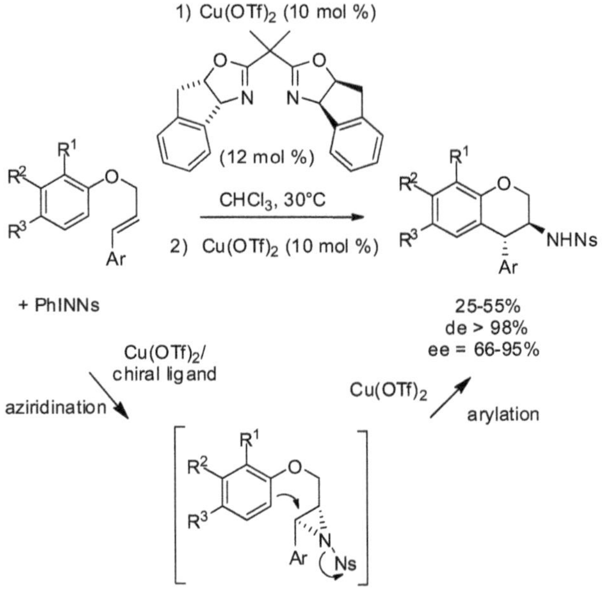

Scheme 3.13 Tandem aziridoarylation reaction catalysed by two copper catalysts.

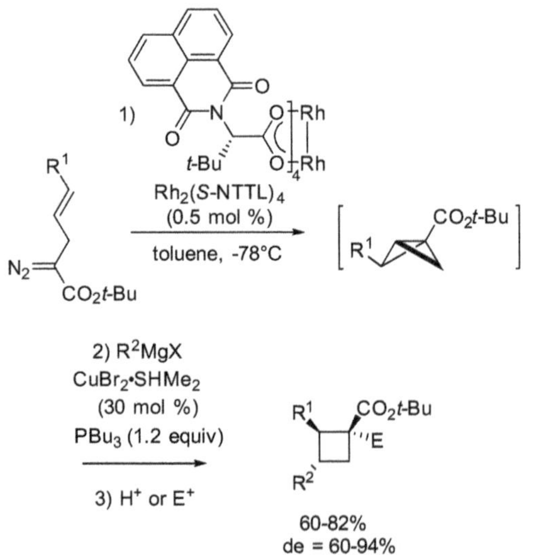

Scheme 3.14 Three-component tandem bicyclobutanation–homoconjugate
addition–enolate trapping reaction catalysed by a combination of
chiral rhodium catalysis and copper catalysis.

References

1. (a) D. E. Fogg and E. N. dos Santos, *Coord. Chem. Rev.*, 2004, **248**, 2365–
   2379; (b) S. Matsunaga and M. Shibasaki, *Bull. Chem. Soc. Jpn.*, 2008, **81**,

60–75; (c) M. Shibasaki, M. Kanai, S. Matsunaga and N. Kumagai, *Acc. Chem. Res.*, 2009, **42**, 1117–1127; (d) M. Shibasaki, M. Kanai, S. Matsunaga and N. Kumagai, *Top. Organomet. Chem.*, 2011, **37**, 1–30; (e) Z.-Y. Han, C. Wang and L.-Z. Gong, in *Science of Synthesis, Asymmetric Organocatalysis*, ed. B. List and K. Maruoka, Georg Thieme Verlag, Stuttgart, 2011, section 2.3.6; (f) J. Park and S. Hong, *Chem. Soc. Rev.*, 2012, **41**, 6931–6943.

2. K. Oisaki, D. Zhao, M. Kanai and M. Shibasaki, *J. Am. Chem. Soc.*, 2007, **129**, 7439–7443.

3. X.-Y. Guan, L.-P. Yang and W. Hu, *Angew. Chem., Int. Ed.*, 2010, **49**, 2190–2192.

4. F. Nahra, Y. Macé, D. Lambin and O. Riant, *Angew. Chem., Int. Ed.*, 2013, **52**, 3208–3212.

5. T. Doi, M. Kokubo, K. Yamamoto and T. Takahashi, *J. Org. Chem.*, 1998, **63**, 428–429.

6. H. Suga, S. Higuchi, M. Ohtsuka, D. Ishimoto, T. Arikawa, Y. Hashimoto, S. Misawa, T. Tsuchida, A. Kakehi and T. Baba, *Tetrahedron*, 2010, **66**, 3070–3089.

7. T.-S. Mei, E. W. Werner, A. J. Burckle and M. S. Sigman, *J. Am. Chem. Soc.*, 2013, **135**, 6830–6833.

8. T. Shimada, K. Mukaide, A. Shinohara, J. W. Han and T. Hayashi, *J. Am. Chem. Soc.*, 2002, **124**, 1584–1585.

9. T. Nemoto, H. Kakei, V. Gnanadesikan, S.-y. Tosaki, T. Oshima and M. Shibasaki, *J. Am. Chem. Soc.*, 2002, **124**, 14544–14545.

10. E. W. Dijk, L. Panella, P. Pinho, R. Naasz, A. Meetsma, A. J. Minaard and B. L. Feringa, *Tetrahedron*, 2004, **60**, 9687–9693.

11. B. M. Trost, M. R. Machacek and B. D. Faulk, *J. Am. Chem. Soc.*, 2006, **128**, 6745–6754.

12. G. Onodera, Y. Nishibayashi and S. Uemura, *Angew. Chem., Int. Ed.*, 2006, **45**, 3819–3822.

13. Y. Shimada, Y. Miyake, H. Matsuzawa and Y. Nishibayashi, *Chem. – Asian J.*, 2007, **2**, 393–396.

14. S. Hajra and D. Sinha, *J. Org. Chem.*, 2011, **76**, 7334–7340.

15. (a) E. M. Vogl, H. Gröger and M. Shibasaki, *Angew. Chem., Int. Ed.*, 1999, **38**, 1570–1577; (b) V. H. Grant and B. Liu, *Tetrahedron Lett.*, 2005, **46**, 1237–1239; (c) K. Ding, *Chem. Commun.*, 2008, 909–921; (d) K. Ishida, H. Kusama and N. Iwasawa, *J. Am. Chem. Soc.*, 2010, **132**, 8842–8843; (e) M. Weber, S. Jautze, W. Frey and R. Peters, *J. Am. Chem. Soc.*, 2010, **132**, 12222–12225.

16. R. Panish, S. R. Chintala, D. T. Boruta, Y. Fang, M. T. Taylor and J. M. Fox, *J. Am. Chem. Soc.*, 2013, **135**, 9283–9286.

CHAPTER 4

Multienzyme-Catalysed Reactions

4.1 Introduction

Enzymes are perfect machines that nature uses to obtain products with excellent enantiopurity. In organic chemistry they can be useful alternatives to newly synthesised catalysts and chemists have studied their applicability in various transformations.[1] The synthetic potential to conduct tandem processes in an asymmetric fashion may conveniently be achieved by making use of the unparalleled chemo-, stereo-, and enantioselectivity of enzymes.[2] Moreover, a highly interesting approach in the application of tandem, domino, and cascade reactions is the use of a multienzyme cocktail to catalyse different reactions under the same mild conditions.[3] Indeed, enzymatic catalysis constitutes a powerful tool for enantioselective green synthesis owing to its high selectivity and non-toxicity. It is now evident that the multienzyme synthesis of natural products has passed from feasibility to practical reality and that there is no limit to the number of enzymes that can be combined in a single reactor to produce a chiral complex structure in good yield and in a one-pot fashion. What is truly remarkable is the lack of product/substrate inhibition, which is probably due to the irreversible nature of many of the later steps in a given sequence. However, owing to the fact that enzymes often have no appropriate active sites for non-natural substrates, the suitable enzyme for each step of a cascade reaction must be carefully optimised. In light of this, there are only limited examples of tandem biocatalysis.

RSC Catalysis Series No. 20
Enantioselective Multicatalysed Tandem Reactions
By Hélène Pellissier
© Hélène Pellissier 2014
Published by the Royal Society of Chemistry, www.rsc.org

4.2   Multienzymatic Synthesis of Chiral Alcohols

For a long time, kinetic resolution of alcohols *via* enantioselective oxidation or *via* acyl transfer employing, for example, lipases along with dynamic kinetic resolution have been the biocatalytic methods of choice for the preparation of chiral alcohols. In recent years, however, impressive progress has been made in the use of alcohol dehydrogenases[4] (ADHs) and ketoreductases (KREDs) for the asymmetric synthesis of alcohols by stereoselective reduction of the corresponding ketones. Furthermore, recent remarkable multienzymatic systems have been successfully applied to the deracemisation of alcohols *via* stereoinversion based on an enantioselective oxidation followed by an asymmetric reduction.

4.2.1   Deracemisation of Alcohols *via* Stereoinversion

The specificity of oxidoreductases towards certain substrates and cofactors allows the possibility of combining reduction and oxidation steps in a one-pot process to be achieved, which is not feasible using classical redox chemistry. Therefore, two stereocomplementary alcohol dehydrogenases (ADHs) can be combined to achieve deracemisation and stereoinversion of alcohols *via* oxidation of one substrate enantiomer to the corresponding ketone and stereoselective reduction of the latter to the alcohol enantiomer of opposite configuration.[5] It has been known for more than 10 years that various microorganisms possessing stereocomplementary *sec*-ADHs are able to perform these transformations, but only recently have they been achieved employing *in vitro* multienzyme systems. A simple combination of two stereocomplementary ADHs in one pot with a cofactor cycling between them does not suffice since it allows equilibration to the thermodynamically most stable product, which is the racemic alcohol. However, decoupling of the cofactor regeneration for the two enzymes provides control over the oxidation–reduction equilibria and therefore over the stereochemical outcome of the process. In 2008, Kroutil *et al.* reported the first example of deracemisation of secondary alcohols through a concurrent tandem biooxidation and bioreduction.[6] They observed that the use of *Alcaligenes faecalis* DSM 13975 as catalyst could use oxygen as oxidant to selectively oxidise the *R*-enantiomer of a racemic secondary alcohol, such as 2-octanol, to form the corresponding intermediate ketone. The combination of this powerful *R*-enantioselective biooxidation reaction with *S*-enantioselective bioreduction of this intermediate ketone catalysed by *S*-selective alcohol dehydrogenase ADA-"A" from *Rhodococcus rubber* DSM 44541 with cofactor recycling provided an easy method for the synthesis of the corresponding enantiomerically pure *S*-2-octanol, which was achieved in 99% yield and 99% ee (Scheme 4.1). This tandem biocatalytic deracemisation of secondary alcohols could be performed on a preparative scale, for example, 0.5 mL of racemic 4-phenyl-2-butanol could be transformed into pure (*S*)-4-phenyl-2-butanol in 91% yield and enantioselectivity of >99% ee. However, while

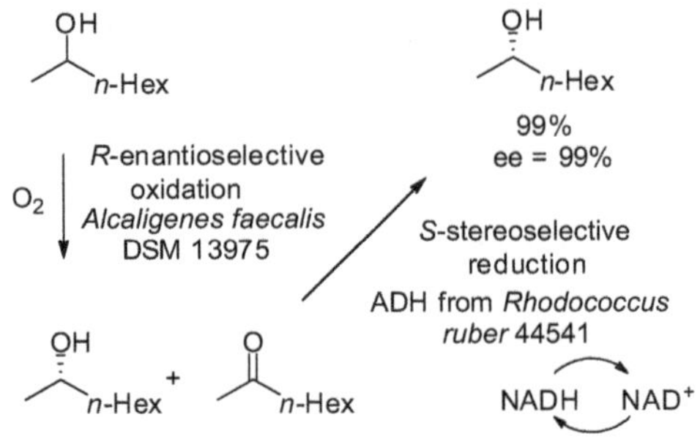

Scheme 4.1 Synthesis of (*S*)-2-octanol.

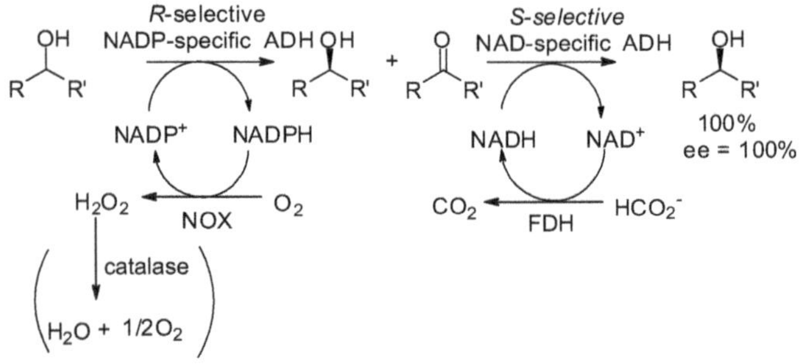

Scheme 4.2 Synthesis of chiral alcohols catalysed by two ADHs.

excellent enantioselectivities were reached in this study using *Prelog* ADHs for the reduction step, the *anti-Prelog* system led to only moderate enantio-meric enrichment. This limitation was overcome by the same authors in a second study using an even more complex multienzyme system.[7]

Control over the reaction equilibria was achieved by using two ADHs with different cofactor specificity (NADH and NADPH) and two independent cofactor regeneration systems (one for NADPH oxidation and the other for NAD^+ reduction). Thus, two ADHs complementary with respect to stereo-selectivity and cofactor specificity were combined with an NADPH-oxidase from *Bacillus subtilis* and an (NAD^+-specific) formate dehydrogenase in one pot without any compartmentalisation (Scheme 4.2). Applying this protocol, different racemic secondary alcohols were transformed into the corres-ponding enantiomerically pure stereoisomers in quantitative yields. A similar system employing whole microbial cells was developed by Xu *et al.*, in 2010.[8] In this study, resting cells of *Microbacterium oxydans* ECU2010 possessing an NAD^+-dependent (*R*)-selective alcohol dehydrogenase and *Rhodotorula* sp. AS2.2241 expressing an NADPH-dependent (*S*)-selective ketoreductase were combined in a one-pot process. The reported yields (43–84%) were lower than those using isolated enzymes, however, the enantioselectivities were excellent (ee >99%) and the use of whole cells rendered external cofactor recycling unnecessary.

4.2.2 Multienzymatic Cascade Reactions

Due to their high stereoselectivity, alcohol dehydrogenases are attractive biocatalysts of processes involving optically active intermediates which are further converted in a linear cascade fashion. As an example, α-chloro ketones may be stereoselectively reduced to the corresponding chloro alcohols by an ADH, followed by ring-closure to yield optically active epoxides. Considerable effort has been made to combine the chloro ketone reduction and epoxide formation in a one-pot process. In the simplest case, ring-closure is achieved by addition of base, but it can also be achieved enzymatically using halohydrin dehalogenases. Thus, enantiomerically enriched epoxides can be obtained from prochiral halo ketones through a two-enzyme cascade. The practicability of this concept was demonstrated by Seisser *et al.* using ADHs from *Rhodococcus brevis* as well as a non-stereoselective halohydrin dehalogenase from *Mycobacterium* sp. GP1.[9] Unfortunately, the unfavourable equilibrium of the ring-closure reaction limited the yield in this biocatalytic cascade. This limitation could be overcome in a following study and therefore drive the reaction to completion.[10] On the other hand, a combination of alcohol dehydrogenases with halohydrin dehalogenases was used by Sheldon *et al.* to prepare enantiopure β-hydroxy nitriles starting from the corresponding α-chloro ketones on the basis of a biocatalytic cascade incorporating three enzymatic processes induced by two enzymes.[11] In the first reaction, the ketone was stereoselectively reduced to the corresponding β-halo alcohol by an alcohol dehydrogenase. In this process, NADPH was used as a cofactor. The recycling of NADPH was provided by the same enzyme in a reaction which converted isopropyl alcohol into acetone. The β-halo alcohol served as a substrate for halohydrin dehalogenase, which catalysed the ring-closure towards the corresponding epoxide, and subsequent ring-opening of this epoxide by cyanide anion, forming an enantiopure β-hydroxy nitrile, which was the final product of the biocatalytic cascade. In addition, these authors applied this cascade methodology to another nucleophilic epoxide ring-opening step using azides as nucleophiles, producing enantiopure β-azido alcohols from the corresponding chloro ketones in a three-step, two-enzyme, and one-pot process (Scheme 4.3).[12]

An alternative approach towards β-hydroxy nitriles is the direct reduction of the corresponding keto nitriles, and this reaction can be coupled with nitrilase-catalysed hydrolysis to give β-hydroxy acids as the final products. Thus, employing a ketone reductase from *Candida magnolia* or an ADH from *Saccharomyces cerevisiae* as well as nitrilases from *Synechocystis* sp. and *Bradyrhizobium japonicum*, in addition to glucose dehydrogenase (GDH), both enantiomers of 1-hydroxy-1-phenylpropanoic acid and some related compounds have been synthesised in enantiopure form with yields of up to 27%, higher than those obtained through a sequential reaction (Scheme 4.4)[13] Previously, Yamada and co-workers have reported the combination of glucose dehydrogenase and an aldehyde dehydrogenase.[14] Indeed, the combination of these two-enzymatic actions in one pot has

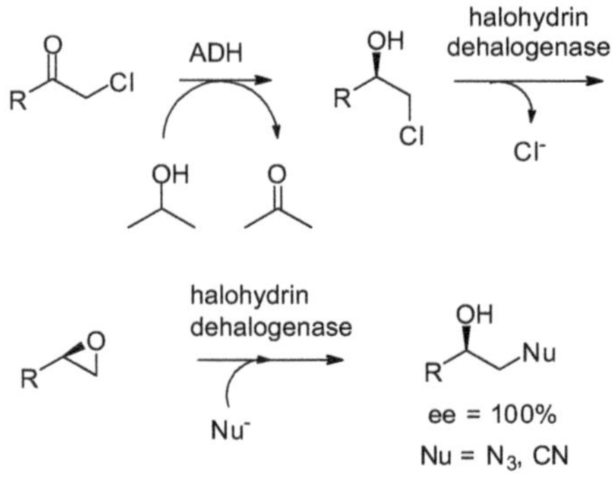

Scheme 4.3   Synthesis of β-azido alcohols and β-hydroxy nitriles catalysed by two enzymes.

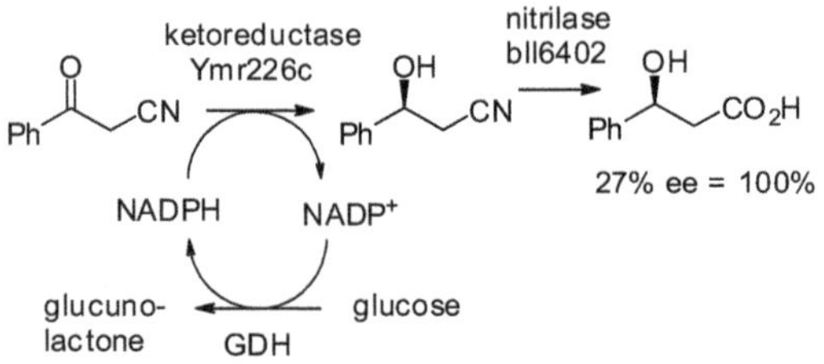

Scheme 4.4   Synthesis of β-hydroxy acids catalysed by three enzymes.

allowed the enantioselective reduction of acetoacetate ester into (*R*)-3-hydroxybutanoate ester, using D-glucose as the reducing agent.

4.3   Multienzymatic Synthesis of Chiral Amines and Amino Acids

4.3.1   Synthesis of Chiral Amines with ω-Transaminases

While the preparation of optically active alcohols using lipases or alcohol dehydrogenases can be considered as an established technology, the biocatalytic asymmetric synthesis of chiral amines has long been problematic. Only recently, the development of efficient cascade systems involving α- and ω-transaminases has provided an efficient biocatalytic entry to α-chiral primary amines, many of which are important for the pharmaceutical industry.[15] Transaminases are often classified into two groups according to their substrate scope; α-transaminases which act on the α-amino group in amino acids, and ω-transaminases which can convert amines that lack a vicinal carboxylic acid functionality. A difficulty in the application of both classes lies in the fact that the amino group transfer is usually reversible, resulting in incomplete conversion. Therefore, transaminases have been a target for the development of cascade processes which allow for shifting the equilibrium to the product side. As an example, Kroutil and co-workers have demonstrated that kinetic resolution and asymmetric synthesis using ω-transaminases of opposite stereopreference could be combined in order to establish a biocatalytic deracemisation system for secondary amines (Scheme 4.5).[16] However, interference of the two stereocomplementary transaminases was observed when the reaction was carried out in a one-pot one-step fashion, leading to decreased enantioselectivity. This problem was overcome by deactivating the first transaminase *via* heat treatment after the kinetic resolution was complete and carrying out the asymmetric synthesis afterwards. Alternatively, the use of immobilised enzymes allowed separation by filtration. This concept has been applied to a variety of α-chiral primary amines, including the pharmaceutically relevant mexiletine which was produced in enantioselectivity of >99% ee and conversion of 55%.[17]

Interestingly, α- and ω-transaminases have been coupled in a parallel cascade system for the concurrent production of (*S*)-amino acids and (*R*)-amines.[18] In this approach, amination of a keto acid with a suitable amino donor afforded the corresponding amino acid and a keto acid by-product, which served as the amino acceptor in the transaminase-catalysed kinetic resolution of an amine (Scheme 4.6). As an example, (*S*)-α-aminobutyrate

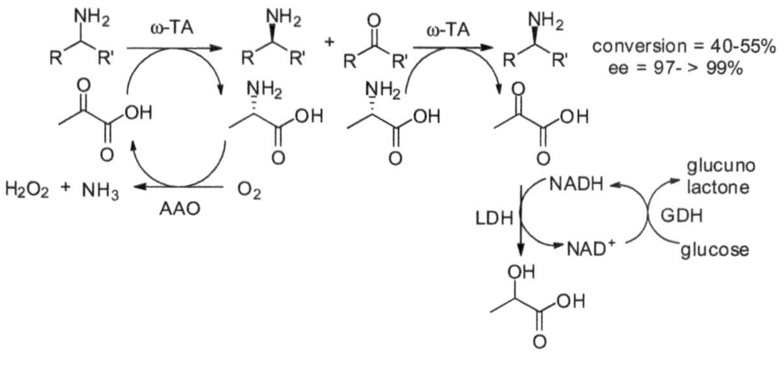

Scheme 4.5 Deracemisation of amines catalysed by five enzymes.

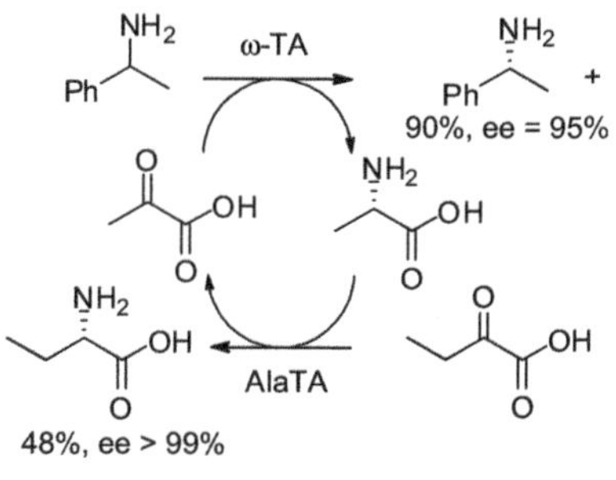

Scheme 4.6 Kinetic resolution of α-methylbenzylamine and synthesis of (*S*)-α-aminobutyric acid catalysed by two enzymes.

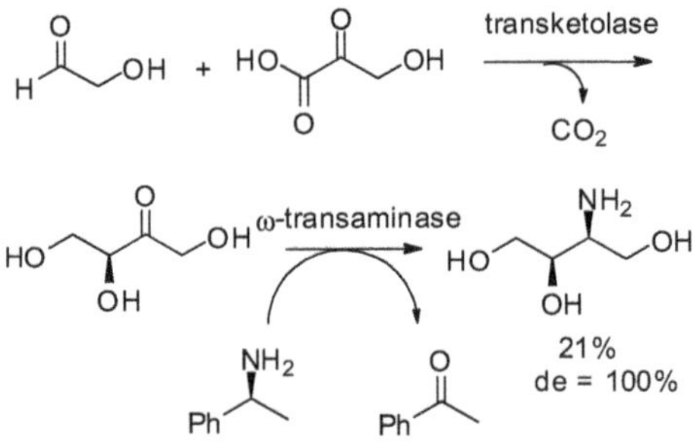

Scheme 4.7 Synthesis of chiral 2-amino-1,3,4-butanetriol catalysed by two enzymes.

and (S)-α-methylbenzylamine could be obtained concurrently on the basis of this strategy, allowing conversions of 48% associated with an enantioselectivity of >99% ee, and 90% combined with an enantioselectivity of 95% ee, to be achieved, respectively. Alanine served as an amino group "shuttle" in this reaction system. However, since equimolar amounts of the substrates, keto acid and racemic amine were used in this study, the cascade was not catalytic in alanine, and indeed one equivalent was added.

On the other hand, a combination of a transketolase from *E. coli* and a transaminase from *Pseudomonas aeruginosa* PAO2 overexpressed in a single host was used for the biocatalytic synthesis of enantiopure 2-amino-1,3,4-butanetriol from hydroxypyruvate, glycolaldehyde and α-methylbenzylamine as an amino donor (Scheme 4.7).[19] Thus, two stereogenic centres were generated *via* consecutive C–C bond formation and transamination in a one-pot process, giving exclusively the enantiopure *syn*-diastereomer in 21% yield.

4.3.2 Synthesis of Chiral Amino Acids with α-Transaminases

On the other hand, α-transaminases have been used extensively in the production of amino acids through kinetic resolution and asymmetric synthesis. While many studies rely on the use of an excess of cosubstrate to drive the reaction to completion, some multienzymatic approaches have been developed as well. As an example, aspartate has been used as an amino donor in a multienzymatic synthesis of L-2-aminobutyrate from L-threonine (Scheme 4.8).[20] The rather complex multistep sequence started with the *in situ* formation of 2-ketobutyrate from L-threonine catalysed by threonine deaminase (ThrDA) from *E. coli*. A tyrosine transaminase (TyrAT) from *E. coli* converted 2-ketobutyrate and L-aspartic acid to L-2-aminobutyrate and oxaloacetate, which spontaneously decarboxylated to give pyruvate. Since the

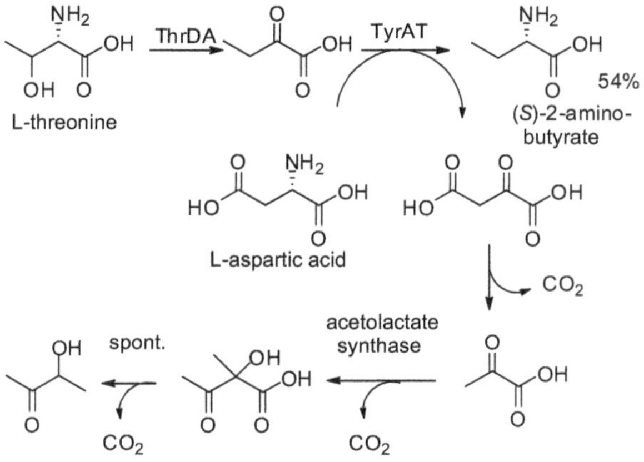

Scheme 4.8 Synthesis of (S)-2-aminobutyrate catalysed by three enzymes.

latter compound could be accepted as a substrate by the transaminase, which would result in incomplete conversion, a third enzyme was employed for pyruvate removal. Acetolactate synthase from *Bacillus subtilis* consumed two molecules of pyruvate to form acetolactate, which again spontaneously decarboxylated to give acetoin as the final product. All the three enzymes were expressed separately in *E. coli* and used in the form of fresh whole cells, allowing simple adjustment of the appropriate relative activities *via* the cell mass. Full conversion was achieved within 24 h at a 500 mM scale; however, the recovery of (S)-2-aminobutyrate was only 54% because of metabolic consumption by the living microorganisms.

Interestingly, α-transaminases have also been coupled with amino acid dehydrogenases in a linear way, achieving a system for deracemisation of amino acids.[21] Branched amino acid transaminase (BCAAT) from *Sinorhizobium meliloti* ATCC 51124 was cloned and overexpressed in *E. coli*. Additionally, D-amino acid dehydrogenase DadA present in the host organism was induced by adding L-alanine to the growth medium, thus giving a whole-cell biocatalyst capable of deracemising α-amino acids by an oxidation–reduction sequence (Scheme 4.9). The use of resting whole cells in this study assured a supply of the needed cofactors and cosubstrates for the reactions. L-4-Chlorophenylalanine was obtained from its racemate in enantiomerically pure form and quantitative yield within 48 h at a 0.5 g L^{-1} scale. A very similar concept has been applied for the deracemisation of naphthylalanine to the corresponding L-enantiomer.[22] In this study, the isolated enzymes D-amino acid oxidase (D-AAO) from *Rhodotorula gracilis* and L-aspartate transaminase (L-AspAT) from *E. coli* were used. Cysteinesulfinic acid served as the amino donor for the transamination step, assuring irreversibility through spontaneous decomposition of the β-keto sulfinic acid by-product (Scheme 4.9).

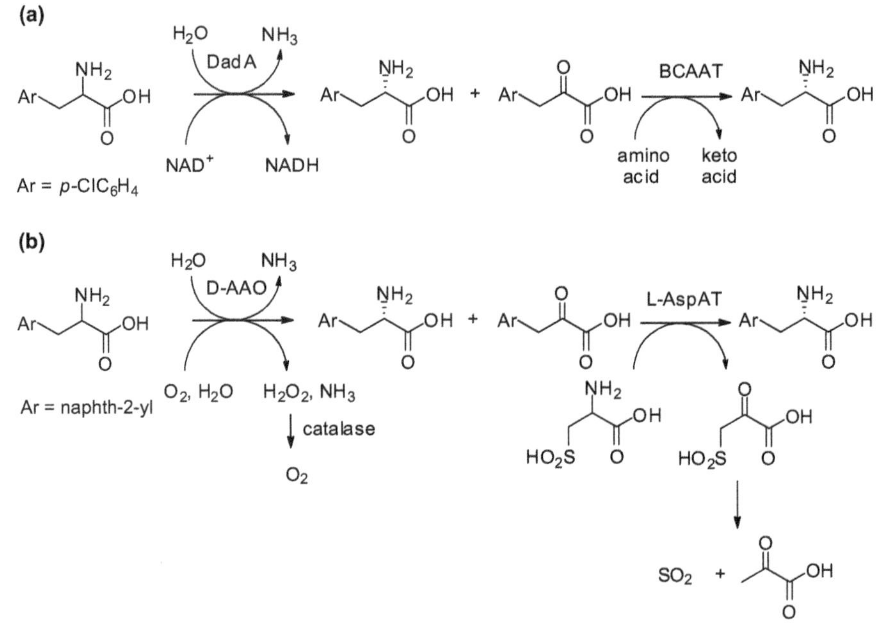

Scheme 4.9   Deracemisations of α-amino acids using (a) a D-selective amino acid dehydrogenase and an L-selective α-transaminase, or (b) a D-selective amino acid oxidase and an L-selective α-transaminase.

4.3.3   Synthesis of Chiral Amino Acids with Other Enzymes

In addition to α-transaminases, amino acid dehydrogenases (AADHs) are also traditionally employed for the production of chiral amino acids. Amino acid dehydrogenases (AADHs) are $NAD(P)^+$-dependent enzymes which convert amino acids into the corresponding keto acids and ammonia. Since this reaction is reversible, they are also useful biocatalysts for the asymmetric synthesis of amino acids *via* reductive amination. Early studies employing, for instance, phenylalanine dehydrogenase were limited by the narrow substrate specificity of the enzyme. Leucine dehydrogenase offers a broader substrate scope and several aliphatic L-amino acids have been obtained in enantiomerically pure form using this enzyme in combination with FDH for cofactor recycling. As an alternative, internal cofactor regeneration may be established *via* enzymatic oxidation of a α-hydroxy acid, thus providing both the substrate and the reduced nicotinamide cofactor for the AADH (Scheme 4.10).[23] Recently, this redox-neutral cascade has been extended by employing mandelate racemase in combination with D-mandelate dehydrogenase (D-MDH) and different AADHs for the conversion of racemic mandelic acid into enantiopure L-phenylglycine in a novel type of "dynamic kinetic asymmetric transformation" system.[24] As shown in Scheme 4.10, employing three enzymes, such as mandelate racemase, mandelate dehydrogenase (D-MDH), and amino acid dehydrogenase (L-AADH), the cascade

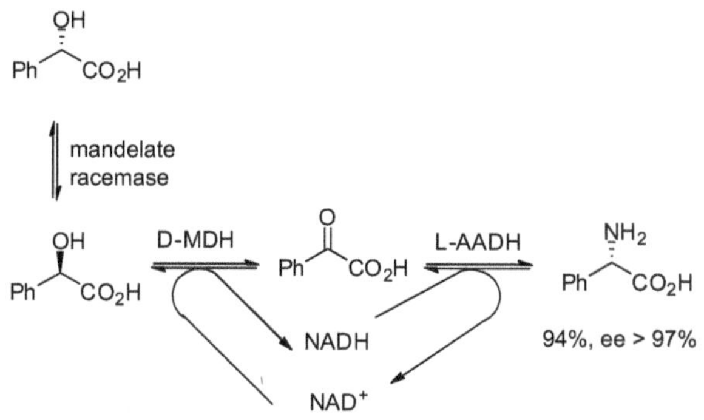

Scheme 4.10   Synthesis of L-phenylglycine catalysed by three enzymes.

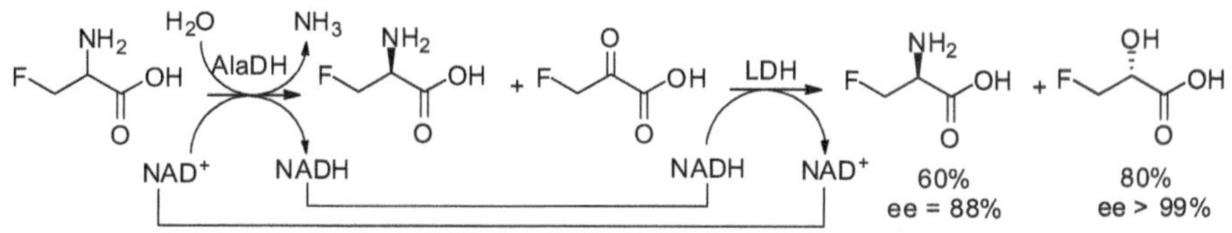

Scheme 4.11   Kinetic resolution of 3-fluoroalanine using alanine dehydrogenase and lactate dehydrogenase.

began with the interconversion of the substrate enantiomers by the enzyme mandelate racemase. Then, D-mandelic acid was oxidised by the D-selective mandelate dehydrogenase to give the corresponding α-oxo acid, consuming NAD^+ and giving NADH. The following step consisted of the reductive amination of this α-oxo acid by an L-selective amino acid dehydrogenase to give the corresponding α-amino acid. As shown in Scheme 4.10, L-phenylglycine was achieved in high yield of up to 94% and excellent enantioselectivity of >97% ee.

Alanine dehydrogenase (AlaDH) has been coupled with lactate dehydrogenase (LDH) for the kinetic resolution of 3-fluoroalanine to the D-enantiomer and L-3-fluorolactic acid (Scheme 4.11).[25] This approach represented the inverse counterpart to the LDH–AlaDH-cascade for the asymmetric synthesis of alanine from lactate as described above. As in the reverse system, internal cofactor regeneration was achieved, resulting in a redox-neutral process. D-3-Fluoroalanine and L-3-fluorolactic acid were obtained in 60% (ee = 88%) and 80% (ee > 99%) yields, respectively.

Other combinations of two dehydrogenases, such as amino acid dehydrogenases and formate dehydrogenase, were described by several groups to achieve the reductive amination of α-keto acids into chiral amino acids in high yields and enantioselectivities.[26] Formate dehydrogenase has also been combined by Bolte and co-workers with an amino acid aminotransferase to achieve enantioselective reductive amination of α-keto acids with small amino acids as the amination agent.[27]

On the other hand, in a very elegant approach, deracemisation of amino acids was achieved by combining D-amino acid oxidase (D-AAO) for enantioselective oxidation and leucine dehydrogenase (LeuDH) for asymmetric reductive amination.[28] Cofactor regeneration for the latter enzyme was supplied by formate dehydrogenase (FDH). Furthermore, catalase was added for hydrogen peroxide degradation in order to prevent oxidation enzyme inactivation (Scheme 4.12). Various α-amino acids were deracemised using this process, affording the L-enantiomers in yields higher than 95% and enantioselectivities higher than 99% ee. On the other hand, Sheldon *et al.* have reported enantioselectivities of 100% ee in a one-pot synthesis of (*S*)-α-*N*-acylamino esters from the corresponding α-amino esters by using a combination of a racemase and a lipase.[29]

In 2010, Janssen and co-workers reported that the kinetic resolution of β-phenylalanine catalysed by a tandem biocatalytic system composed of phenylalanine aminomutase (PAM) and phenylalanine ammonia lyase (PAL) yielded the corresponding enantiopure (*S*)-β-phenylalanine in good yield (48%) and excellent enantiomeric excess of >99% ee (Scheme 4.13).[30] The process was based upon the PAM-catalysed, reversible, enantioselective transformation of (*R*)-β-phenylalanine to (*S*)-α-phenylalanine. The latter one was transformed in a PAL-catalysed regioselective process into (*E*)-cinnamic acid, with liberation of ammonia. This constituted an example of a tandem biocatalytic, kinetic resolution in which one enzyme catalysed the equilibration between the substrate and reaction intermediate, while the other shifted this equilibrium between the substrate towards the final product.

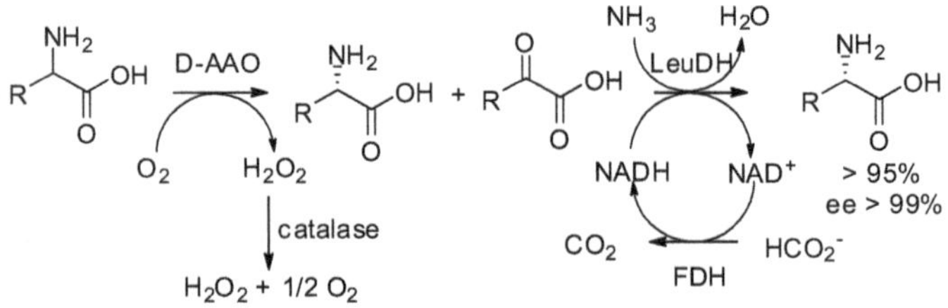

Scheme 4.12 Deracemisation of amino acids using four enzymes.

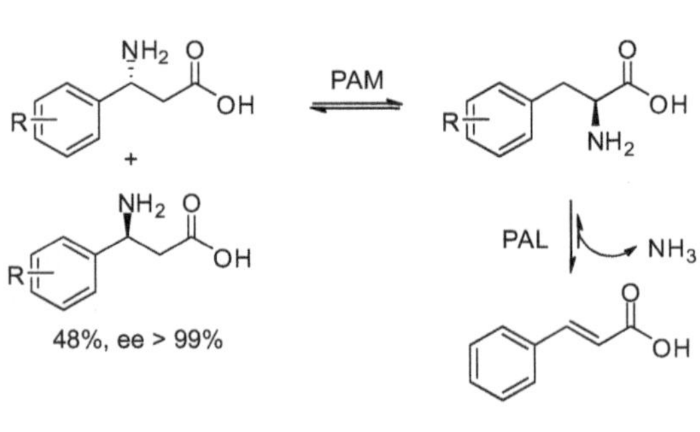

Scheme 4.13 Deracemisation of β-phenylalanine using two enzymes.

4.4 Other Multienzymatic Reactions

The concept of coupling the synthetic power of carbon–carbon bond-forming enzymes with other enzymes in a single reactor truly imitates nature in generating complex chiral molecules from simple substrates. Aldolases are the most widely applied C–C bond-forming enzymes in organic synthesis. They catalyse an aldol-type condensation reaction of a donor compound to an acceptor compound, leading to the formation of one or two novel stereogenic centres. They constitute an attractive alternative to chemical methods as they do not require protecting groups. Most of them show high specificity towards a donor substrate (usually a ketone) but tolerate a broad range of acceptor substrates. Aldolases are classified according to the nature of donor substrates. *N*-Acetylneuraminic acid aldolases (NeuAc aldolases) use pyruvate as the donor substrate and catalyse the aldol reaction between pyruvate and mannose or mannose derivatives. They have gained a lot of interest as the produced sialic acids find use in cancer therapy and as anti-infectives. The reaction is reversible with an equilibrium constant close to one. The equilibrium can be pulled towards aldol formation by adding an excess of pyruvate, but the latter seriously interferes with product isolation. A possible solution is the application of multienzyme one-pot reactions.[31] In this context, Ichikawa *et al.* have developed the synthesis of a sialyl trisaccharide through a nine-enzyme one-pot procedure starting from *N*-acetylmannosamine.[32] The reaction started with the NeuAc aldolase-mediated conversion of *N*-acetylmannosamine to *N*-acetylneuraminic acid in the presence of pyruvic acid (Scheme 4.14a). The produced *N*-acetylneuraminic acid then reacted to cytidine 5′-monophospho-*N*-acetyl-neuraminic acid synthetase (CMP-sialic acid synthetase), thus impeding the aldolase back reaction. Further enzymatic steps led to the formation of the sialyl trisaccharide *N*-acetylid-α-(2,6)-galactose-β-(1,4)-*N*-acetylglucosamine in an overall yield of 22%. Yu *et al.* have investigated three different CMP-sialyl acid synthetases with different substrates and could synthesise various CMP-sialyl acid derivatives.[33] In a similar approach, Cao *et al.* have more recently used C-5-hydroxy-substituted mannose derivatives to produce a variety of C-5-substituted sialosides employing a CMP-sialic acid synthetase and a sialyl-transferase.[34] Taking advantage of the two-way reaction of NeuAc aldolases, Miyazaki *et al.* have developed an elegant [3-^{13}C]-labelling method for *N*-acetylneuraminic acid analogues (Scheme 4.14b).[35] In a first step, *N*-acetylneuraminic acid analogues were degraded to *N*-acetylmannosamine and pyruvate by NeuAc aldolase. The reaction equilibrium was shifted towards degradation by addition of lactate dehydrogenase which removed pyruvic acid from the reaction mixture. After the reaction was completed, the cofactor for lactate dehydrogenase was degraded by the addition of nucleotide pyrophosphatase. Then, [3-^{13}C]-pyruvic acid was added and incorporated into *N*-acetylneuraminic acid by the NeuAc aldolase. This one-pot method allowed [3-^{13}C]-labelling in satisfactory yield (46–76%) and with an excellent degree of ^{13}C-enrichment. It must be noted that

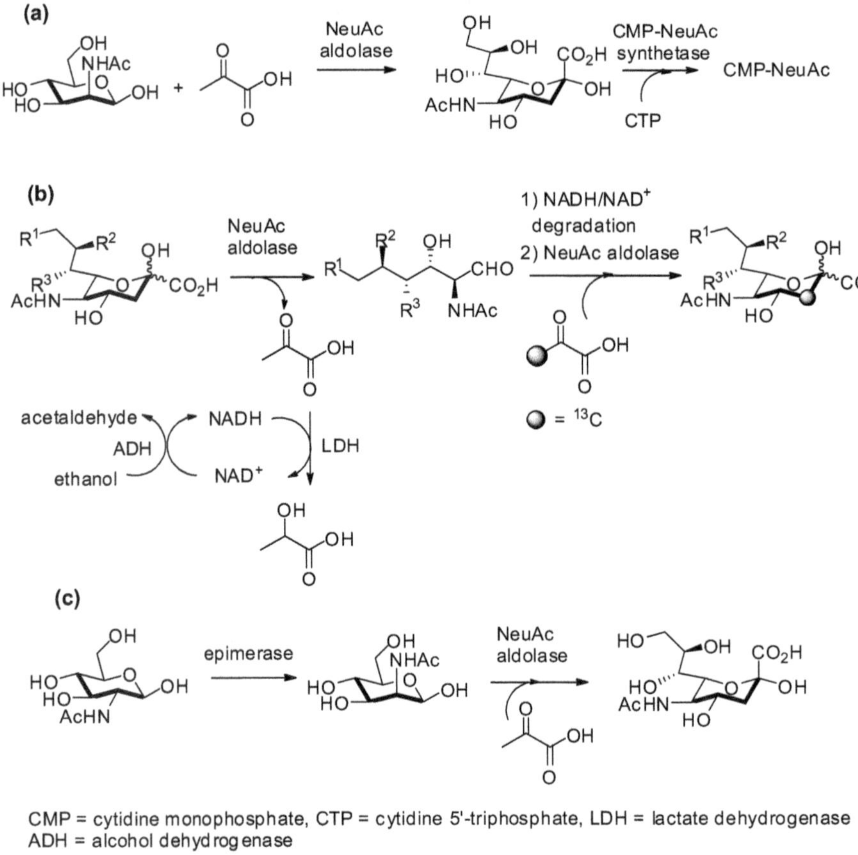

CMP = cytidine monophosphate, CTP = cytidine 5'-triphosphate, LDH = lactate dehydrogenase
ADH = alcohol dehydrogenase

Scheme 4.14 Multienzymatic reactions with NeuAc aldolase.

N-acetylmannosamine, the substrate for NeuAc aldolases, is rather expensive and difficult to prepare on a large scale. A possibility to circumvent this is the *in situ* formation of *N*-acetylmannosamine from *N*-acetylglucosamine by *N*-acetylglucosamine 2-epimerase (Scheme 4.14c). In this context, Kragl *et al.* have achieved an enzymatic membrane reactor containing free *N*-acetylglucosamine 2-epimerase and NeuAc aldolase.[36] After an initiation phase, *N*-acetylneuraminic acid could be produced continuously over several hours with a space–time yield of 109 g L^{-1} d^{-1}. In 2009, *N*-acetyl-D -gluco-samine 2-epimerase and *N*-acetyl-D-neuraminic acid aldolase were over-expressed as double-tagged gene fusions, thus greatly facilitating enzyme isolation.[37]

2-Deoxy-D-ribose 5-phosphate aldolases (DERAs) belong to the class of acetaldehyde-dependent aldolases. In contrast to dihydroxyacetone phos-phate (DHAP) aldolases, the donor substrate specificity is not as strict, allowing for chain elongation by two or three carbon atoms. DERA aldolases have been successfully applied to the production of nucleosides. Horinouchi

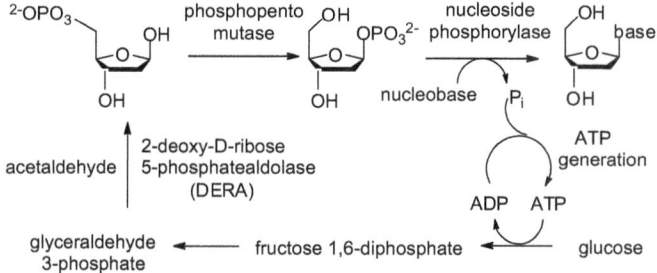

Scheme 4.15 Synthesis of 2′-deoxyribonucleosides catalysed by three enzymes.

et al. have described the production of 2′-deoxyribonucleosides from glucose, acetaldehyde and a nucleobase through a multienzymatic sequential reaction (Scheme 4.15).[38] They used permeabilised yeast cells in a first step to produce D-glyceraldehyde 3-phosphate from glucose through the glycolytic pathway. In a second step, condensation of D-glyceraldehyde 3-phosphate and acetaldehyde catalysed by a recombinant DERA aldolase from *Klebsiella pneumonia* led to the formation of 2-deoxyribose 5-phosphate. In a third step, a recombinant phosphopentomutase catalysed the isomerisation to 2-deoxyribose 1-phosphate. In a last step, a commercial nucleoside phosphorylase catalysed the nucleobase transfer yielding final 2′-deoxynucleosides. The overall yield of this process was, however, low (18% for example for the synthesis of 2′-deoxyinosine from glucose, acetaldehyde and adenine) due to inhibition effects of phosphate and the phosphorylated glycolysis intermediates. Coupling the different reaction steps in a one-pot cascade allowed a 3-fold increase of 2′-deoxyribonucleoside production to be achieved.[39]

In 2007, Griengl *et al.* developed the synthesis of chiral aromatic 1,2-amino alcohols on the basis of a bienzymatic dynamic kinetic asymmetric transformation process.[40] The reaction occurred between a benzaldehyde derivative and glycine in the presence of L-threonine aldolase from *Pseudomonas putida* and L-tyrosine decarboxylase from either *Enterococcus faecalis* or two genes from *Enterococcus faecium*. The best results were obtained for the production of (*S*)-octopamine (99%, ee = 81%), and (*S*)-noradrenaline (76%, ee = 79%), as shown in Scheme 4.16.

In 1995, Gijsen and Wong reported a domino aldol reaction catalysed by the aldolases, 2-deoxyribose 5-phosphate aldolase (DERA) and fructose 1,6-diphosphate aldolase (RAMA). This multienzyme system was used to induce a ternary crossed aldol reaction between a α-substituted acetaldehyde derivative, acetaldehyde, and dihydroxyacetone phosphate.[41] At the same time, these authors developed an enzymatic synthesis of enantiomerically pure L-fructose from dihydroxyacetone phosphate (DHAP) and L-glyceraldehyde, carried out by a multienzyme system comprising rhamnulose-1-phosphate aldolase (RhaD) and acid phosphatase (AP), using a stereospecific aldol addition reaction by this aldolase.[42] This latter methodology suffered,

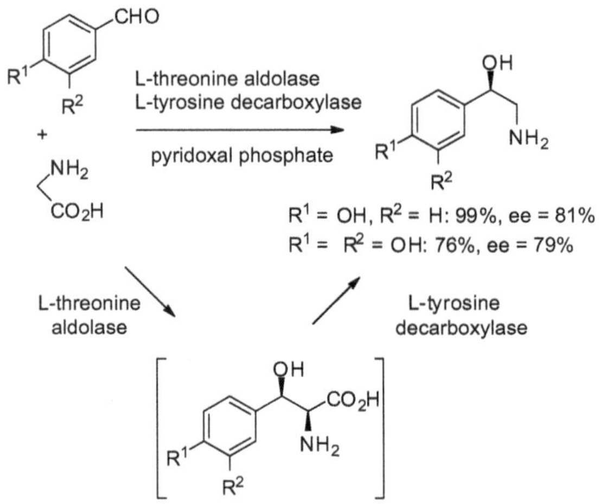

Scheme 4.16 Synthesis of chiral 1,2-amino alcohols catalysed by two enzymes.

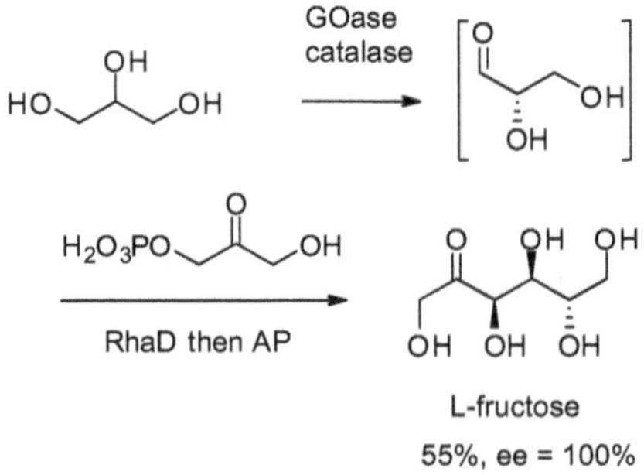

Scheme 4.17 Synthesis of L-fructose catalysed by four enzymes.

however, from two limitations. Firstly, L-glyceraldehyde is not commercially available and, secondly, this starting material is known to be thermodynamically metastable and decomposes easily. In this way, enantiopure L-fructose could be produced in one pot in 55% yield from glycerol in the presence of GOase (galactose oxidase), catalase, RhaD (rhamnulose-1-phosphate aldolase), and AP (acid phosphatase), as shown in Scheme 4.17.[43]

Hydroxynitrile lyases are powerful tools for the synthesis of chiral cyanohydrins which constitute versatile building blocks in organic synthesis. Hydroxynitrile lyases appear in a wide variety of higher plants and are classified into non-FAD- and FAD-containing enzymes.[44] In 2006, Mateo *et al.* employed hydroxynitrile lyases in combination with an unspecific nitrilase immobilised as a cross-linked enzyme aggregate (CLEA) for the production of enantiopure (*S*)-mandelic acid (Scheme 4.18a).[45] The process was based on the enantioselective hydrocyanation of benzaldehyde to mandelonitrile by hydroxynitrile lyase followed by enzymatic hydrolysis into mandelic acid in the presence of a non-selective nitrilase. The hydroxynitrile

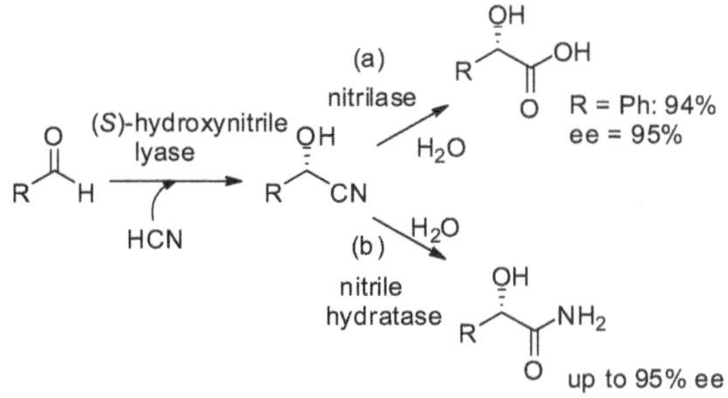

Scheme 4.18 Bienzymatic reactions using hydroxynitrile lyase.

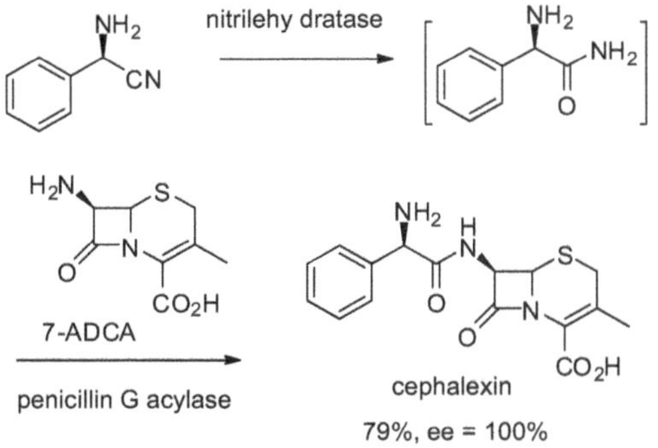

Scheme 4.19 Synthesis of cephalexin catalysed by two enzymes.

lyase determined the configuration of the stereogenic centre in the molecule. The main challenge lay in the different reaction optima of the enzymes used. While hydroxynitrile lyases have a pH optimum of approximately 5, nitrilases are usually most effective at close to neutral pH values. Additionally, reactions should be performed at low pHs and in the presence of organic solvents in order to suppress non-enzymatic hydrocyanation. Therefore, a recombinant nitrilase from *Pseudomonas fluorescens* EBC 191 was employed which still possessed significant activity at pH 5.5. Under optimised conditions, (S)-mandelic acid was produced with high enantioselectivity of up to 95% ee. In a similar process, a hydroxynitrile lyase and a novel nitrile hydratase from *Nitriliruptor alkaliphilus*, both immobilised as CLEAs, were coupled for the synthesis of enantiopure aliphatic α-hydroxycarboxylic amides (Scheme 4.18b).[46] The application of CLEAs improved the stability of the nitrile hydratase in cell-free preparations.

In 2002, Sheldon *et al.* reported a two-step, one-pot enzymatic synthesis of cephalexin from D-phenylglycine nitrile.[47] The nitrile hydratase-catalysed hydration of D-phenylglycine nitrile to the corresponding amide was combined with the penicillin G acylase-catalysed acylation of 7-aminodesacetoxycephalosporanic acid (7-ADCA) to afford, in one pot, enantiopure cephalexin in 79% yield, as shown in Scheme 4.19.

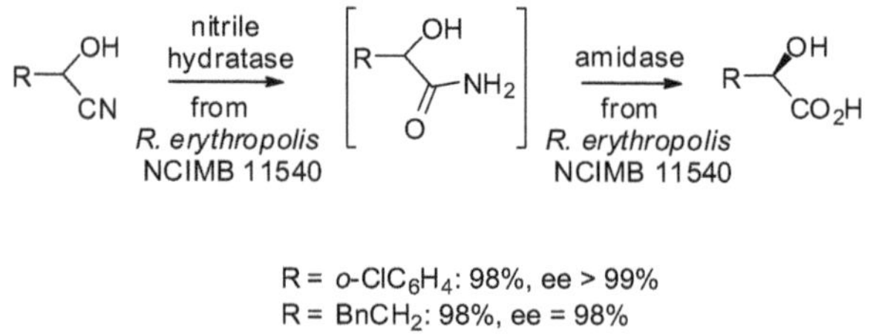

R = o-ClC$_6$H$_4$: 98%, ee > 99%
R = BnCH$_2$: 98%, ee = 98%

Scheme 4.20   Synthesis of (*R*)-2-chloromandelic acid and (*R*)-2-hydroxy-4-phenyl-butyric acid catalysed by two enzymes.

In 2003, Griengl *et al.* reported the hydrolysis of cyanohydrins by treatment with bacterial cells of *Rhodococcus erythropolis* NCIMB 11540, which have a highly active nitrile hydratase–amidase enzyme system.[48] In this manner, two important pharmaceutical intermediates, such as (*R*)-2-chloromandelic acid and (*R*)-2-hydroxy-4-phenylbutyric acid, could be prepared in high optical and chemical yields after short reaction times (3 and 1.5 h, respectively), as shown in Scheme 4.20.

In another context, Långström *et al.* have described the multienzymatic synthesis of carboxy-^{11}C-labelled L-tyrosine, L-DOPA, L-tryptophan and 5-hydroxy-L-tryptophan, starting from racemic [1-^{11}C]alanine with enantiomeric purities higher than 99% ee.[49] The enzymatic reactions were performed using, simultaneously, D-amino acid oxidase, catalase, glutamic-pyruvic transaminase, and β-tyrosinase (for L-tyrosine and L-DOPA), or tryptophanase (for L-tryptophan and 5-hydroxy-L-tryptophan), in a one-pot reaction. In 1991, Gygax *et al.* described the synthesis of β-D-glucuronides on the basis of a one-pot multienzyme system with *in situ* regeneration of uridine 5′-diphosphoglucuronic acid.[50] This stereoselective simple reaction involved the use of glucose-1-phosphate as a donor of the glucuronic acid moiety, and phosphoenolpyruvate and NAD (nicotinamide adenine dinucleotide) as cosubstrates. On the other hand, Thiem and Weimann have shown that galactosyltransferase catalysed the galactosylation of oligosaccharides terminated by glucose and by 2-acetamido-2-deoxy-glucopyranose, respectively.[51] The glycosyl donor, uridine-5′-diphosphogalactose, was generated *in situ* by the treatment of UDP-glucose with UDP-galactose-4-epimerase. In the presence of a glycosyl acceptor and galactosyltransferase, the corresponding galactosylated oligosaccharide was obtained (Scheme 4.21). Later, Wong *et al.* reported a multienzyme system for a one-pot synthesis of sialyl oligosaccharides on the basis of a combined use of β-galactosidase and α-(2,6)-sialyltransferase coupled with *in situ* regeneration of CMP-sialic acid.[52] Thus, the synthesis of sialyl oligosaccharides was achieved with a β-galactosidase-catalysed galactosylation of an acceptor followed by a sialyltransferase-catalysed sialylation with *in situ* regeneration of CMP-sialic acid.

On the other hand, Kren and Thiem have developed a sequential multienzyme one-pot system with cofactor regeneration in order to prepare rather complicated hetero-oligoglycosides such as the sialylated antigen T-epitope

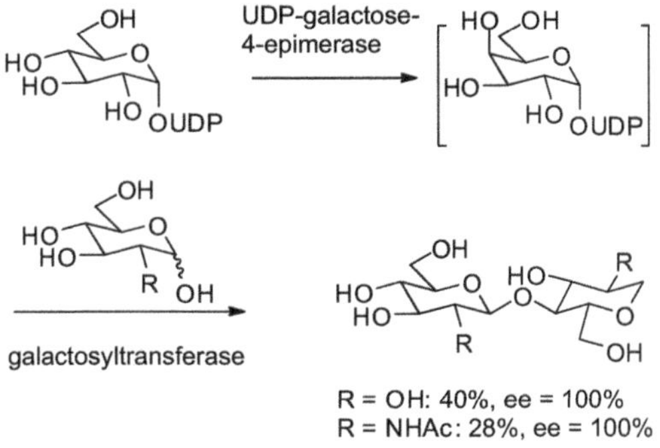

Scheme 4.21 Synthesis of galactose-terminated oligosaccharides catalysed by two enzymes.

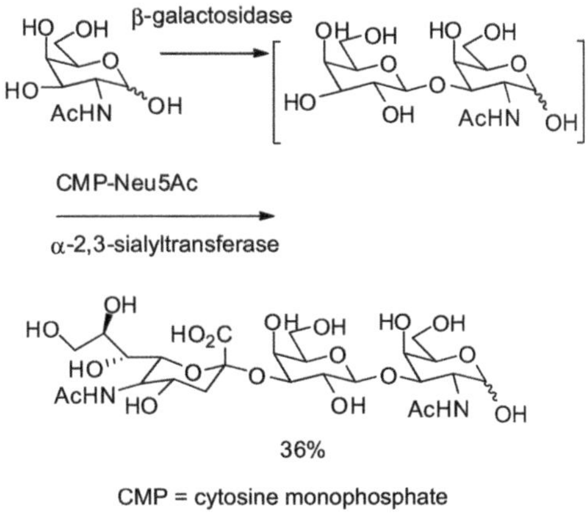

Scheme 4.22 Synthesis of fluorinated T- and ST-antigens catalysed by two enzymes.

(Scheme 4.22).[53] This was achieved in 36% yield through the sequential action of a glycosidase and a glycosyltransferase although they have maximum activities at very different pH values.

Overexpression of tumor-associated carbohydrate antigens (TACAs) on cell surfaces is a common phenomenon for cancer progression. The Thomsen–Friedenreich antigen (T-antigen) is one of the most common TACAs. The synthesis of fluorinated oligosaccharides including T-antigens is challenging for vaccine development. In this context, Cao and co-workers have recently described the synthesis of fluorinated Thomsen–Friedenreich (T) antigens from fluorinated monosaccharides based on a highly efficient one-pot two-enzyme approach involving a recombinant galactokinase (EcGalk) and a D-galactosyltransferase (BiGalHexNAcP), as shown in Scheme 4.23.[54] The formed fluorinated T-antigens were obtained in high yields (80–88%) and further sialylated to form the corresponding fluorinated ST-antigens in excellent yields (92–99%) using another one-pot two-enzyme system

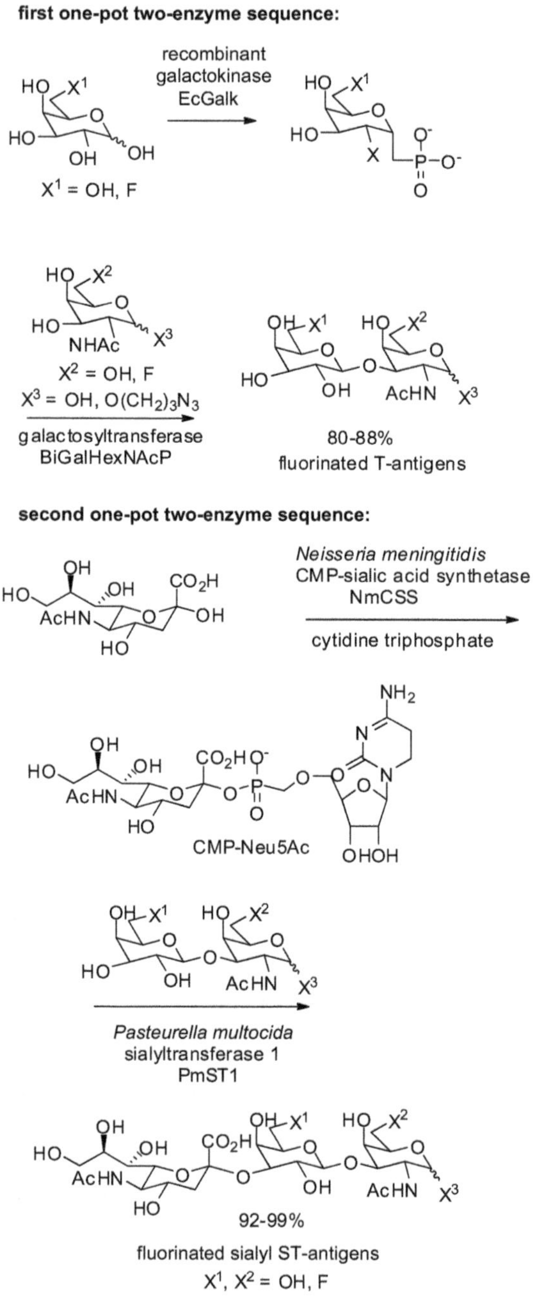

first one-pot two-enzyme sequence:

second one-pot two-enzyme sequence:

Scheme 4.23 Synthesis of fluorinated T-antigens and sialyl ST-antigens catalysed by two enzymes.

containing a CMP-sialic acid synthetase (NmCSS) and a sialyltransferase 1 (PmST1). It must be noted that the enzymes were sequentially added to accommodate their distinct pH preferences.

In another context, another remarkable multienzyme cocktail was used for the one-pot synthesis of precorrin-5, starting from δ-aminolevulinic acid. In this transformation, up to eight different enzymes were used, including 5-aminolevulinic acid dehydratase to form porphobilinogen followed by porphobilinogen deaminase, uroporphyringen III synthase, uroporphyringen III methyltransferase, precorrin-2 methyltransferase, precorrin-3 oxidase, precorrin-3 hydroxylactone methyltransferase, and precorrin-4 methyltransferase.[55] The desired enantiopure precorrin-5 was achieved in 30% overall yield, as shown in Scheme 4.24.

On the other hand, Duggan *et al.* supplied erythrose-4-phosphate and the unnatural substrate, 3-fluorophosphoenolpyruvate, to enzymes of the shikimate biosynthetic pathway to produce unnatural (*R*)- and (*S*)-6-fluoro analogues of shikimic acid, potentially useful as antibiotics.[56] In 1999, Sung *et al.* reported the production of aromatic D-amino acids from α-ketoacids and ammonia by the coupling of four enzyme reactions.[57] The multienzyme system composed of glutamate racemase, thermostable D-amino acid aminotransferase, glutamate dehydrogenase and formate dehydrogenase was employed for the synthesis of the enantiomerically pure D-amino acids, D-phenylalanine and D-tyrosine, from the corresponding α-ketoacids, phenylpyruvate and hydroxyphenylpyruvate, respectively (Scheme 4.25).

Enantiomerically pure epoxides are of great importance in synthetic chemistry. In this context, Sello *et al.* have developed a biocatalytic synthesis of chiral epoxide derivatives on the basis of the cascade use of two enzymatic

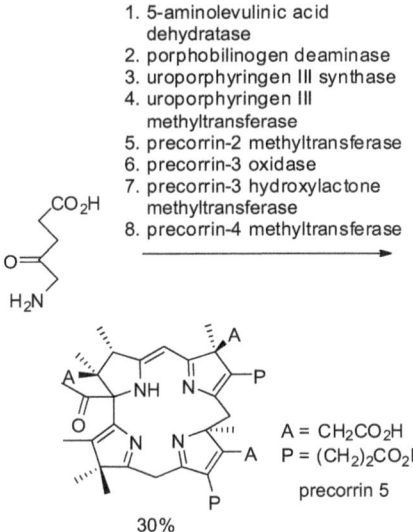

1. 5-aminolevulinic acid dehydratase
2. porphobilinogen deaminase
3. uroporphyringen III synthase
4. uroporphyringen III methyltransferase
5. precorrin-2 methyltransferase
6. precorrin-3 oxidase
7. precorrin-3 hydroxylactone methyltransferase
8. precorrin-4 methyltransferase

A = CH₂CO₂H
P = (CH₂)₂CO₂H

precorrin 5

30%

Scheme 4.24 Synthesis of precorrin-5 catalysed by eight enzymes.

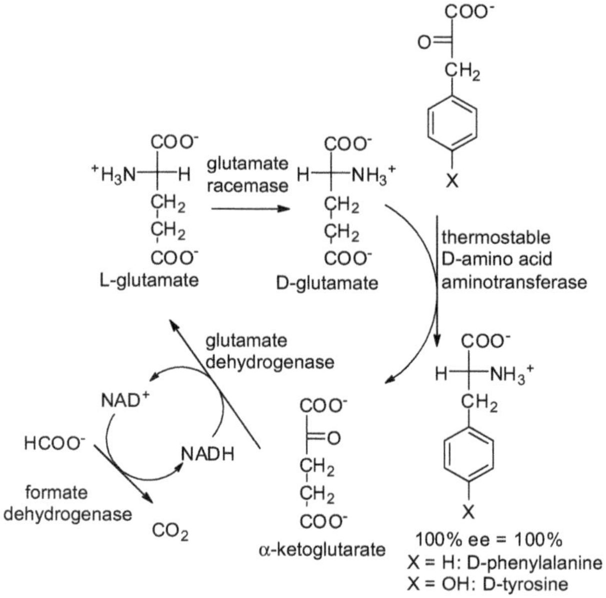

Scheme 4.25   Synthesis of D-amino acids catalysed by four enzymes.

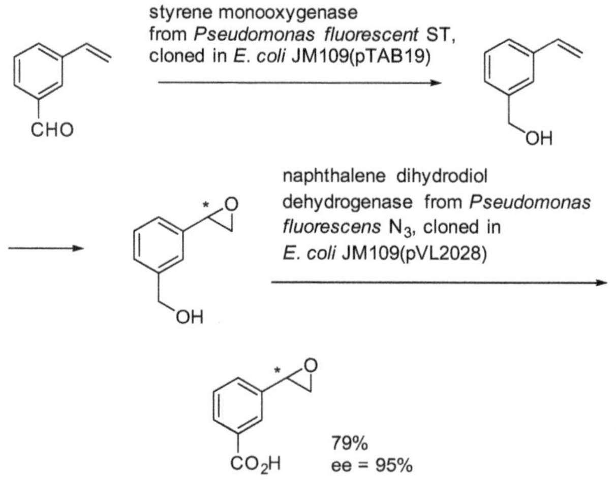

Scheme 4.26   Synthesis of a chiral epoxide catalysed by two enzymes.

activities in one pot.[58] As shown in Scheme 4.26, using whole-cell biocata-lysts, (−)-3-(oxiran-2-yl) benzoic acid could be synthesised in high yield and enantiomeric purity of up to 95% ee through a one-pot three-step procedure. The enzymes relevant to these transformations are styrene monooxygenase and naphthalene dihydrodiol dehydrogenase; the first comes from *Pseudo-monas fluorescent* ST, the second from *Pseudomonas fluorescens* N$_3$. They both have been cloned in *E. coli* JM109.

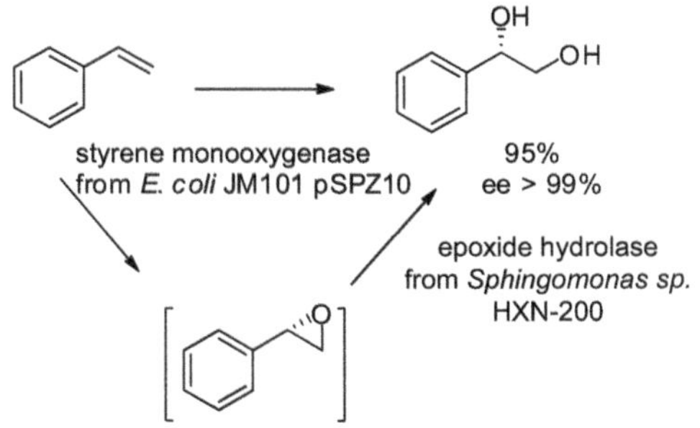

Scheme 4.27 Synthesis of (*S*)-styrene diol catalysed by two enzymes.

In 2009, Li *et al.* reported the preparation of chiral aryl vicinal diols through a tandem epoxidation and hydrolysis reaction with the combination of two biocatalysts (Scheme 4.27).[59] Thus, the enantioselective epoxidation of styrene with *E. coli* JM101 (pSPZ10) occurred efficiently to generate (*S*)-styrene oxide in enantioselectivity of >99% ee, which was hydrolysed into the corresponding (*S*)-diol with excellent regioselectivity induced by the ly-ophilised cell-free extract of *Sphingomonas sp.* HXN-200. This tandem re-action was carried out in a two-phase system, and styrene mainly remained in the organic phase, which allowed a facile product separation, and reduced the inhibition of styrene and styrene oxide to the enzyme. As such, the use of tandem biocatalysts could achieve higher conversion for the epoxidation than the use of only styrene monooxygenase, and more efficient synthesis of enantiopure vicinal diols than the use of dioxygenase, which afforded mix-tures of diol and cyclohexadiene-*cis*-diol with low to moderate yields and enantioselectivities.

References

1. (a) G. M. Whitesides and C.-H. Wong, *Angew. Chem., Int. Ed. Engl.*, 1985, **24**, 617–638; (b) C.-H. Wong and G. M. Whitesides, *Enzymes in Synthetic Organic Chemistry*, Elsevier, Oxford, 1994.
2. K. Faber, *Biotransformations in Organic Chemistry*, Springer Verlag, Heidelberg, 4th edn, 2000.
3. (a) A. Bruggink, R. Schoevaart and T. Kieboom, *Org. Process Res. Dev.*, 2003, **7**, 622–640; (b) T. Kieboom in *Catalysis for Renewables*, ed. G. Centi and R. A. van Santen, Wiley-VCH, Weinheim, 2007; (c) E. Ricca, B. Brucher and J. H. Schrittwieser, *Adv. Synth. Catal.*, 2011, **353**, 2239–2262; (d) P. A. Santacoloma, G. Sin, K. V. Gernaey and J. M. Woodley, *Org. Process Res. Dev.*, 2011, **15**, 203–212; (e) B. T. Ueberbacher, M. Hall and K. Faber, *Nat. Prod. Rep.*, 2012, **29**, 337–350.
4. W. Hummel and M.-R. Kula, *Eur. J. Biochem.*, 1989, **184**, 1–13.
5. (a) C. V. Voss, C. C. Gruber and W. Kroutil, *Synlett*, 2010, 991–998; (b) R. Wohlgemuth, *Curr. Opin. Microbiol.*, 2010, **13**, 283–292.

6. C. V. Voss, C. C. Gruber and W. Kroutil, *Angew. Chem., Int. Ed.*, 2008, **47**, 741–745.
7. C. V. Voss, C. C. Gruber, K. Faber, T. Knaus, P. Macheroux and W. Kroutil, *J. Am. Chem. Soc.*, 2008, **130**, 13969–13972.
8. Y. L. Li, J. H. Xu and Y. Xu, *J. Mol. Catal. B: Enzym.*, 2010, **64**, 48–52.
9. B. Seisser, I. Lavandera, K. Faber, J. H. Lutje Spelberg and W. Kroutil, *Adv. Synth. Catal.*, 2007, **349**, 1399–1404.
10. J. H. Schrittwieser, I. Lavandera, B. Seisser, B. Mautner, J. H. Lutje Spelberg and W. Kroutil, *Tetrahedron: Asymmetry*, 2009, **20**, 483–488.
11. S. K. Ma, J. Gruber, C. Davis, L. Newman, D. Gray, A. Wang, J. Grate, G. W. Huisman and R. A. Sheldon, *Green Chem.*, 2010, **12**, 81–86.
12. J. H. Schrittwieser, I. Lavandera, B. Seisser, B. Mautner and W. Kroutil, *Eur. J. Org. Chem.*, 2009, 2293–2298.
13. H. Ankati, D. N. Zhu, Y. Yang, E. R. Biehl and L. Hua, *J. Org. Chem.*, 2009, **74**, 1658–1662.
14. S. Shimizu, M. Kataoka, M. Katoh, T. Miyoshi and H. Yamada, *Appl. Environ. Microbiol.*, 1990, **56**, 2303–2310.
15. C. K. Savile, J. M. Janey, E. C. Mundorff, J. C. Moore, S. Tam, W. R. Jarvis, J. C. Colbeck, A. Krebber, F. J. Fleitz, J. Brands, P. N. Devine, G. W. Huisman and G. J. Hughes, *Science*, 2010, **329**, 305–309.
16. D. Koszelewski, D. Clay, D. Rozzell and W. Kroutil, *Eur. J. Org. Chem.*, 2009, 2289–2292.
17. D. Koszelewski, D. Pressnitz, D. Clay and W. Kroutil, *Org. Lett.*, 2009, **11**, 4810–4812.
18. B. K. Cho, H. J. Cho, S. H. Park, H. Yun and B. G. Kim, *Biotechnol. Bioeng.*, 2003, **81**, 783–789.
19. C. U. Ingram, M. Bommer, M. E. B. Smith, P. A. Dalby, J. M. Ward, H. C. Hailes and G. J. Lye, *Biotechnol. Bioeng.*, 2007, **96**, 559–569.
20. I. G. Fotheringham, N. Grinter, D. P. Pantaleone, R. F. Senkpeil and P. P. Taylor, *Bioorg. Med. Chem.*, 1999, **7**, 2209–2213.
21. D. I. Dato, K. Miyamoto and H. Ohta, *Biocatal. Biotransform.*, 2005, **23**, 375–379.
22. A. Caligiuri, P. D'Arrigo, T. Gefflaut, G. Molla, L. Pollegioni, E. Rosini, C. Rossi and S. Servi, *Biocatal. Biotransform.*, 2006, **24**, 409–413.
23. (a) C. Wandrey, in *Enzymes as catalysts in organic synthesis*, ed. M. P. Schreiner, Springer, Berlin, vol. 1, 1986, pp. 263–284; (b) E. Schmidt, Đ. Vasić-Rački and C. Wandrey, *Appl. Microbiol. Biotechnol.*, 1987, **26**, 42–48; (c) C. Wandrey, E. Fiolitakis, U. Wichmann and M. R. Kula, *Ann. N. Y. Acad. Sci.*, 1984, **434**, 91–94.
24. V. Resch, W. M. F. Fabian and W. Kroutil, *Adv. Synth. Catal.*, 2010, **352**, 993–997.
25. L. P. B. Gonçalves, O. A. C. Antunes, G. F. Pinto and E. G. Oestreicher, *Tetrahedron: Asymmetry*, 2000, **11**, 1465–1468.
26. (a) A. S. Bommarius, K. Drautz, W. Hummel, M. R. Kula and C. Wandrey, *Biocatalysis*, 1994, **10**, 37–47; (b) K. Seelbach and U. Kragl, *Enzyme*

Microb. Technol., 1997, **20**, 389–392; (c) R. N. Patel, *Adv. Synth. Catal.*, 2001, **343**, 527–546.

27. V. Hélaine, J. Rossi, T. Gefflaut, S. Alaux and J. Bolte, *Adv. Synth. Catal.*, 2001, **343**, 692–697.
28. N. Nakajima, N. Esaki and K. Soda, *J. Chem. Soc., Chem. Commun.*, 1990, 947–948.
29. M. A. Wegman, M. A. P. J. Hacking, J. Rops, P. Pereira, F. van Rantwijk and R. A. Sheldon, *Tetrahedron: Asymmetry*, 1999, **10**, 1739–1750.
30. B. Wu, W. Szymanski, S. de Wildeman, G. J. Poelarends, B. L. Feringa and D. B. Janssen, *Adv. Synth. Catal.*, 2010, **352**, 1409–1412.
31. K. M. Koeller and C.-H. Wong, *Chem. Rev.*, 2000, **100**, 4465–4494.
32. Y. Ishikawa, J. L. C. Liu, G. J. Shen and C. H. Wong, *J. Am. Chem. Soc.*, 1991, **113**, 6300–6302.
33. H. Yu, R. Karpel and X. Chen, *Bioorg. Med. Chem.*, 2004, **12**, 6427–6435.
34. H. Z. Cao, Y. H. Li, K. Lau, S. Muthana, H. Yu, J. S. Cheng, H. A. Chokhawala, G. Sugiarto, L. Zhang and X. Chen, *Org. Biomol. Chem.*, 2009, **7**, 5137–5145.
35. T. Miyazaki, H. Sato, T. Sakakibara and Y. Kajihara, *J. Am. Chem. Soc.*, 2000, **122**, 5678–5694.
36. U. Kragl, D. Gygax, O. Ghisalba and C. Wandrey, *Angew. Chem., Int. Ed. Engl.*, 1991, **30**, 827–828.
37. T. H. Wang, Y. Y. Chen, H. H. Pan, F. P. Wang, C. H. Cheng and W. C. Lee, *BMC Biotechnol.*, 2009, **9**, 63.
38. N. Horinouchi, J. Ogawa, T. Kawano, T. Sakai, K. Saito, S. Matsumoto, M. Sasaki, Y. Mikami and S. Shimizu, *Appl. Microbiol. Biotechnol.*, 2006, **71**, 615–621.
39. N. Horinouchi, J. Ogawa, T. Kawano, T. Sakai, K. Saito, S. Matsumoto, M. Sasaki, Y. Mikami and S. Shimizu, *Biotechnol. Lett.*, 2006, **28**, 877–881.
40. (a) J. Steinreiber, M. Schürmann, F. van Assema, M. Wolberg, K. Fesko, C. Reisinger, D. Mink and H. Griengl, *Adv. Synth. Catal.*, 2007, **349**, 1379–1386; (b) J. Steinreiber, M. Schürmann, M. Wolberg, F. van Assema, C. Reisinger, K. Fesko, D. Mink and H. Griengl, *Angew. Chem., Int. Ed.*, 2007, **46**, 1624–1626.
41. H. J. M. Gijsen and C.-H. Wong, *J. Am. Chem. Soc.*, 1995, **117**, 2947–2948.
42. (a) R. Alajarin, E. Garcia-Junceda and C.-H. Wong, *J. Org. Chem.*, 1995, **60**, 4294–4295; (b) I. Henderson, K. B. Sharpless and C.-H. Wong, *J. Am. Chem. Soc.*, 1994, **116**, 558–561.
43. D. Franke, T. Machajewski, C.-C. Hsu and C.-H. Wong, *J. Org. Chem.*, 2003, **68**, 6828–6831.
44. A. Hickel, M. Hasslacher and H. Griengl, *Physiol. Plant.*, 1996, **98**, 891–898.
45. C. Mateo, A. Chmura, S. Rustler, F. van Rantwijk, A. Stolz and R. A. Sheldon, *Tetrahedron: Asymmetry*, 2006, **17**, 320–323.
46. S. van Pelt, F. van Rantwijk and R. A. Sheldon, *Adv. Synth. Catal.*, 2009, **351**, 397–404.

47. M. A. Wegman, L. M. van Langen, F. van Rantwijk and R. A. Sheldon, *Biotechnol. Bioeng.*, 2002, **79**, 356–361.
48. I. Osprian, M. H. Fechter and H. Griengl, *J. Mol. Catal. B: Enzym.*, 2003, **24–25**, 89–98.
49. (a) P. Bjurling, G. Antoni, Y. Watanabe and B. Långström, *Acta Chem. Scand.*, 1990, **44**, 178–182; (b) P. Bjürling, Y. Watanabe, S. Oka, T. Nagasawa, H. Yamada and B. Långström, *Acta Chem. Scand.*, 1990, **44**, 183–188.
50. D. Gygax, P. Spies, T. Winkler and U. Pfaar, *Tetrahedron*, 1991, **47**, 5119–5122.
51. (a) J. Thiem and T. Wiemann, *Synthesis*, 1992, 141–145; (b) J. Thiem and T. Wiemann, *Angew. Chem., Int. Ed. Engl.*, 1991, **30**, 1163–1164.
52. G. F. Herrmann, Y. Ichikawa, C. Wandrey, F. C. A. Gaeta, J. C. Paulson and C.-H. Wong, *Tetrahedron Lett.*, 1993, **34**, 3091–3094.
53. V. Kren and J. Thiem, *Angew. Chem., Int. Ed. Engl.*, 1995, **34**, 893–895.
54. J. Yan, X. Chen, F. Wang and H. Cao, *Org. Biomol. Chem.*, 2013, **11**, 842–848.
55. (a) A. I. Scott, *Synlett*, 1994, 871–883; (b) J. Park, J. Tai, C. A. Roessner and A. I. Scott, *Biorg. Med. Chem.*, 1996, **4**, 2179–2185; (c) C. A. Roessner and A. I. Scott, *Chem. Biol.*, 1996, **3**, 325–330.
56. P. J. Duggan, E. Parker, J. Coggins and C. Abell, *Biorg. Med. Chem. Lett.*, 1995, **5**, 2347–2352.
57. H.-S. Bae, S.-G. Lee, S.-P. Hong, M.-S. Kwak, N. Esaki, K. Soda and M.-H. Sung, *J. Mol. Catal. B: Enzym.*, 1999, **6**, 241–247.
58. G. Sello, S. Bernasconi, F. Orsini and P. Di Gennaro, *Tetrahedron: Asymmetry*, 2009, **20**, 563–565.
59. Y. Xu, X. Jia, S. Panke and Z. Li, *Chem. Commun.*, 2009, 1481–1843.

CHAPTER 5

Conclusions

This first section illustrates how much asymmetric multicatalysis based on the use of catalysts belonging to the same discipline, such as two or three organocatalysts, two metal catalysts, or two or more biocatalysts, has contributed to the development of various types of novel and powerful enantioselective tandem reactions. Indeed, these beautiful one-pot reactions can be considered as one of the most influential reaction classes of the last few years. The combination of asymmetric multicatalysis with the concept of tandem reactions has allowed the easy attainment of high molecular complexity with often high levels of stereocontrol, simple operational procedures, and advantages of savings in solvent, time, energy, and costs. Indeed, the combination of catalysts has emerged as a new and powerful strategy for developing new and valuable reactions, attracting increasing attention as it can enable the development of unprecedented transformations that are not possible by use of either of the catalytic systems alone. This section collects all the major progress in the field of enantioselective tandem reactions promoted by multiple organocatalysts, two metal catalysts, or two or more biocatalysts. It demonstrates the power of these elegant one-pot processes of two or more bond-forming reactions, occurring with minimum workup or change in conditions in which the subsequent transformation takes place at the functionalities obtained in the former transformation, following the same principles that are found in biosynthesis in nature. These fascinating reactions have rapidly become one of the most current fields in organic chemistry. As the concept of combined catalysis expands, the number of novel transformations, new catalyst combinations, and new applications in particular in enantioselective tandem reactions will continue to increase.

RSC Catalysis Series No. 20
Enantioselective Multicatalysed Tandem Reactions
By Hélène Pellissier
© Hélène Pellissier 2014
Published by the Royal Society of Chemistry, www.rsc.org

SECTION II
Asymmetric Tandem Reactions Catalysed by Multiple Catalysts from Different Disciplines

CHAPTER 6

Introduction

The combination of catalysts from different disciplines, such as that of organocatalysts with metal catalysts, metal catalysts with biocatalysts, and organocatalysts with biocatalysts, has attracted increasing attention in the synthetic community in the past ten years, allowing unprecedented transformations not currently possible by use of any catalysis alone, and making current synthetic methods more economical and practical.[1] Indeed, the combination of different types of catalysts is fruitful for the development of new synthetic methods of complex molecules with good chemo- and stereoselectivity inaccessible through the use of single specific catalytic systems. This section, including all the major progress in the field of enantioselective tandem reactions promoted by multiple catalysts belonging to different disciplines, is divided into five chapters, dealing successively after an introduction (Chapter 6) with reactions catalysed by a combination of metals and organocatalysts (Chapter 7), reactions catalysed by a combination of metals and enzymes (Chapter 8), and reactions catalysed by a combination of organocatalysts and enzymes (Chapter 9), followed by conclusions (Chapter 10). Chapter 7, dedicated to combined organocatalysis with transition metal catalysis, is divided into three sections which successively deal with cooperative, relay or sequential catalysis. Chapter 8 collects enantioselective chemoenzymatic tandem reactions, not based on DKR in the first section, and based on DKR in the second section. Finally, Chapter 9 collects the first examples appearing in the literature of enantioselective tandem reactions catalysed by combinations of organocatalysts and enzymes.

References

1. (a) J.-A. Ma and D. Cahard, *Angew. Chem., Int. Ed.*, 2004, **43**, 4566–1508; (b) M. Kanai, N. Kato, E. Ichikawa and M. Shibasaki, *Synlett*, 2005, 1491–1508; (c) D. H. Paull, C. J. Abraham, M. T. Scerba, E. Alden-Danforth and T. Leckta, *Acc. Chem. Res.*, 2008, **41**, 655–663; (d) Z. Shao and H. Zhang, *Chem. Soc. Rev.*, 2009, **38**, 2745–2755; (e) C. Zhong and X. Shi, *Eur. J. Org. Chem.*, 2010, 2999–3025; (f) M. Rueping, R. M. Koenigs and I. Atodiresei, *Chem. – Eur. J.*, 2010, **16**, 9350–9365; (g) J. Zhou, *Chem. – Asian J.*, 2010, **5**,

422–434; (h) L. M. Ambrosini and T. H. Lambert, *ChemCatChem*, 2010, **2**, 1373–1380; (i) S. Piovesana, D. M. Scarpino Schietroma and M. Bella, *Angew. Chem., Int. Ed.*, 2011, **50**, 6216–6232; (j) N. T. Patil, V. S. Shinde and B. Gajula, *Org. Biomol. Chem.*, 2012, **10**, 211–224; (k) A. E. Allen and D. W. C. MacMillan, *Chem. Sci.*, 2012, **3**, 633–658; (l) Z. Du and Z. Shao, *Chem. Soc. Rev.*, 2013, **42**, 1337–1378.

CHAPTER 7

Reactions Catalysed by a Combination of Metals and Organocatalysts

7.1 Introduction

Recently, the combination of organocatalysts and transition metal catalysts has evolved as a new strategy to carry out enantioselective transformations that could not be performed in a traditional way by simply employing one of the two catalysts. These transformations not only demonstrate the potential of this merged catalytic approach, but they also show that there are more options to render a reaction highly enantioselective than testing different chiral metal–ligand complexes, organocatalysts, or additives. By using appropriate combinations of an organocatalyst and an achiral or chiral transition metal catalyst, facile ways for reaction optimisation can be achieved by simply varying one of the two existing catalysts. The first example of combining transition metal and organocatalyst was reported by Ito *et al.* in 1986, dealing with a remarkable enantioselective domino aldol–cyclisation reaction of aldehydes with methyl isocyanoacetate catalysed by a combination of a gold complex and a chiral tertiary amine as organocatalyst, allowing diastereo- and enantioselectivities of up to 100% de and 97% ee, respectively, to be achieved in combination with yields of 83–100%.[1] Further studies by Togni and Pastor contributed to elucidating the cooperative nature of the system, where the pendant terminal tertiary amine deprotonates the isocyanoacetate that has been previously activated *via* complexation with the gold cation.[2] In 2003, Krische and co-workers reported an effective dual catalytic process achieved with a starting

RSC Catalysis Series No. 20
Enantioselective Multicatalysed Tandem Reactions
By Hélène Pellissier
© Hélène Pellissier 2014
Published by the Royal Society of Chemistry, www.rsc.org

enone-tethered terminal allyl carbonate, through both the use of tributyl-phosphane as the Lewis base and of palladium-activated Tsuji–Trost π-allyl as the electrophile.[3] The corresponding allylated cyclic enones were produced in excellent yields. Importantly, this work initiated the field of combining transition metal and Lewis base catalysis. Although the combination of transition metal catalysis with organocatalysis has allowed a range of novel and useful reactions to be achieved,[4] the development of tandem reactions induced by a combination of two types of catalysts still remains a challenge. While the organocatalysis is dominated by Lewis base catalysts, such as amines, carbenes, and tertiary phosphines, a metal catalyst usually has an empty coordination site to interact and activate a substrate. The challenge in combining an organocatalyst and a metal catalyst is in part to avoid the deactivation of catalyst by Lewis acid–base interaction. Even in the absence of a catalyst poison, the presence of a Lewis base can erode the chiral environment of a chiral metal complex. Consequently, the success of tandem catalysis will need a fine tuning of the hardness and softness of the metal catalyst and the organocatalyst to increase their compatibility. Moreover, it will be possible to add the two catalysts successively to the reaction media (sequential catalysis), or to use phase separation techniques to partition incompatible catalysts. In spite of these difficulties, there are, however, many beautiful examples of asymmetric tandem reactions based on the use of combinations of organocatalysts of several types with a range of metals, acting through cooperative, relay as well as sequential catalysis. Cooperative catalysis means that both the organocatalyst and the metal catalyst share the same catalytic cycle working cooperatively to form a product. On the other hand, relay and sequential catalysis require both the organocatalyst and the metal catalyst to perform two distinct catalytic cycles for the consecutive reactions, whereby the substrates first react to form an intermediate in the first catalytic cycle, which can either be the organocatalytic cycle or the transition metal catalytic cycle. Subsequently, this intermediate is converted to the final product by another independent catalyst. Whereas relay catalysis involves no change in the reaction conditions and consequently both the two catalysts are present at the beginning of the reaction, in sequential catalysis the second catalyst is added after completion of the first catalytic cycle. Transition metals have been already combined with almost all types of organocatalysts, such as aminocatalysts including secondary and primary amines, phosphoric acids, (thio)ureas, N-heterocyclic carbenes, phase transfer catalysts, Brønsted base catalysts, and Lewis base catalysts. On the other hand, a wide variety of transition metals such as titanium, vanadium, manganese, iron, cobalt, nickel, copper, zinc, ytterbium, niobium, ruthenium, rhodium, palladium, silver, indium, iridium, platinum, and gold have been found effective catalysts in combination with organocatalysts. Among them, it must be noted that the combination of gold catalysis and organocatalysis has known a particularly rapid growth in recent years.[5]

7.2 Cooperative Catalysis

A combination of an aluminium catalyst, such as $AlEt_2Cl$, with a chiral phosphoric acid has been reported by Zhu *et al.* to promote the addition of α-isocyanoacetamides to aldehydes, affording after subsequent cyclisation the corresponding Passerini-type products in high yields and enantio-selectivities of up to 95% ee, as shown in Scheme 7.1.[6]

In the last few years, several groups have developed enantioselective domino reactions catalysed by combinations of organocatalysts with palladium complexes. As an example, Murkherjee and List have reported a domino synthesis of β-all carbon quaternary amines on the basis of a highly enantioselective α-alkylation of α-branched aldehydes, involving an achiral palladium catalyst and a chiral phosphoric acid.[7] Under the catalysis of phosphoric acid, a secondary allylamine reacted with an α-branched aldehyde to form an enammonium phosphate salt, which upon reaction with palladium catalyst afforded a cationic π-allyl palladium complex (Scheme 7.2). This intermediate resulted in the formation of an α-allylated iminium ion, which could be reduced to the corresponding final chiral amine in high yield and excellent enantioselectivity of 97% ee. The synthetic utility of this transformation was also demonstrated by a formal synthesis of (+)-cuparene.

Later, a combination of a chiral cinchona alkaloid Lewis base and a palladium catalyst was employed by Lectka *et al.* to induce the domino fluorination–amidation reaction of acid chlorides.[8] The process employed *N*-fluorobenzenesulfonimide (NFSI) as fluorinating agent and was quenched by subsequent addition of an aryl amine, providing the corresponding enantiopure α-fluorinated amides in moderate to good yields and generally almost complete enantioselectivity of >99% ee in all cases of substrates studied, as shown in Scheme 7.3. The authors demonstrated that adding a Lewis acidic lithium salt, such as $LiClO_4$, to specifically coordinate NFSI by

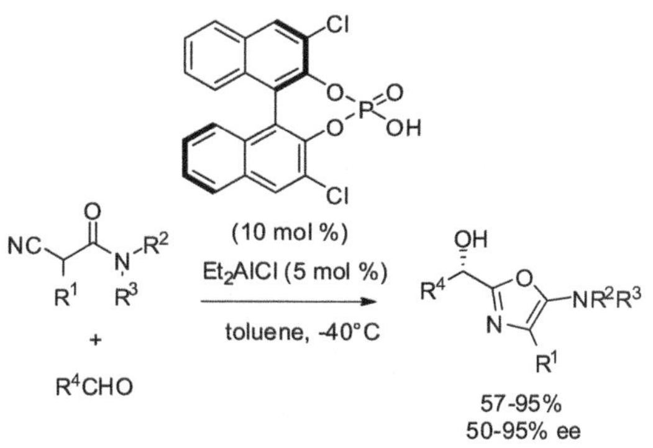

Scheme 7.1 Passerini-type reaction catalysed by chiral phosphoric acid catalysis and aluminium catalysis.

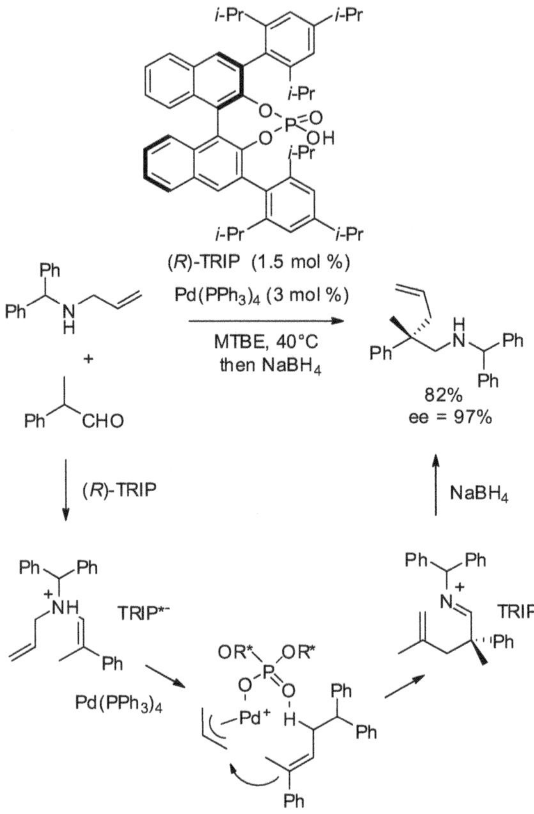

Scheme 7.2 Domino nucleophilic addition–dehydration–reduction reaction catalysed by chiral phosphoric acid catalysis and palladium catalysis.

increasing its electrophilicity to the dually activated system, led to better
yields while maintaining the excellent enantioselectivity.

In 2010, Cordova *et al.* developed a highly enantioselective domino
oxa-Michael–carbocyclisation reaction between propargyl alcohols and
α,β-unsaturated aldehydes induced by a combination of PdCl₂ and a chiral
L-proline-derived catalyst, such as chiral diphenylprolinol triethylsilyl ether.[9]
The process provided the corresponding chiral dihydrofurans in moderate
to good yields and good to excellent enantioselectivities of up to 99% ee,
albeit with moderate diastereoselectivities (≤48% de), as shown in
Scheme 7.4. The proposed mechanism consisted of a formal dynamic kinetic
asymmetric transformation which began with the formation of an iminium
intermediate **22** from the enal and the chiral amine catalyst (Scheme 7.4).
Next, the oxa-Michael addition of the propargylic alcohol to this iminium
intermediate rendered the corresponding enamine intermediates **23** and
23′. Subsequent hydrolysis of **23** and **23′** could give **24**; however, retro-
Michael reaction and racemisation was favoured, giving back the starting
materials. Next, stereoselective *Re*-facial oxidative cycloaddition of the less

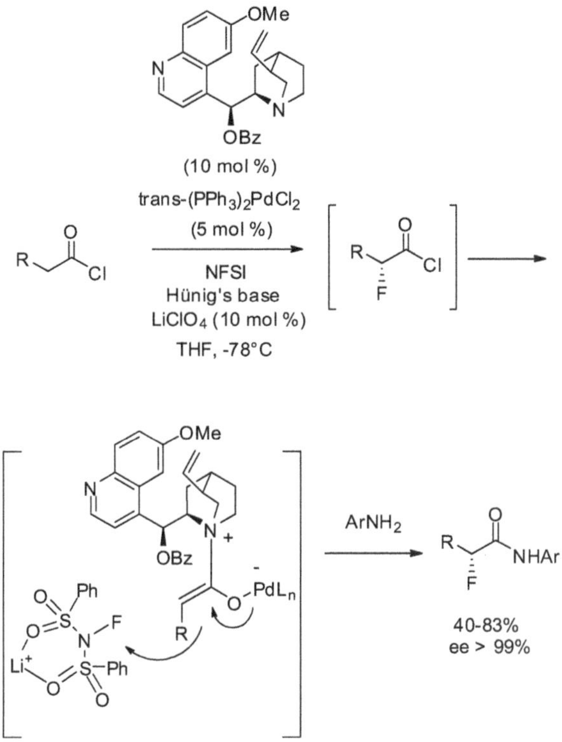

Scheme 7.3   Domino α-fluorination–amidation reaction catalysed by chiral cinchona catalysis and palladium catalysis.

sterically hindered chiral enamine **25** as compared to **25'** afforded bicyclic Pd(ɪᴠ) intermediate **26**. β-Elimination and protonation of palladium gave iminium intermediate **27**. Subsequent reductive elimination released Pd(ɪɪ), which could then take part in its catalytic cycle again, and rendered iminium intermediate **28**. Hydrolysis of this intermediate **28** released the chiral amine catalyst and afforded the final dihydrofuran. Indeed, the simultaneous cooperative catalysis of the chiral amine catalyst and the metal was essential to achieve product formation.

    In the same context, these authors have also developed a highly enantioselective domino Michael–carbocyclisation reaction of carbon nucleophiles, such as dialkyl propargylmalonates, with α,β-unsaturated aldehydes, providing the corresponding functionalised chiral cyclopentenes in good yields and generally excellent enantioselectivities ranging from 93 to 99% ee, as shown in Scheme 7.5.[10] In this case, the best results for this formal dynamic kinetic asymmetric transformation were reached through cooperative catalysis with chiral diphenylprolinol trimethylsilyl ether and Pd(PPh₃)₄. This reaction was previously investigated by Wang *et al.*, employing a combination of PdCl₂ with the (*R*)-enantiomer of the same organocatalyst in the presence of benzoic acid as an additive in dichloromethane as the solvent at

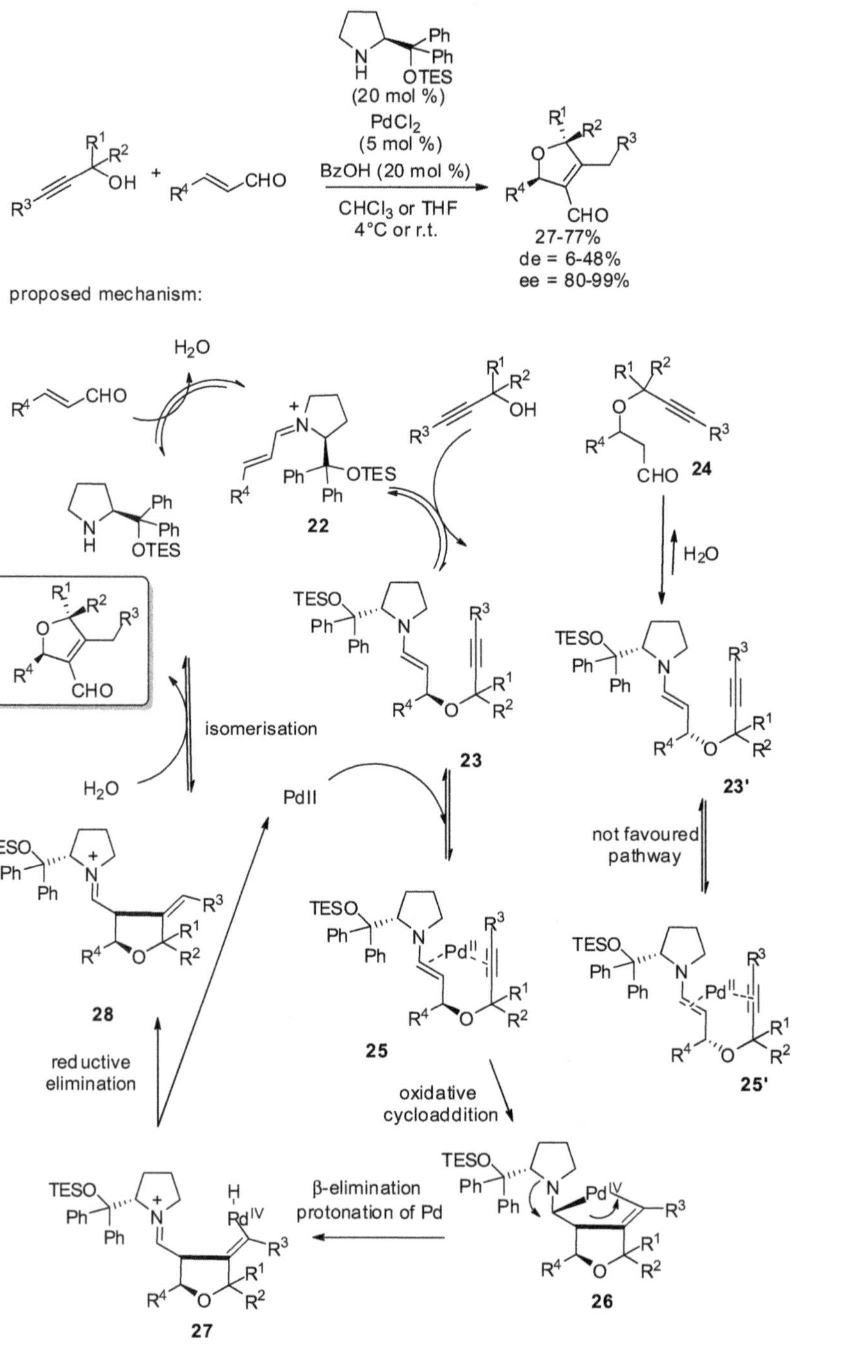

Scheme 7.4 Domino oxa-Michael–carbocyclisation reaction catalysed by chiral amine catalysis and palladium catalysis.

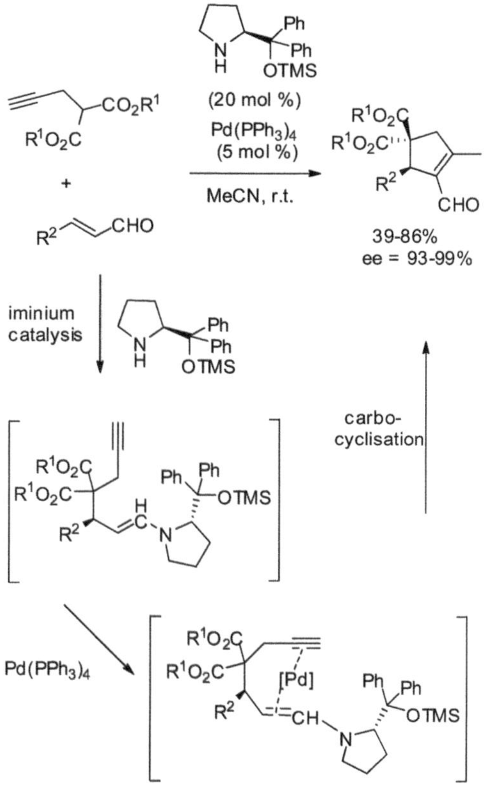

Scheme 7.5 Domino Michael–carbocyclisation reaction catalysed by chiral amine catalysis and palladium catalysis.

room temperature.[11] Under these conditions, the corresponding (S)-enantiomeric cyclopentenes were achieved in better yields of 61 to 91% and enantioselectivities ranging from 89 to 99% ee.

In 2011, another catalyst system consisting of chiral diphenylprolinol triethylsilyl ether and PdCl$_2$ was applied by Wang *et al.* to promote the related enantioselective domino aza-Michael–carbocyclisation reaction between α,β-unsaturated aldehydes and N-tosyl propargylamines, affording the corresponding chiral 2,5-dihydropyrroles in 41 to 87% yields, diastereoselectivities of 88 to >90% de, and enantioselectivities of 73 to >99% ee, as shown in Scheme 7.6.[12] In this case, the best results were obtained by using NaOAc, H$_2$O and DMAP as additives, which had a beneficial effect on the acceleration of the reaction without decreasing the enantioselectivities. A plausible mechanism is proposed in Scheme 7.6, involving an iminium activation of the aldehyde by the chiral diphenylprolinol triethylsilyl ether catalyst in the aza-Michael first step, followed successively by an enamine-metal cooperative activation in the second carbocyclisation step.

On the other hand, the catalytic asymmetric borane Michael addition has been proved to be feasible through the merger of transition metal and

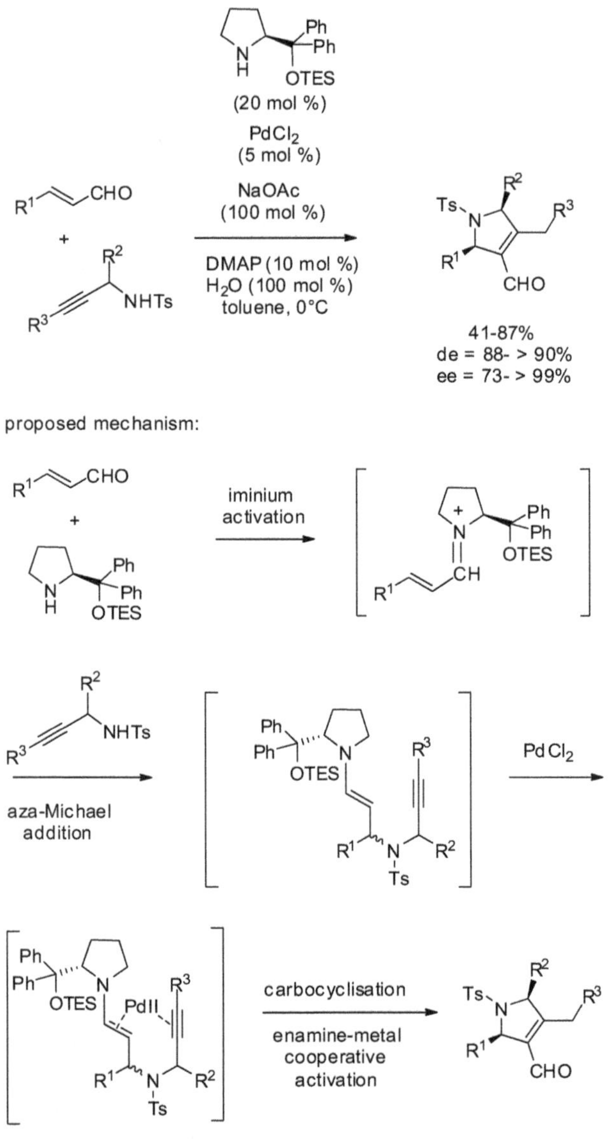

Scheme 7.6 Domino aza-Michael–carbocyclisation reaction catalysed by chiral amine catalysis and palladium catalysis.

iminium catalysis. Indeed, Cordova *et al.* have developed this type of reaction, using chiral diphenylprolinol trimethylsilyl ether in combination with $Cu(OTf)_2$ as catalyst system and providing Michael adducts, which were subsequently trapped by a Wittig reaction to afford the corresponding chiral homoallylboronates, as shown in Scheme 7.7.[13] The process tolerated α,β-unsaturated aldehydes with both aryl and aliphatic substituents at the β-position. Moreover, cinnamic α,β-unsaturated aldehydes with an

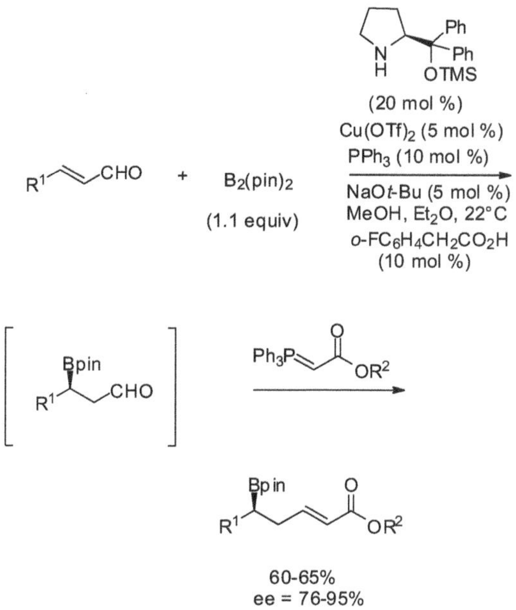

Scheme 7.7 Domino Michael–Wittig reaction catalysed by chiral amine catalysis and copper catalysis.

electron-withdrawing group at the *para*, *ortho*, or *meta*-position exhibited a higher reactivity in the β-boration step. On the other hand, α,β-unsaturated aldehydes bearing heterocyclic substituents at the β-position were not suitable substrates for this three-component reaction. Although the multi-catalysed step of this methodology consisted of a simple Michael reaction, it was decided to include these results because the global process is a tandem Michael–Wittig reaction.

Copper was also employed by Jang and co-workers in a highly enantiose-lective synthesis of α,β-disubstituted aldehydes on the basis of an iminium–copper catalysis.[14] As shown in Scheme 7.8, the β-substitution took place through iminium-catalysed Michael addition of nitromethane or diethyl-malonate as nucleophile to the α,β-unsaturated aldehyde, followed by copper-assisted addition of 2,2,6,6-tetramethylpiperidin-1-yloxyl (TEMPO) at the aldehyde α-position. The reaction employed a combination of chiral diphenylprolinol trimethylsilyl ether as organocatalyst and CuCl in the case of nitromethane addition or Cu(OTf)$_2$ in the case of ethylmalonate addition. The domino products were achieved in generally very high yields, diastereo- and enantioselectivities, as shown in Scheme 7.8. The authors have proposed the catalytic cycle depicted in Scheme 7.8 to explain the results, which began with the formation of the iminium salt from the enal and the organocatalyst. The subsequent addition of the nucleophile to its β-position provided the corresponding enamine intermediate, which underwent either hydrolysis or copper-TEMPO addition. In the presence of TEMPO and the

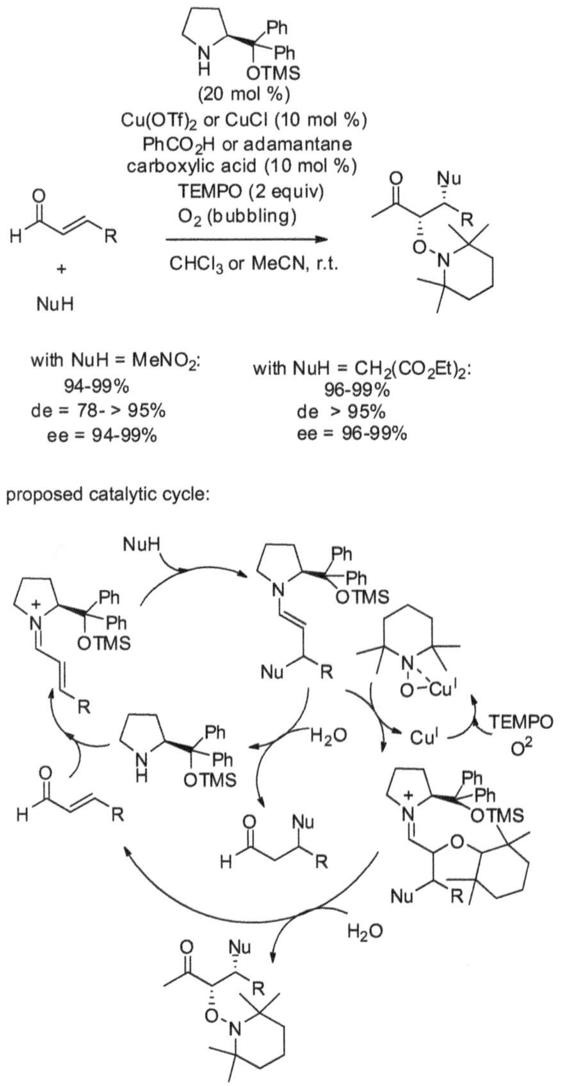

with NuH = MeNO$_2$:
94-99%
de = 78- > 95%
ee = 94-99%

with NuH = CH$_2$(CO$_2$Et)$_2$:
96-99%
de > 95%
ee = 96-99%

proposed catalytic cycle:

Scheme 7.8 Domino Michael–TEMPO addition reaction catalysed by chiral amine catalysis and copper catalysis.

copper complex, this enamine reacted with the copper-TEMPO complex to afford the final α,β-disubstituted aldehyde.

Cooperative catalysis using cinchona alkaloid derivatives in combination with metals such as silver have also been widely developed.[15] On the basis of this concept, Escolano *et al.* have disclosed an enantioselective domino Michael–cyclisation reaction.[16] This formal [3 + 2] cycloaddition occurred between isocyanoacetates and enones in the presence of a combination of a chiral bifunctional cinchona alkaloid, such as cupreine, and AgNO$_3$ to provide the corresponding chiral 2,3-dihydropyrroles in low to high yields and

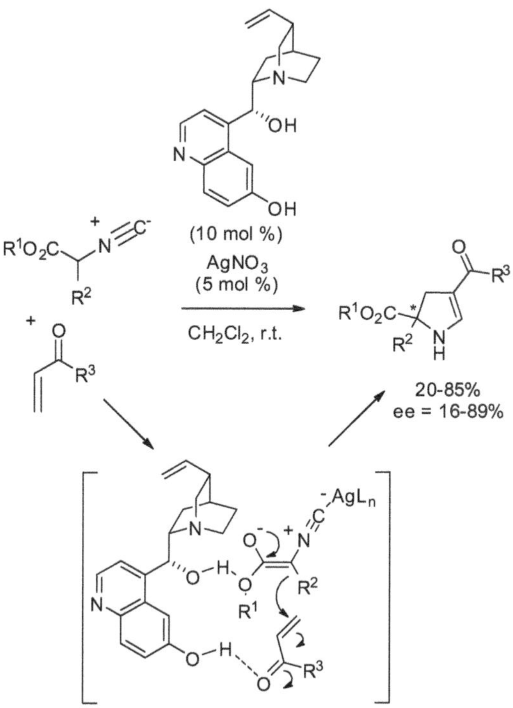

Scheme 7.9 Domino Michael–cyclisation reaction catalysed by chiral cinchona alkaloid catalysis and silver catalysis.

enantioselectivities of up to 89% ee, as shown in Scheme 7.9. In this process, AgNO$_3$ was used to increase the acidity of the pronucleophile while the bifunctional cupreine catalyst was responsible for the dual activation through hydrogen-bonding interactions, as shown in Scheme 7.9.

In the same area, Shi *et al.* reported a cinchona alkaloid-derived squaramide–AgSbF$_6$ cooperative catalytic system for the highly diastereo- and enantioselective formal [3 + 2] cycloaddition of α-aryl isocyanoacetates with *N*-aryl-substituted maleimides.[17] As shown in Scheme 7.10, a range of chiral 1,3a,4,5,6,6a-hexahydropyrrolo[3,4-c]pyrrole derivatives was prepared in high yields, excellent general diastereoselectivities of >90% de, combined with good to high enantioselectivities of up to 92% ee. To explain the stereoselectivity of the reaction, the authors have proposed the transition-state model depicted in Scheme 7.10, in which one carbonyl group of the maleimide was hydrogen-bonded to the squaramide motif, while the α-proton of the isocyanoacetate was easily deprotonated by the quinuclidine nitrogen of cinchona catalyst due to the activation of Ag(ı) chelating to the terminal carbon of the isocyano group. A single hydrogen bond was then formed between the OH group of the enolised isocyanoacetate and the tertiary amine of the cinchona alkaloid. A weak hydrogen bond between the OR^2 group of the enolised isocyanoacetate and the NH group in the

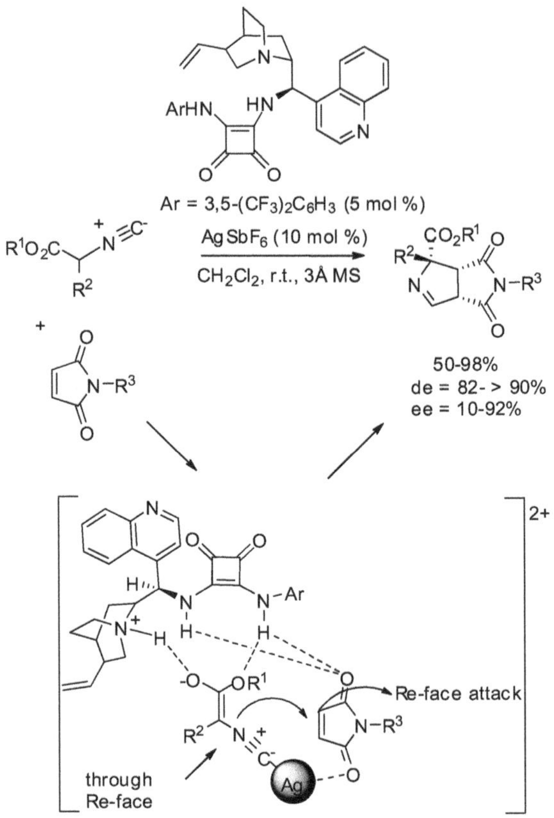

Scheme 7.10   Domino Michael–cyclisation reaction catalysed by chiral cinchona alkaloid catalysis and silver catalysis.

squaramide moiety, as well as an interaction between Ag(I) and the other carbonyl group of the maleimide could be formed concurrently, thus forcing the isocyanoacetate enolate to attack the maleimide from the *Re*-face, thereby leading to the formation of two newly generated stereocentres with (*R,R*)-configuration. Subsequently, a 5-*endo-dig* cyclisation took place assisted by electrophilic silver isocyanide activation. The third stereocentre was formed as *S*-configuration after the cyclisation step.

In another context, Dixon *et al.* have developed a new class of chiral aminophosphine precatalysts derived from 9-amino(9-deoxy) epicinchona alkaloids, which have been successfully applied in combination with Ag(I) salts, such as Ag$_2$O, to induce a related domino aldol–cyclisation reaction of branched aliphatic as well as aromatic aldehydes with various iso-cyanoacetates including α-substituted ones.[18] The corresponding chiral oxazolines were obtained in 50 to 93% yields, good to high diastereo-selectivities of up to 98% de, and good to excellent enantioselectivities of up to 98% ee, as shown in Scheme 7.11. This protocol could be performed by mixing together the ligand and Ag$_2$O, without the need to preform the active

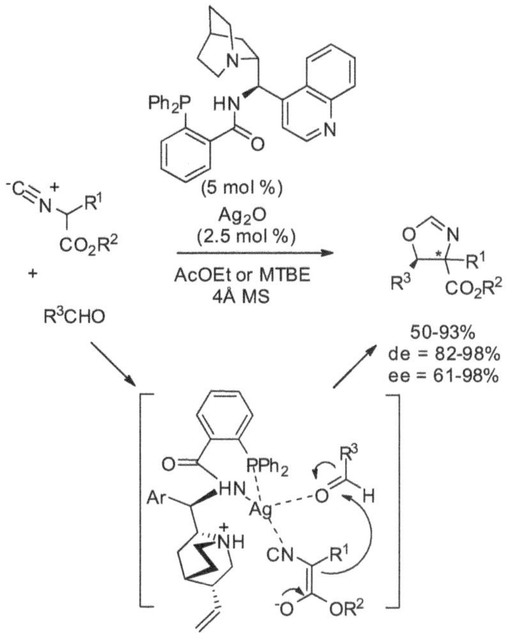

Scheme 7.11 Domino aldol–cyclisation reaction catalysed by chiral cinchona alkaloid catalysis and silver catalysis.

catalytic species. Moreover, the possibility of lowering the catalyst loading to 2 mol% of chiral ligand and 0.5 mol% of Ag_2O was demonstrated since, under these conditions, yields of 44 to 90% were obtained in combination with diastereo- and enantioselectivities of up to 86% de and 94% ee, respectively. Interestingly, when α-substituted isocyanoacetates were used, the opposite facial selectivity in the nucleophilic component was observed.

On the other hand, another cooperative catalysis approach was developed by Oh and Kim with a highly diastereo- and enantioselective domino aldol–cyclisation reaction occurring between aldehydes and methyl α-isocyanoacetate.[19] The process employed a combination of a chiral cobalt complex derived from brucine amino diol and an achiral thiourea. The reaction was applicable to a range of aliphatic, aromatic and heteroaromatic aldehydes, providing the corresponding chiral oxazolines in good yields and diastereoselectivities of up to >90% de combined with good to excellent enantioselectivities of up to 98% ee, as shown in Scheme 7.12.

Although a large number of transformations exist for *N*-heterocyclic carbene catalysis, until recently there was no knowledge about their compatibility with metals as catalysts and their workability in presence of each other.[20] In 2010, Scheidt *et al.* reported the first enantioselective cooperative catalytic system, consisting of $Mg(Ot\text{-}Bu)_2$ and a chiral *N*-heterocyclic carbene, which was applied in the presence of a base, such as 1,5,7-triazabicyclo[4.4.0]dec-5-ene (TBD), to a diastereo- and enantioselective synthesis

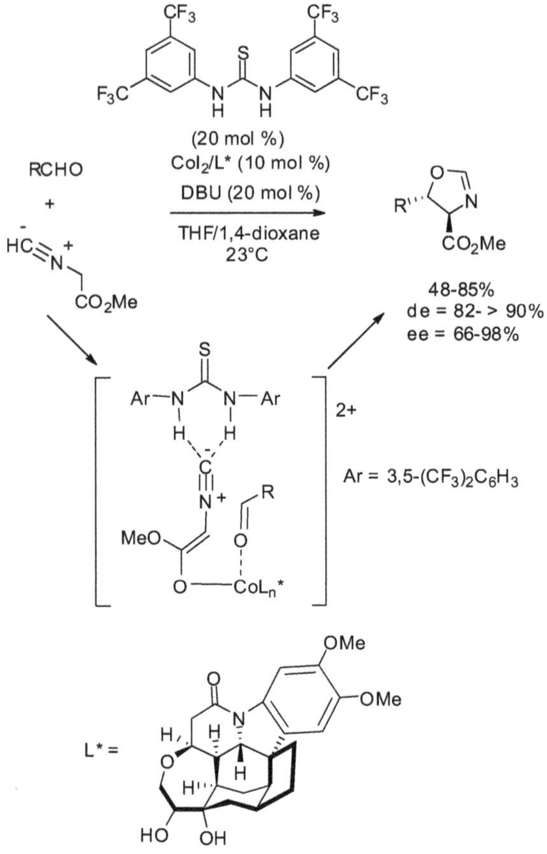

Scheme 7.12 Domino aldol–cyclisation reaction catalysed by thiourea catalysis and chiral cobalt catalysis.

of chiral γ-lactams from the reaction of *N*-acyl hydrazones with α,β-unsaturated aldehydes (Scheme 7.13).[21] The key behind the success was the reversible magnesium–*N*-heterocyclic carbene interaction, allowing high levels of diastereo- and enantioselectivities of up to 90% de, and 98% ee, respectively, to be achieved, as shown in Scheme 7.13.

The same authors have also used this chiral *N*-heterocyclic carbene catalyst in combination with Ti(O*i*-Pr)$_4$ to promote a highly diastereo- and enantioselective annulation reaction between α,β-unsaturated aldehydes and enones, which provided the corresponding chiral substituted *cis*-cyclopentenes in good yields and remarkable diastereo- and enantioselectivities of 90% de and 98 to 99% ee, respectively, as shown in Scheme 7.14.[22]

The scope of this protocol was extended by these authors to homoenolate addition to β,γ-unsaturated α-ketoesters, which afforded the corresponding chiral highly substituted and functionalisable cyclopentanols in good yields, moderate to high diastereoselectivities of up to 90% de, and excellent enantioselectivities of up to 98% ee, as shown in Scheme 7.15.[23]

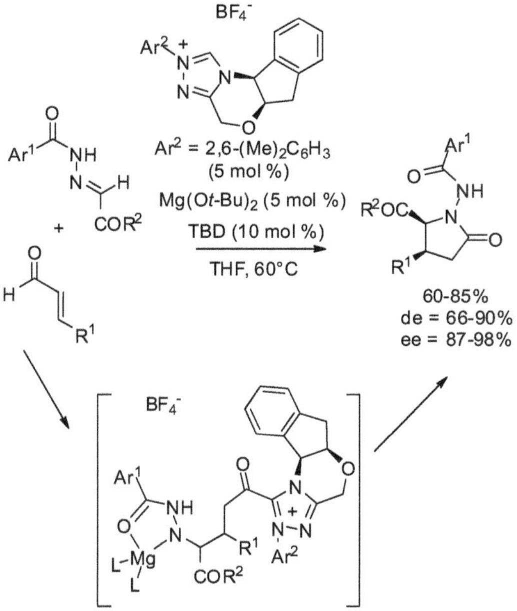

Scheme 7.13 Formal [3 + 2] cycloaddition catalysed by chiral *N*-heterocyclic carbene catalysis and magnesium catalysis.

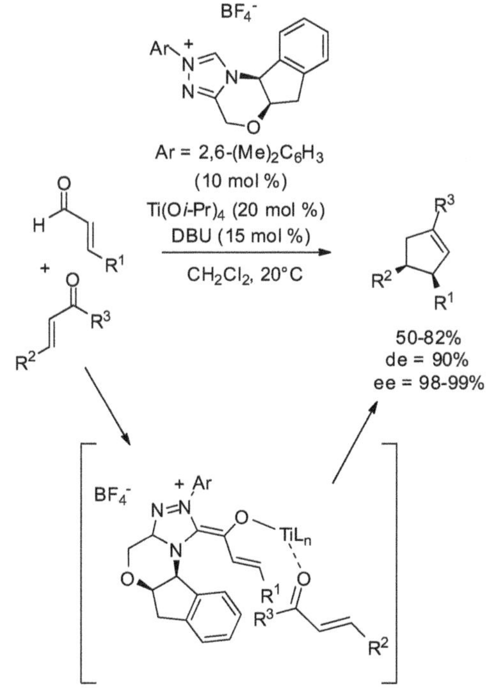

Scheme 7.14 Annulation reaction catalysed by chiral *N*-heterocyclic carbene catalysis and titanium catalysis.

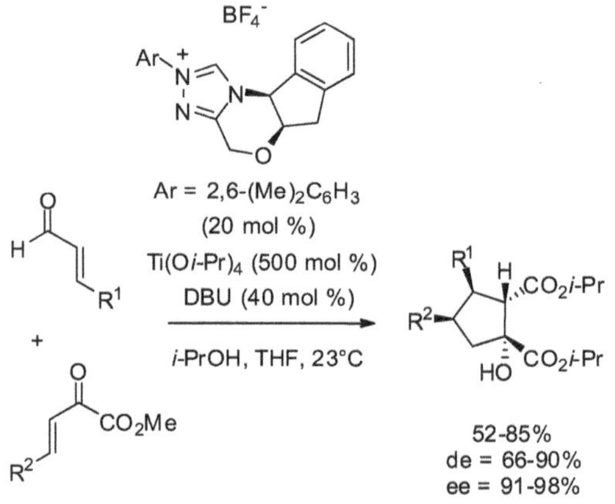

Scheme 7.15 Annulation reaction catalysed by chiral *N*-heterocyclic carbene catalysis and titanium catalysis.

In addition, these authors have used a closely related chiral *N*-heterocyclic carbene in association with Ti(O*i*-Pr)$_4$ to induce, in the presence of a base such as 1,5,7-triazabicyclo[4.4.0]dec-5-ene (TBD), a highly diastereo- and enantioselective dimerisation of α,β-unsaturated aldehydes to yield the corresponding chiral *cis*-cyclopentenes in good yields and high diastereo- and enantioselectivities of >90% de, and up to 90% ee, respectively, as shown in Scheme 7.16.[24] To explain the results, the authors have proposed the mechanism depicted in this scheme, beginning with the initial co-ordination of the enal to the titanium(ɪᴠ) complex which induced the formation of the extended Breslow intermediate **29**. The coordination of a second enal to the aldehyde–titanium(ɪᴠ) complex increased its reactivity towards conjugate addition and organised the reactants in space as shown in **30** to promote 1,4-addition over 1,2-addition. Both the homoenolate and the enal were thus poised to react through an s-*cis* conformation, ensuring high *cis*-diastereoselectivity in the products. Following the C–C bond formation to give intermediate **31**, the bis-enolate underwent protonation, tautomerisation and intramolecular aldol to afford intermediate **32**. The *N*-heterocyclic carbene catalyst was regenerated from acylation of the acyl azolium to give intermediate **33** and the final chiral cyclopentene was formed after base-promoted elimination.

In addition, Moberg *et al.* have reported a synthesis of enantioenriched *O*-acylated cyanohydrins cooperatively catalysed by a chiral titanium complex and an achiral base such as TEA.[25] The reaction occurred between acyl cyanides and prochiral aldehydes in dichloromethane at −40 °C, providing the corresponding *O*-acylated cyanohydrins in good to high yields (64–93%) and moderate to high enantioselectivities ranging from 20 to 96% ee. As shown in Scheme 7.17, it was supposed that the Lewis acid was able to activate both the acyl cyanide and the aldehyde which was verified by ^{13}C NMR spectroscopy, and the Lewis base simultaneously served to liberate the

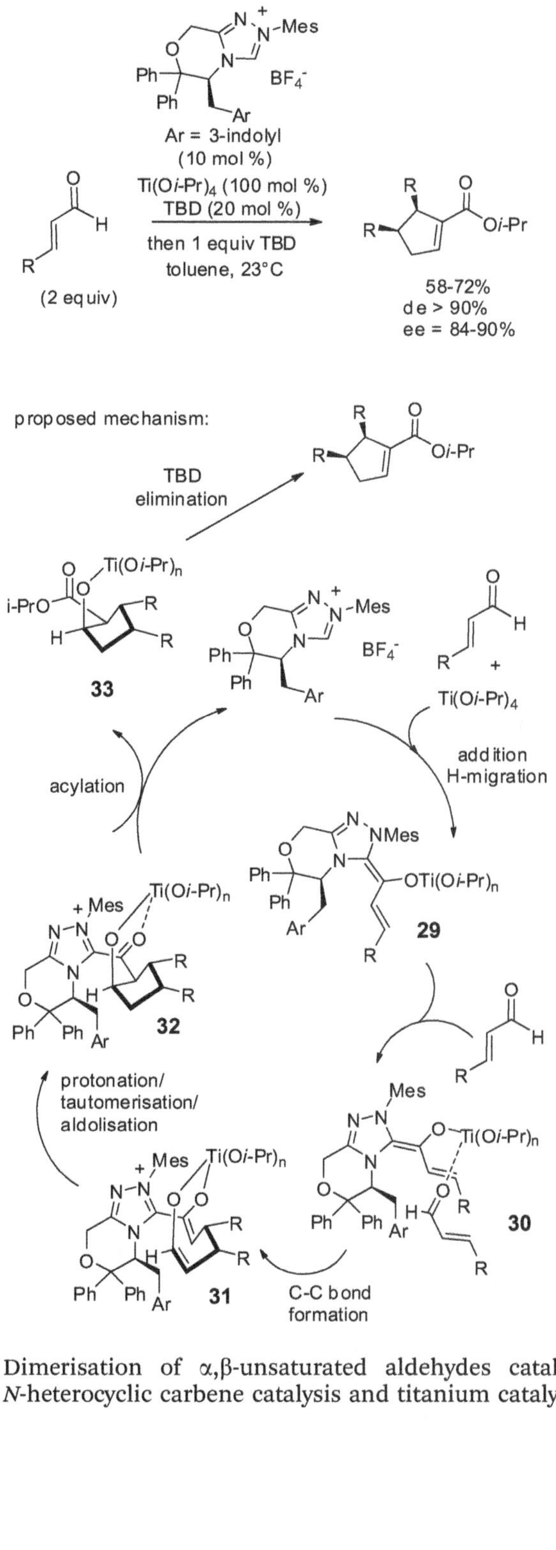

Scheme 7.16 Dimerisation of α,β-unsaturated aldehydes catalysed by chiral *N*-heterocyclic carbene catalysis and titanium catalysis.

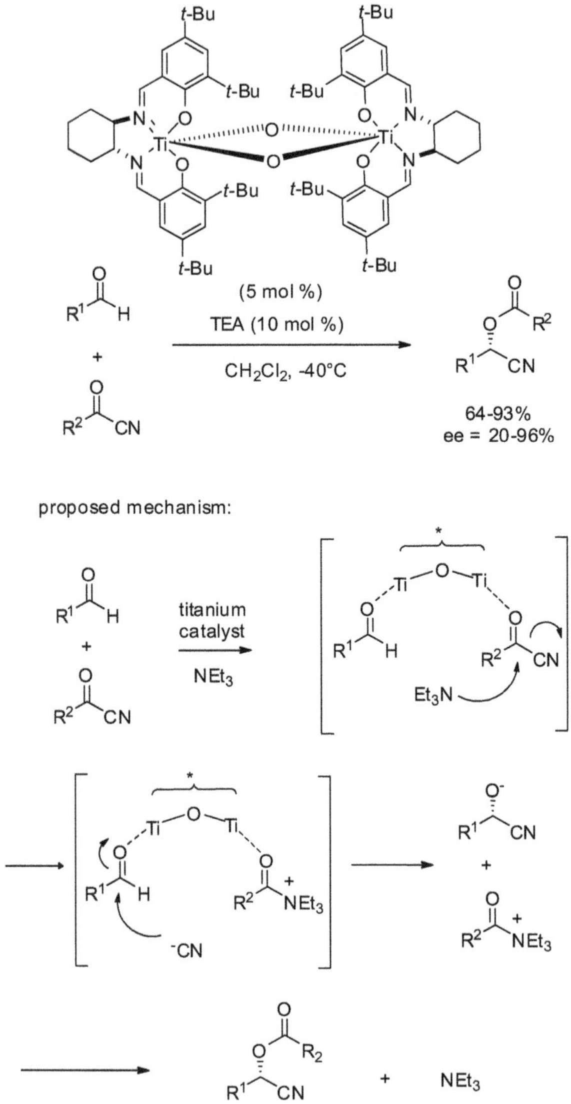

Scheme 7.17   Domino acylcyanation reaction catalysed by Lewis base catalysis and chiral titanium catalysis.

nucleophile CN⁻ and to form a potent acylating agent. Indeed, the authors assumed that the reaction proceeded *via* a route where both carbonyl compounds were bound to a dimeric titanium complex, the structure of which was, however, not known. Interestingly, the authors found that the reaction rate was increased by the presence of a tertiary amine, but the use of a chiral amine was not required for excellent enantioselectivities. Replacement of TEA with a cinchona-alkaloid had only minimal effect on both the yield and enantioselectivity.

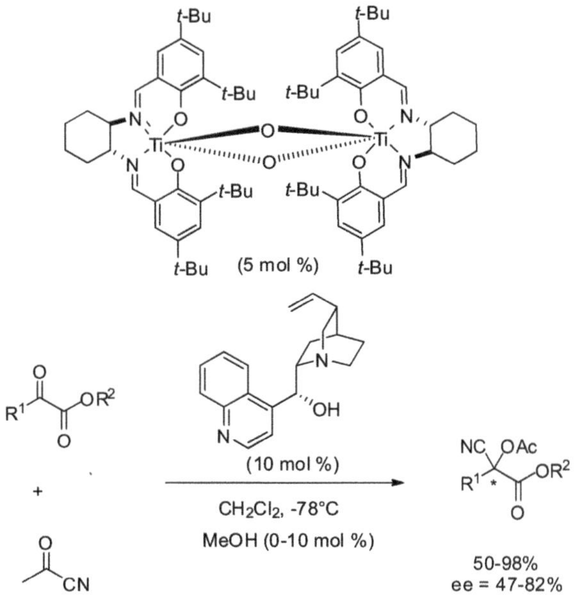

Scheme 7.18 Domino acylcyanation reaction catalysed by chiral cinchona catalysis and chiral titanium catalysis.

Other works by the same group on the cyanation of α-ketoesters with acetyl cyanide revealed that, when using a ketone as substrate, a chiral base such as cinchonidine became responsible for enantioselection instead of the chiral titanium(IV) salen complex (Scheme 7.18), leading to the corresponding tertiary cyanohydrins in good enantioselectivities of up to 82% ee.[26]

In 2010, Shibata and co-workers developed an enantioselective enamine–trifluoropyruvate domino aldol–cyclisation reaction to yield chiral pyrrolidones.[27] Several commercially available derivatives of cinchona alkaloids were screened in combination with Ti(O*i*-Pr)₄. Hydroquinine diether ((DHQD)₂AQN) was found to be the best ligand and afforded the products in high yields and enantioselectivities of up to 92% ee, as shown in Scheme 7.19. Using this system, five cyclic enamines with different protecting groups were screened with remarkable results. The enantiomers of the products were also accessible by applying the pseudoenantiomeric cinchona alkaloid.

In 2009, Xiao and co-workers described a combination of a chiral iridium complex with a chiral phosphoric acid to induce a direct asymmetric reductive amination of ketones with anilines which provided the corresponding chiral amines in good yields of up to 94% and high enantioselectivities of up to 97% ee, as shown in Scheme 7.20.[28] The iridium complex reduced the *in situ* generated iminium cation *via* ionic hydrogenation and phosphoric acid aiding enantioselective hydrogen transfer *via* ion pairing of its conjugate base with iminium ion.

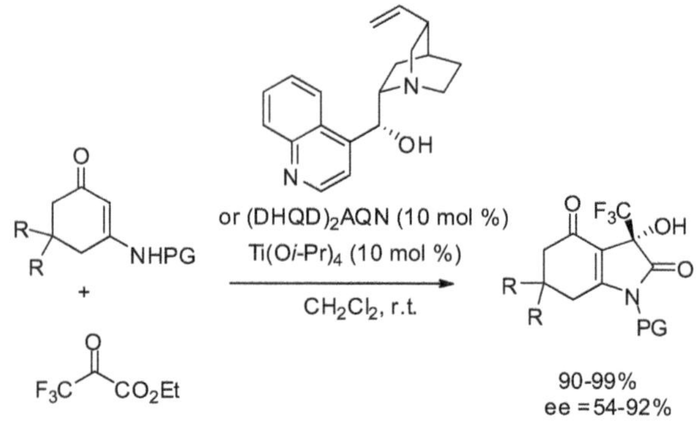

Scheme 7.19 Domino aldol–cyclisation reaction catalysed by chiral cinchona catalysis and titanium catalysis.

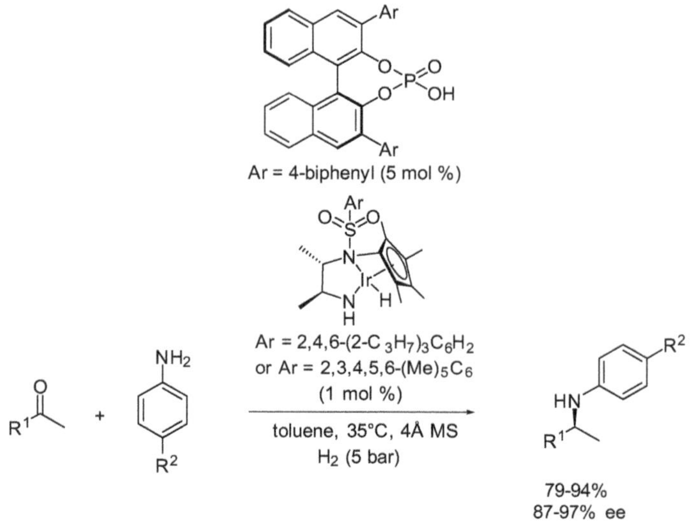

Scheme 7.20 Domino reductive amination catalysed by chiral phosphoric acid catalysis and chiral iridium catalysis.

Iridium–diamine complexes are inactive in hydrogenation reactions; however, upon protonation by a strong Brønsted acid, they turn into highly active hydrogenation catalysts. If a chiral, strongly acidic Brønsted acid is applied, the acid not only activates the catalyst but also renders it chiral. Thus, enantioselectivities depend on the chiral activating acid. In this context, Rueping and Koenigs have combined a racemic iridium complex with a chiral *N*-triflylphosphoramide to induce the asymmetric hydrogenation of quinolones by kinetic discrimination.[29] As shown in Scheme 7.21, a variety of chiral 2-substituted quinolones could be reduced to the corresponding tetrahydroquinolines in good yields and enantioselectivities of up to 82% ee.

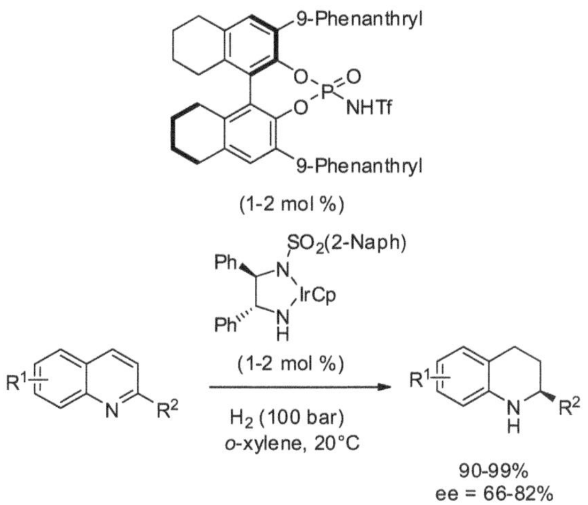

Scheme 7.21 Domino hydrogenation of quinolines catalysed by chiral phosphoramide catalysis and iridium catalysis.

This work constituted the first application of chiral *N*-triflylphosphoramides in combination with a metal catalyst.

Gold catalysts have also been combined with organocatalysts to cooperatively induce enantioselective domino reactions. As a very recent example, an enantioselective cooperative triple catalysis was employed by Patil *et al.* to promote a domino addition–cycloisomerisation–transfer hydrogenation reaction between 2-aminobenzaldehydes and terminal alkynes in the presence of Hantzsch ester.[30] Indeed, this process involved the simultaneous action of three catalysts, such as an achiral gold catalyst, an achiral amine such as *p*-anisidine, and a chiral phosphoric acid, affording the corresponding chiral 2-substituted tetrahydroquinolines in good to very high yields and excellent enantioselectivities of up to 99% ee (Scheme 7.22). The authors have experimentally demonstrated the concerted/simultaneous action of the three catalysts.

Another type of cooperatively catalysed domino reaction was described by Zhou *et al.*, in 2011.[31] It consisted of the decomposition of a diazo compound, such as an α-aryl- or α-alkyl-α-diazoacetate derivative, in the presence of a catalytic amount of $Rh_2(TPA)_4$ to generate a rhodium carbene intermediate **34**, which reacted with $BocNH_2$ to provide ylide **35**. The second catalyst, a chiral spiro phosphoric acid, assisted proton transfer *via* a seven-membered-ring intermediate **36** to form the final N–H insertion product, and regeneration of the two catalysts. It must be noted that this process used a low catalyst loading of only 1 mol% for each catalyst. As shown in Scheme 7.23, a range of chiral alkyl 2-(*tert*-butoxycarbonylamino)-2-arylacetate derivatives were synthesised in both high yields and enantioselectivities of up to 95% ee. It must be noted that the lowest enantioselectivities (50% ee) were obtained in the cases of α-alkyl-α-diazoacetate derivatives

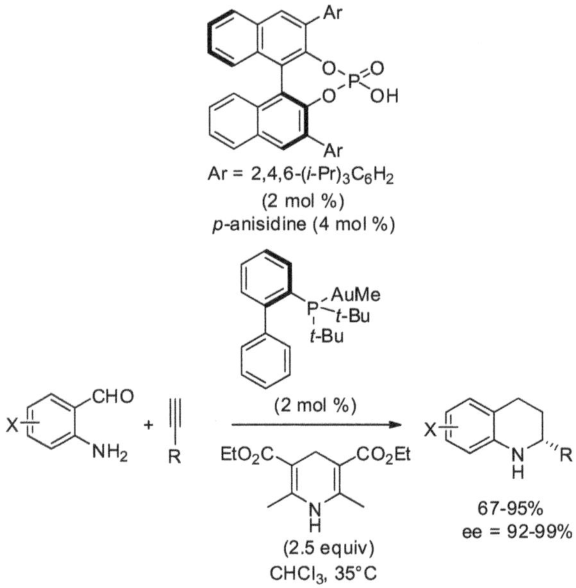

Scheme 7.22 Domino addition–cycloisomerisation–transfer hydrogenation reaction catalysed by chiral phosphoric acid catalysis, amine catalysis, and gold catalysis.

whereas enantioselectivities ranging from 90 to 95% ee were achieved in the cases of α-aryl-α-diazoacetate derivatives.

Combinations of metal catalysts such as rhodium complexes with organocatalysts have also been recently employed to induce enantioselective multicomponent domino reactions. For example, in 2008 Hu and co-workers developed an enantioselective four-component reaction of aryl diazoacetates with alcohols, aldehydes, and amines on the basis of the cooperative catalysis of an achiral rhodium complex and a chiral phosphoric acid to readily achieve the corresponding β-amino-α-hydroxyl acid derivatives with excellent control of chemo-, diastereo- and enantioselectivity, as shown in Scheme 7.24.[32] Under catalysis with the Brønsted acid, benzaldehyde reacted with *p*-methoxyaniline (PMP–NH$_2$) to form the corresponding enammonium phosphate salt, which reacted with the oxonium ylide intermediate generated *in situ* from diazoacetate and alcohol upon catalysis with rhodium catalyst. These two intermediates could be organised through hydrogen-bonding with the bridging chiral anion of the chiral acid to form a three-component intermediate, which provided the final product in high yields, and excellent diastereo- and enantioselectivity of up to >98% de and 97% ee, respectively.

Later, these authors reported the utilisation of a combination of Rh$_2$(OAc)$_4$ with a chiral BINOL-derived phosphoric acid to induce an enantioselective three-component Mannich-type reaction of diazo compounds, carbamates and imines, providing a rapid and efficient access to

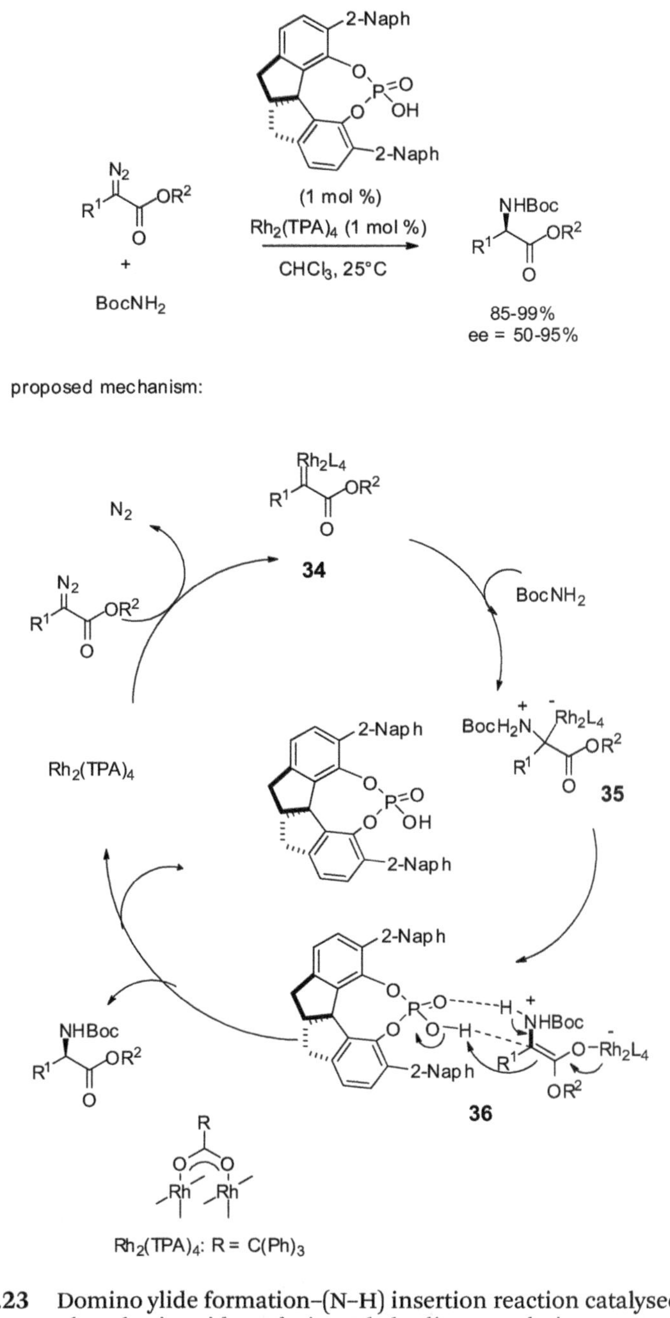

Scheme 7.23 Domino ylide formation–(N–H) insertion reaction catalysed by chiral phosphoric acid catalysis and rhodium catalysis.

both *syn*- and *anti*-α-substituted α,β-diamino acid derivatives with a high level control of chemo-, diastereo-, and enantioselectivity of up to 99% ee.[33] As shown in Scheme 7.25, the use of the (*R*)-enantiomer of a phosphoric acid

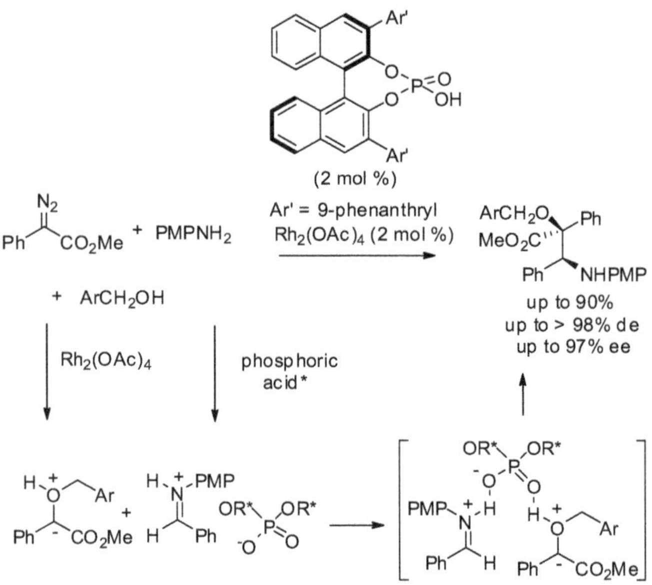

Scheme 7.24   Three-component reaction of aryl diazoacetate with alcohol, aldehyde and amine catalysed by chiral phosphoric acid catalysis and
rhodium catalysis.

stereoselectively led to the corresponding *syn*-products, whereas the in-
volvement of the (*S*)-enantiomer of a closely related phosphoric acid stereoselectively afforded the corresponding *anti*-products. The mechanism of
this process involved the decomposition of the diazo compound in the
presence of a catalytic amount of $Rh_2(OAc)_4$ to generate a rhodium carbene
intermediate **39**, which reacted with the carbamate to provide protic carbamate ammonium ylides **37** and **38**. Then, ylide **38** underwent reaction with
the imine, giving protic carbamate ammonium ylide trapping product *via*
zwitterion **40**. These novel diastereoselectively switchable enantioselective
three-component reactions opened up an avenue to all isomers of α,β-dia-
mino acid derivatives with an α-quaternary carbon centre in an optically
active form.

   As an extension, the same authors have reported a highly enantioselective
trapping of zwitterionic intermediates generated from *N*-aryldiazoamides
in the presence of rhodium by imines with the activation of the same
chiral phosphoric acid.[34] This process provided the corresponding chiral
polyfunctionalised oxindole derivatives in good to high yields, diastereo-,
and enantioselectivities of up to 81%, 98% de, and 98% ee, respectively
(Scheme 7.26). In addition, they reported an extension of the zwitterionic
intermediate trapping process to a three-component reaction between an
α-aryl-α-diazoacetate, an imine and an indole, leading to the corresponding
chiral indole derivatives in high to excellent diastereo- and enantioselectivities of up to >90% de and 98% ee, respectively, as shown in Scheme 7.26.

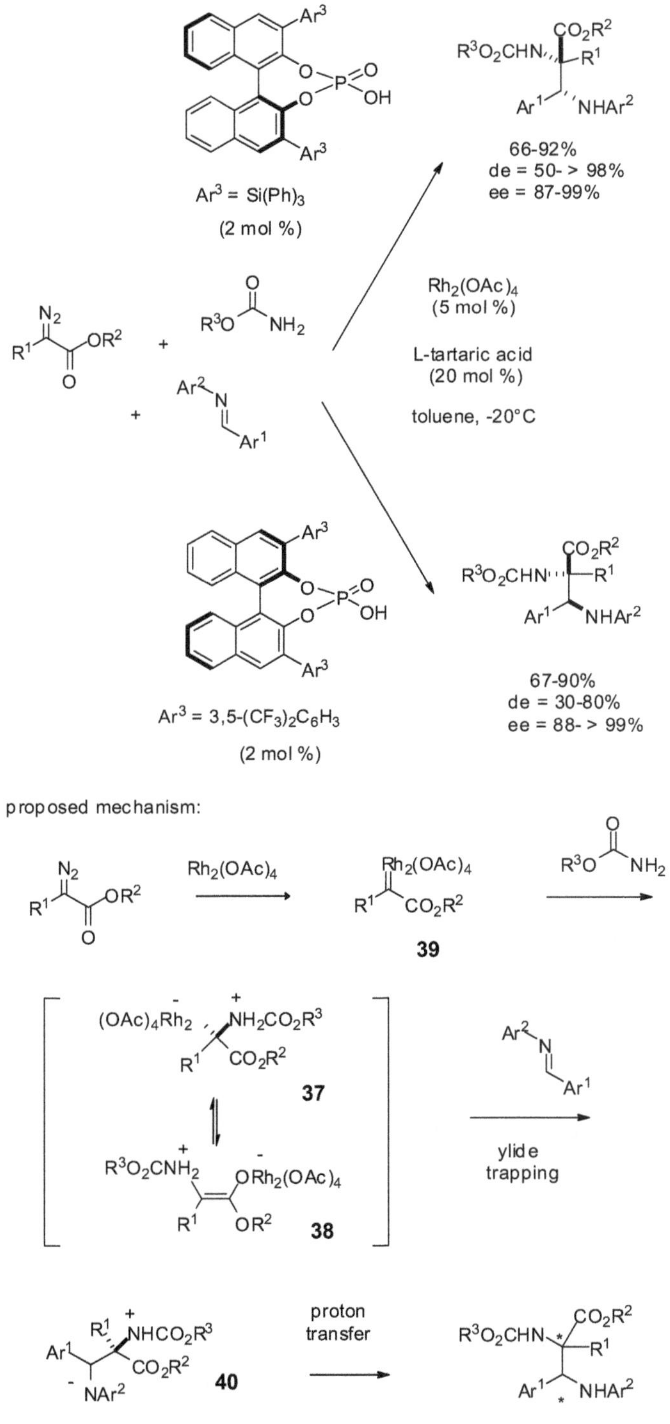

proposed mechanism:

Scheme 7.25 Mannich-type three-component reaction catalysed by chiral phosphoric acid catalysis and rhodium catalysis.

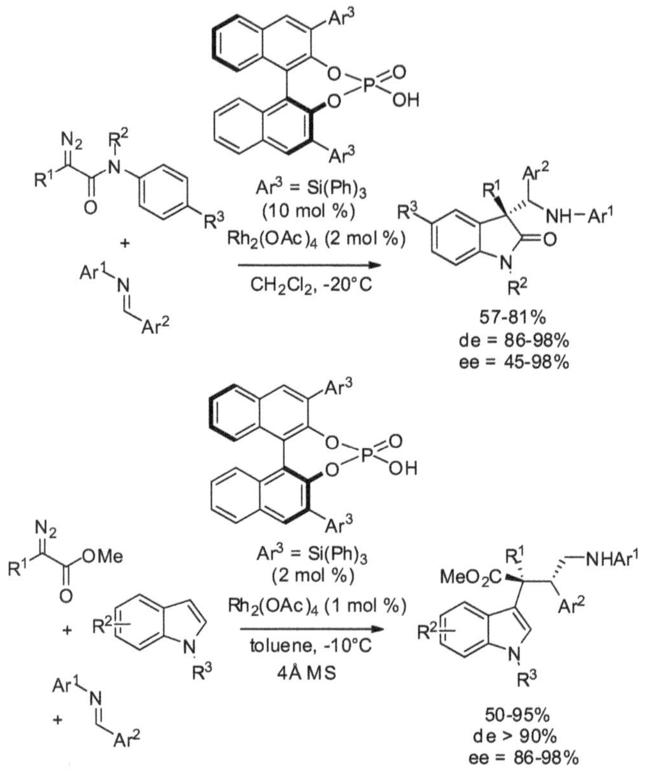

Scheme 7.26 Mannich-type two- and three-component reactions catalysed by chiral phosphoric acid catalysis and rhodium catalysis.

In addition, these authors have developed a related highly diastereo- and enantioselective three-component Mannich-type reaction between diazoacetophenones, 9-anthryl alcohol and imines, which provided the corresponding chiral β-amino-α-hydroxyl ketones.[35] Again, the process was cooperatively catalysed by Rh$_2$(OAc)$_4$ and a chiral BINOL-derived phosphoric acid, allowing the *anti*-products to be obtained in good to high yields, excellent diastereoselectivities of >96% de in almost all the cases studied, and high enantioselectivities ranging from 80 to 98% ee, as shown in Scheme 7.27. In this process, the diazoacetophenone and the alcohol served as an enol equivalent of an α-alkoxy aryl ketone by forming an oxonium ylide in the presence of the rhodium catalyst, and the chiral phosphoric acid was used to control the selectivity of the reaction by activating an imine component. This methodology was applied to an efficient synthesis of a taxol side chain and (−)-*epi*-cytoxazone.[36]

The same authors have described a related enantioselective three-component Mannich-type reaction of diazo compounds, imines and water in the presence of a combination of Rh$_2$(OAc)$_4$ with a chiral BINOL-derived phosphoric acid.[37] Following the same mechanism as above (with water

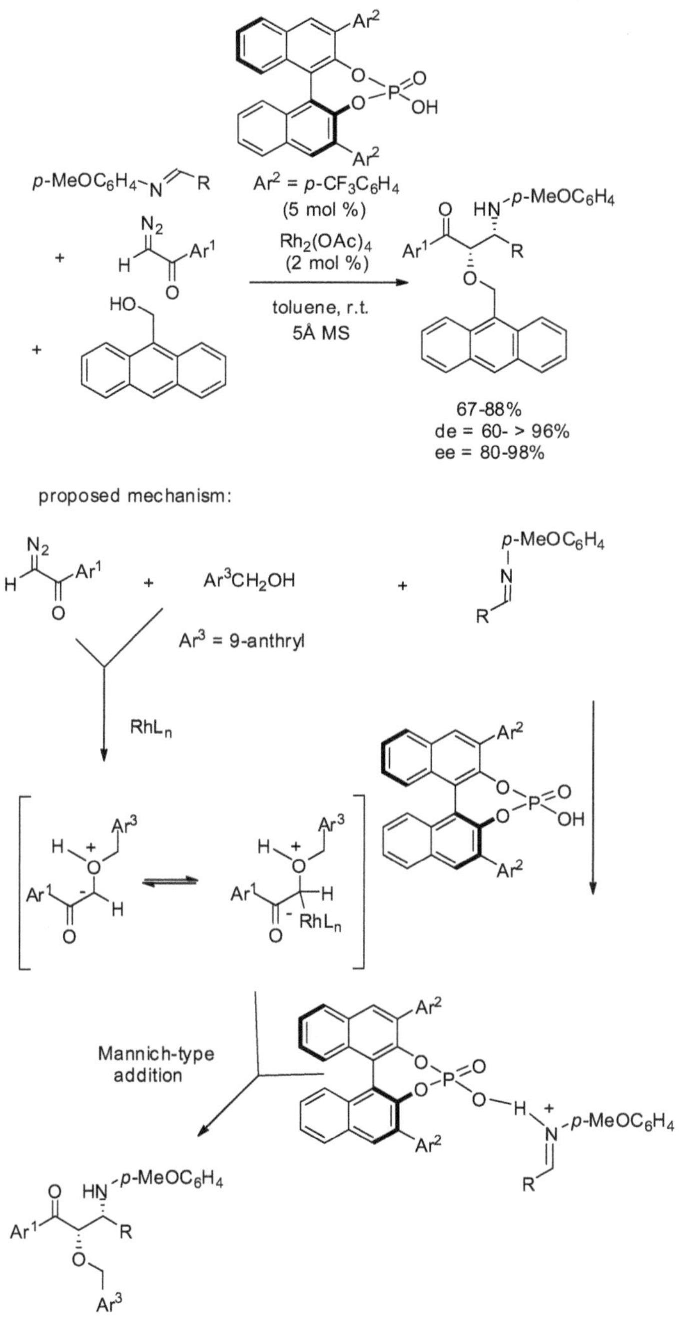

proposed mechanism:

Scheme 7.27 Mannich-type three-component reaction catalysed by chiral phosphoric acid catalysis and rhodium catalysis.

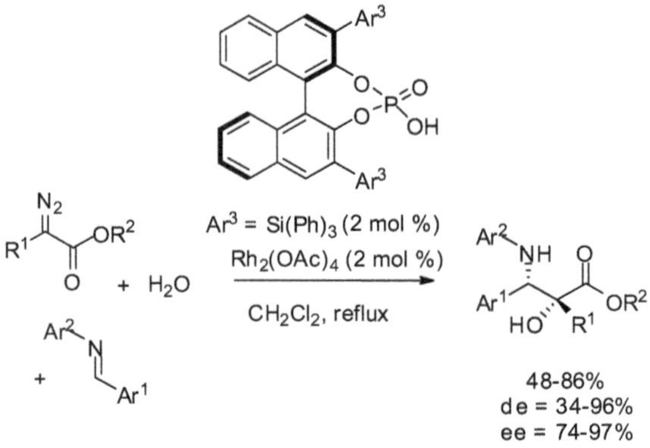

Scheme 7.28   Three-component reaction catalysed by chiral phosphoric acid catalysis and rhodium catalysis.

instead of alcohol), the corresponding chiral β-amino-α-hydroxy acid derivatives were produced in good yields and diastereoselectivities of up to 96% de combined with high enantioselectivities ranging from 74 to 97% ee, as shown in Scheme 7.28.

The Biginelli reaction is a three-component reaction allowing the synthesis of 3,4-dihydropyrimidin-2-(1*H*)-ones or -thiones by reacting urea or thiourea, a 1,3-dicarbonyl derivative, and an aldehyde.[38] The heterocyclic pyrimidinone products are known to exhibit a wide range of important pharmacological properties and make up a large family of medicinally relevant compounds. Consequently, asymmetric catalytic Biginelli reactions have been a long-standing challenge. In this context, Xu *et al.* have recently developed a novel enantioselective Biginelli reaction cooperatively catalysed by a combination of NbCl₅, responsible for the reactivity, and a chiral primary amine derived from quinine, introducing the stereoselectivity of the reaction.[39] As shown in Scheme 7.29, the corresponding chiral dihydropyrimidiones were produced in good to excellent yields and enantioselectivities ranging from 43 to 84% ee. The authors assumed that the reaction began with the formation of the enamine of the 1,3-dicarbonyl substrate by reaction with the chiral primary amine of the organocatalyst. This enamine activated by the metal-based Lewis acid reacted with *N*-acylimine *in situ* generated from aldehyde and urea to provide the final dihydropyrimidione, as shown in Scheme 7.29.

7.3   Relay Catalysis

In recent years, several groups have developed enantioselective tandem reactions based on the combination of gold catalysis and organocatalysis.[5,40] Among them, Gong *et al.* reported that an achiral gold complex compatibly worked with a chiral phosphoric acid to promote a domino intramolecular hydroamination–reduction reaction, readily transforming

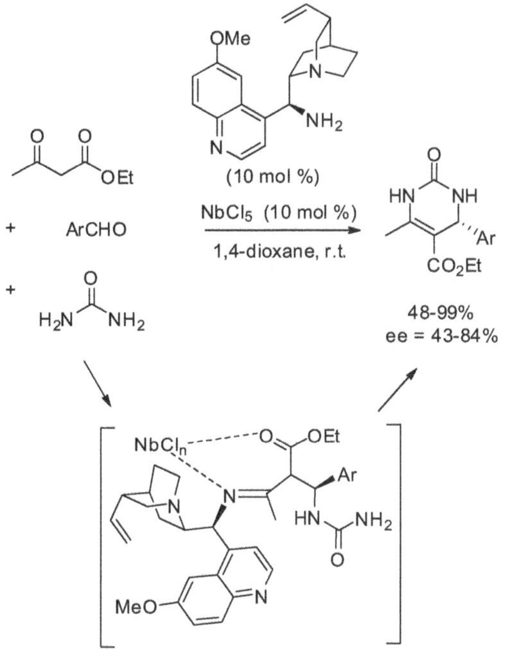

Scheme 7.29 Biginelli reaction catalysed by chiral amine catalysis and niobium catalysis.

2-(2-propynyl)aniline derivatives into the corresponding chiral tetra-hydroquinolines in one operation with excellent enantioselectivities of up to 97% ee combined with almost quantitative yields.[41] This process was initiated by a gold-catalysed intermolecular hydroamination of an alkyne, followed by a chiral phosphoric acid-catalysed enantioselective transfer hydrogenation using Hantzsch ester (Scheme 7.30).

Almost at the same time, Liu and Che independently applied the same strategy to the synthesis of chiral secondary amines through a domino intermolecular hydroamination–transfer hydrogenation of alkynes using a gold(ı) complex in combination with a chiral phosphoric acid.[42] This domino process has a broad substrate scope since a wide variety of aryl, alkenyl, and aliphatic alkynes could be coupled with anilines with different electronic properties to afford chiral amines in excellent enantioselectivities of up to 94% ee, as shown in Scheme 7.31.

In 2013, Gong *et al.* reported a relay catalytic domino hydroamination–redox reaction, which was able to directly assemble tertiary amine substituted 3-en-1-yne derivatives and anilines into the corresponding cyclic aminals by using a gold(ı) complex and an achiral phosphoric acid.[43] By using a chiral phosphoric acid, the authors have developed an enantioselective version of this reaction. As shown in Scheme 7.32, the reaction of 1-(2-ethynylphenyl)pyrrolidine and *p*-anisidine in the presence of a catalytic combination of a chiral phosphoric acid and Ph₃PAuNTf₂ led to the

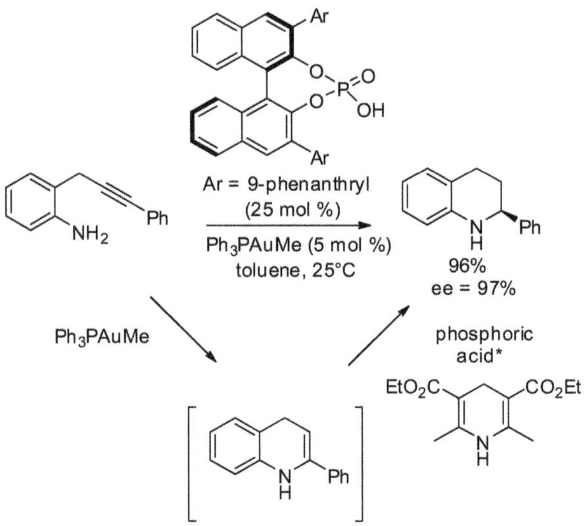

Scheme 7.30 Domino intramolecular hydroamination–reduction reaction catalysed by chiral phosphoric acid catalysis and gold catalysis.

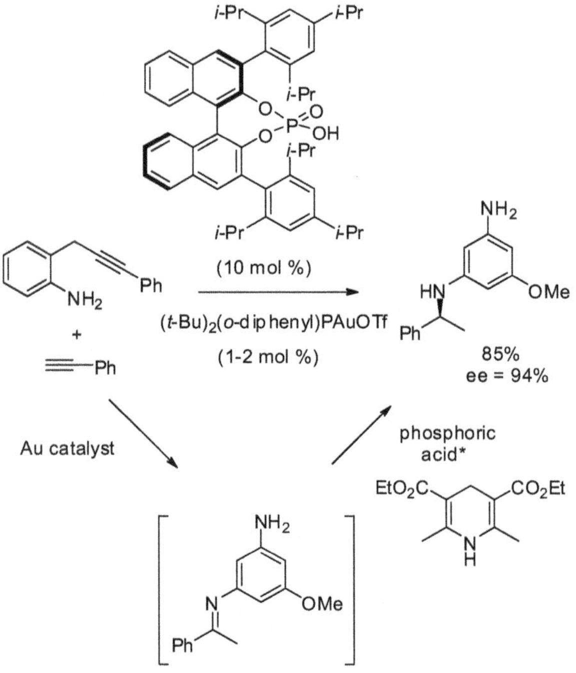

Scheme 7.31 Domino intermolecular hydroamination–reduction reaction catalysed by chiral phosphoric acid catalysis and gold catalysis.

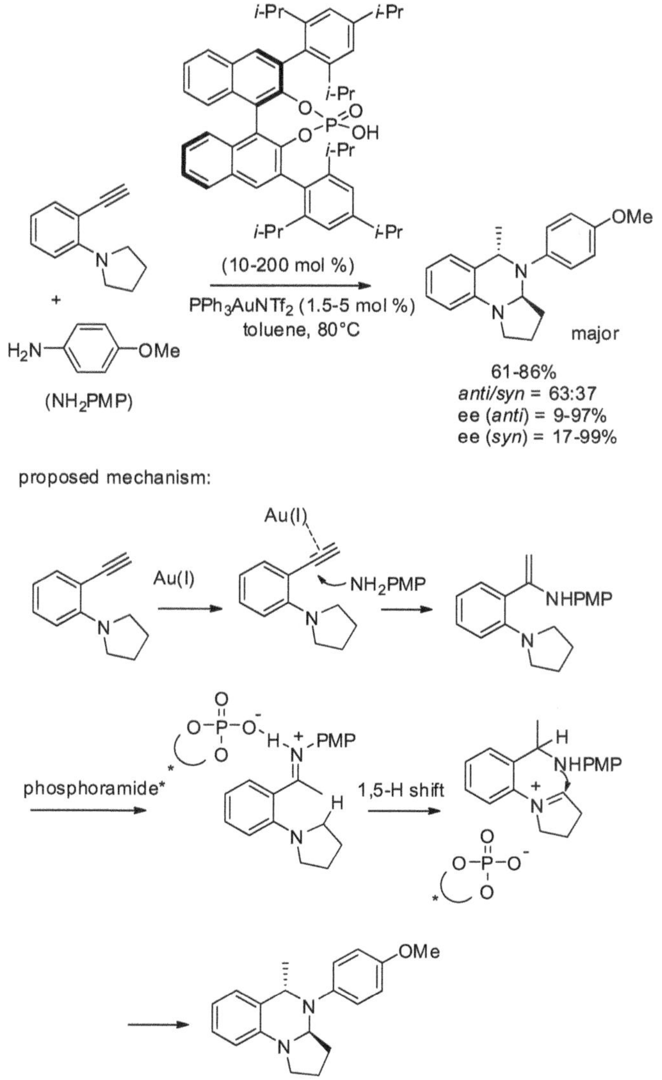

Scheme 7.32 Domino intermolecular hydroamination–redox reaction catalysed by chiral phosphoric acid catalysis and gold catalysis.

corresponding chiral tricyclic domino product in good yield (61–86%) as a 63 : 37 mixture of *anti-* and *syn*-diastereomers, which were obtained with enantioselectivities of up to 97 and 99% ee, respectively. The authors have proposed the mechanism depicted in Scheme 7.32 to explain the results. Basically, in the presence of the gold(ı) complex, the terminal alkyne was able to undergo an intermolecular hydroamination with *p*-anisidine to give the corresponding imine intermediate, which formed an iminium species with the chiral phosphoric acid. The latter species then underwent a

subsequent 1,5-hydride transfer to generate a transient iminium intermediate, which finally underwent a cyclisation reaction to give the product.

In 2010, Jørgensen *et al.* reported a gold/chiral organo orthogonal-relay
catalytic system for the synthesis of chiral cyclopentene carbaldehydes.[44]
These domino products were obtained in good yields and enantioselectivities of up to 96% ee starting from enones and propargylated malononitrile, as shown in Scheme 7.33. Moreover, when this reaction was
catalysed by a combination of CuOTf (5 mol%) with the same chiral organocatalyst derived from L-proline, even higher enantioselectivities ranging
from 90 to 99% ee were reached combined with yields of 49 to 91%. This
relay catalytic system was extended to the domino reaction of enones and
cyanoacetates to give the corresponding chiral cyclopentenes in moderate to
good yields (50–69%), moderate to excellent diastereoselectivities (50–>90%
de), and generally excellent enantioselectivities ranging from 90 to 99% ee.
As depicted in Scheme 7.33, the domino reaction began with the activation
of the enal through iminium ion formation by using the chiral amine
catalyst which induced the Michael addition of the propargylated nucleophile. The intermediate **41** formed underwent a 5-*exo-dig* cyclisation, forming a C–C bond in which both the organocatalyst and the Lewis acid were
involved, followed by double bond isomerisation to give the final chiral
product.

Other chiral organocatalysts, such as chiral phosphoric acids, have been
associated with gold complexes to perform relay catalysis applied to enantioselective tandem reactions. As an example, Gong *et al.* have described
asymmetric cyclisation of alkynol triggered addition of azlactones catalysed
by a combined catalyst system consisting of a gold complex of an achiral
sterically bulky phosphine and a chiral BINOL-derived phosphoric acid as
organocatalyst.[45] As shown in Scheme 7.34, the tandem reaction produced
chiral restricted amino acid precursors bearing vicinal quaternary stereogenic centres in high yields, moderate to good diastereoselectivities combined with low to high enantioselectivities of up to 95% ee (Scheme 7.34).
This process presumably proceeded initially with an intramolecular hydro-
alkoxylation with a suitable gold catalyst to generate enol ether **42**. This enol
ether participated in the addition reaction of the azlactone through two
possible intermediates including (1) an ion pair of a chiral conjugate base
and an oxonium ion **43** formed from protonation of the enol ether with the
phosphoric acid and (2) a chiral oxonium intermediate **43'** generated from
coordination of the chiral gold catalyst to the double bond of the enol ether.
Either intermediate **43** or **43'** was able to participate in the enantioselective
addition of the azlactone *via* the transition states **44** and **44'**, furnishing the
final chiral product.

In 2010, Gong *et al.* reported a highly enantioselective three-component
domino reaction, consisting of an enantioselective aza-Diels–Alder cycloaddition catalysed by a chiral phosphoric acid and a subsequent intramolecular hydroamination catalysed by a gold complex.[46] The domino
reaction occurred between aldehydes, an enamide, and 2-(2-propynyl)aniline

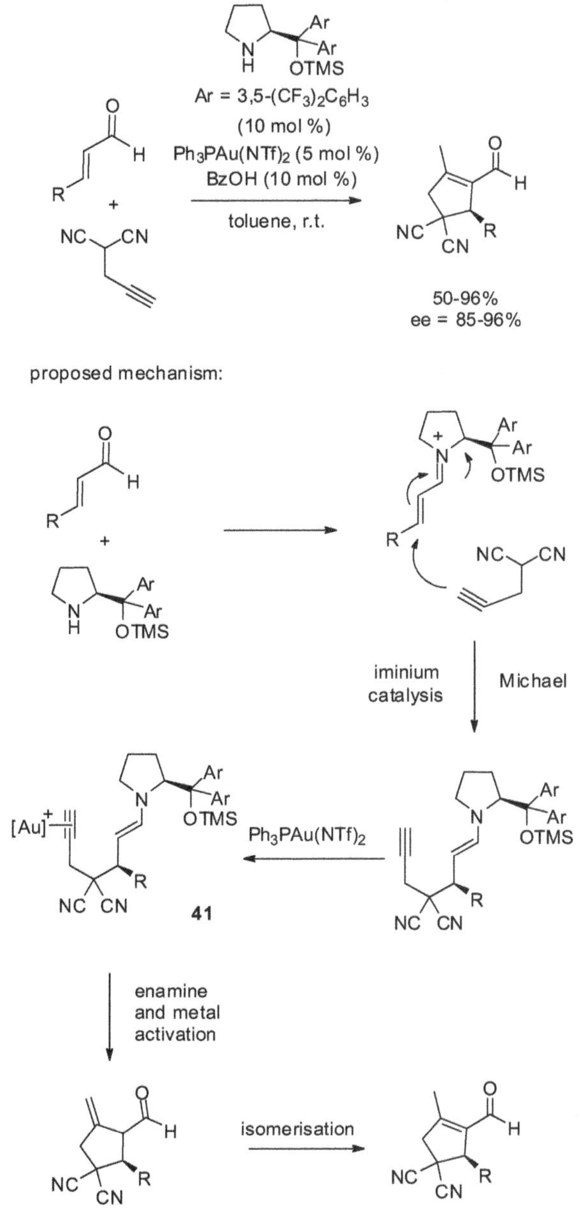

Scheme 7.33 Domino Michael–cyclisation reaction catalysed by chiral amine catalysis and gold catalysis.

derivatives to give, after subsequent reduction by treatment with acetic acid and sodium triacetoxyborohydride added to the reaction media, the corresponding structurally diverse julolidine derivatives in good yields, moderate *cis*-diastereoselectivities, and generally excellent enantioselectivities

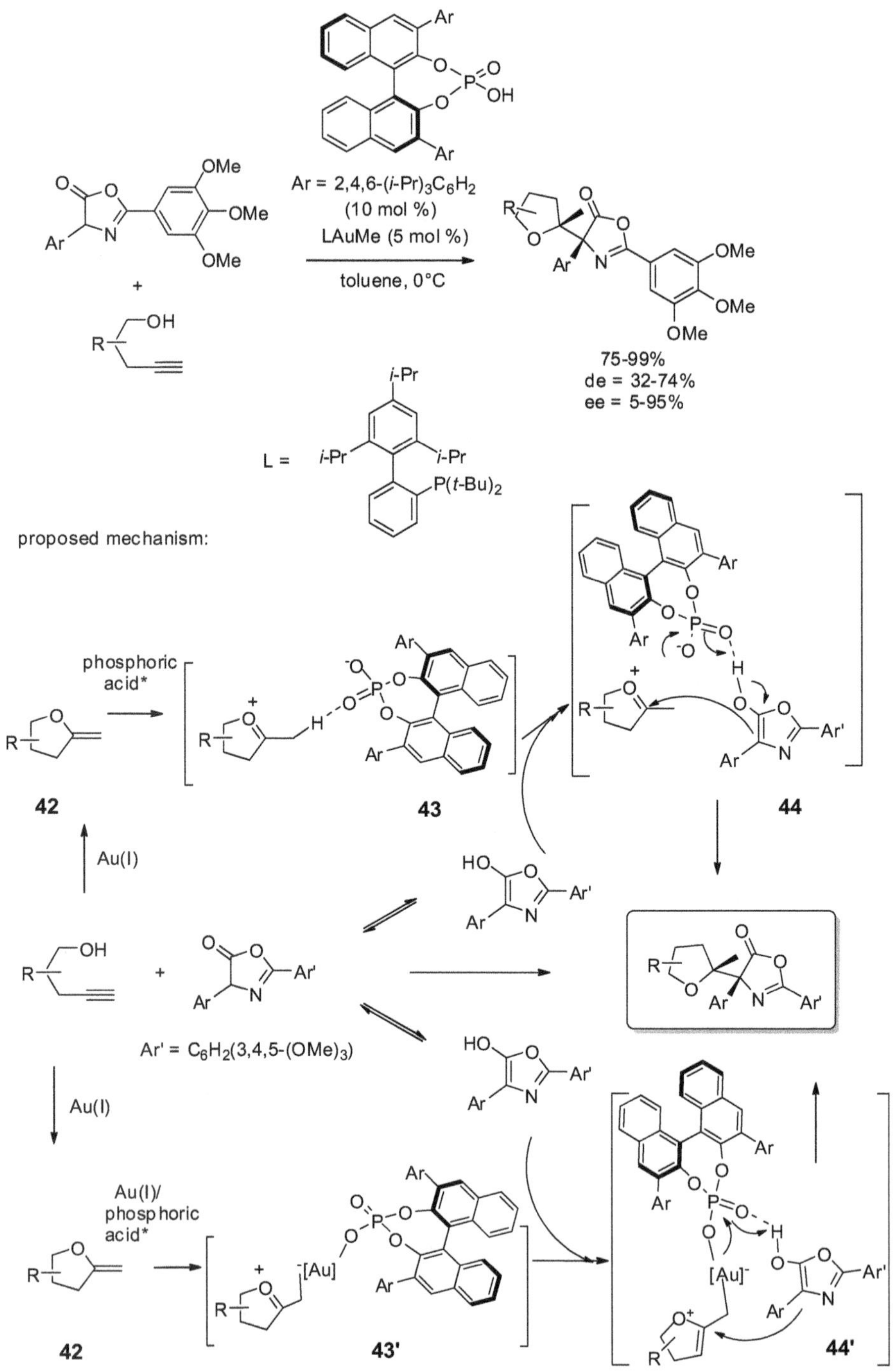

Scheme 7.34 Domino cyclisation–addition reaction catalysed by chiral phosphoric acid catalysis and gold catalysis.

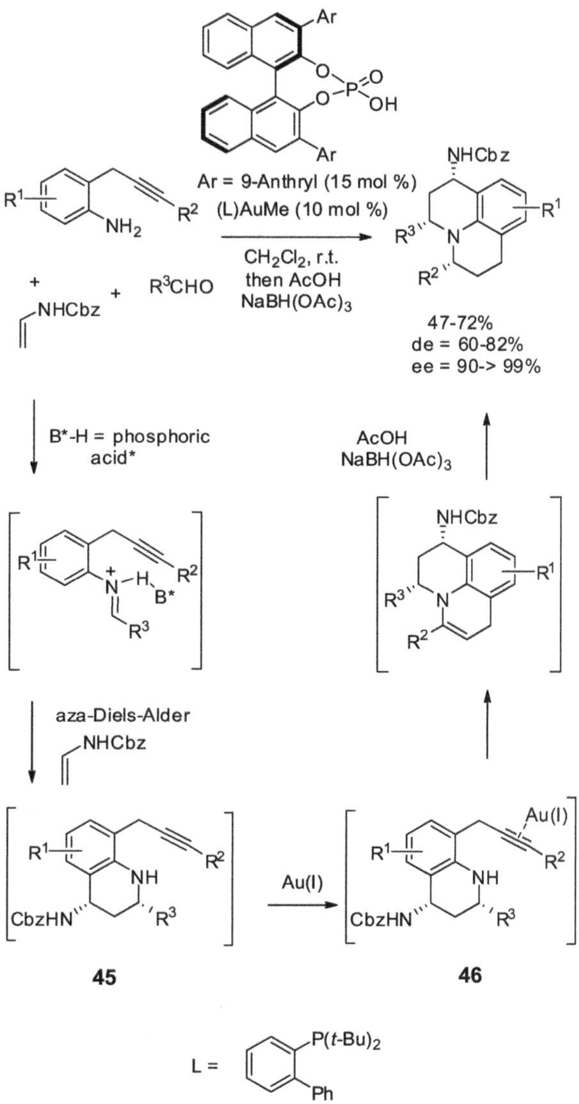

Scheme 7.35 Three-component domino aza-Diels – Alder–hydroamination reaction catalysed by chiral phosphoric acid catalysis and gold catalysis followed by reduction.

ranging from 90 to >99% ee, as shown in Scheme 7.35. A mechanism is proposed in Scheme 7.35, showing that the process began with a three-component inverse electron-demand aza-Diels–Alder reaction (Povarov reaction) of the 2-(2-propynyl)aniline derivative, the aldehyde and the enamide catalysed by the chiral phosphoric acid to give the intermediate **45**, which underwent a subsequent hydroamination reaction under catalysis with the gold complex to give the final chiral product through an active intermediate **46**.

More recently, Wang *et al.* reported the first asymmetric relay catalytic domino intramolecular hydrosiloxylation–Diels–Alder reaction of enynyl silanols with quinones.[47] This remarkable process was promoted by a combination of a gold(i) complex with a chiral BINOL-derived phosphoramide, which enabled these enynes to serve as latent 1,3-silyloxydienes capable of subsequently participating in an asymmetric Diels–Alder reaction with quinones, providing the corresponding polycyclic products as single diastereomers in high yields and generally high enantioselectivities ranging from 87 to 96% ee, as shown in Scheme 7.36.

In 2013, another intramolecular hydrosiloxylation was combined with a Mukaiyama aldol condensation by Gong *et al.* in an enantioselective relay catalytic cascade.[48] This domino reaction occurred between arylacetylene silanols and glyoxylates and was induced by a combination of an achiral gold catalyst and a chiral *N*-triflyl phosphoramide. As shown in Scheme 7.37,

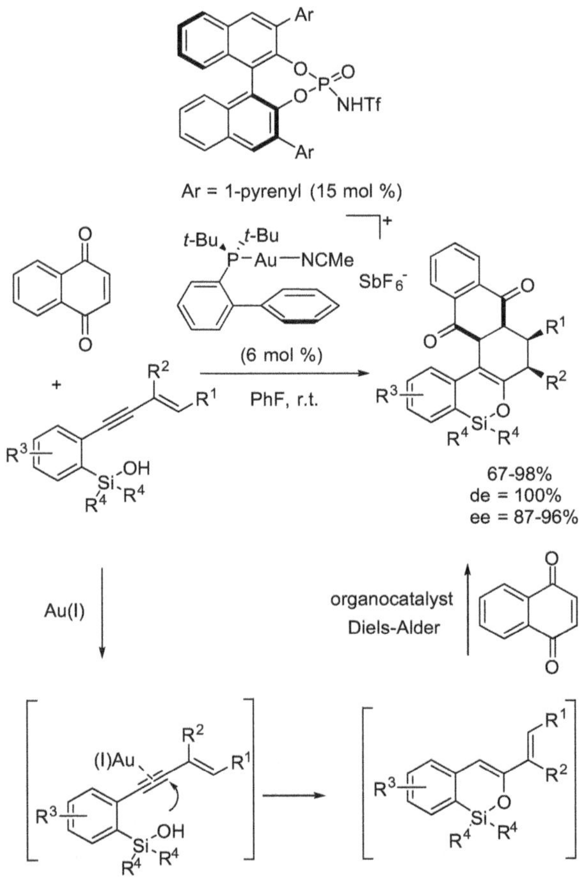

Scheme 7.36 Domino intramolecular hydrosiloxylation–Diels – Alder reaction catalysed by chiral phosphoramide catalysis and gold catalysis.

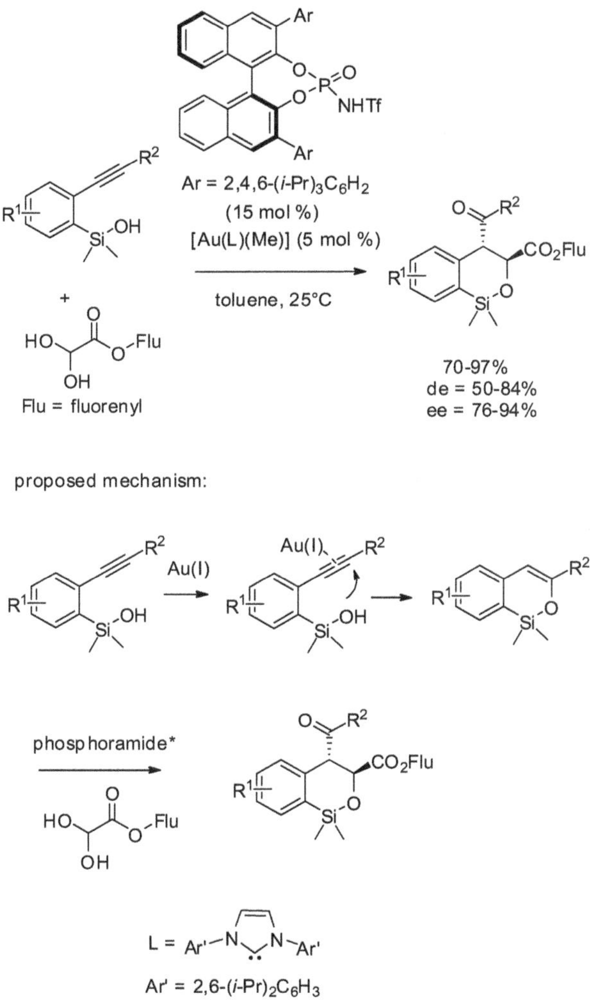

proposed mechanism:

Scheme 7.37 Domino intramolecular hydrosiloxylation–Mukaiyama aldol reaction catalysed by chiral phosphoramide catalysis and gold catalysis.

it afforded the corresponding chiral bicyclic products in good to high yields, and diastereo- and enantioselectivities of up to 84% de, and 94% ee, respectively. The mechanism of the sequence, involving gold-catalysed hydrosiloxylation followed by enantioselective phosphoramide-catalysed Mukaiyama aldol reaction is also depicted in Scheme 7.37.

In another context, a gold(i)-catalysed hydroamination has been recently combined by Patil *et al.* with a chiral phosphoric acid-catalysed enantioselective aminalisation between 2-alkynylbenzaldehydes and 2-aminobenzamides.[49] The key to this methodology was to search for a suitable achiral gold(i) catalyst which should only catalyse the hydroamination and should not take part in the enantioselective aminalisation reaction, resulting in

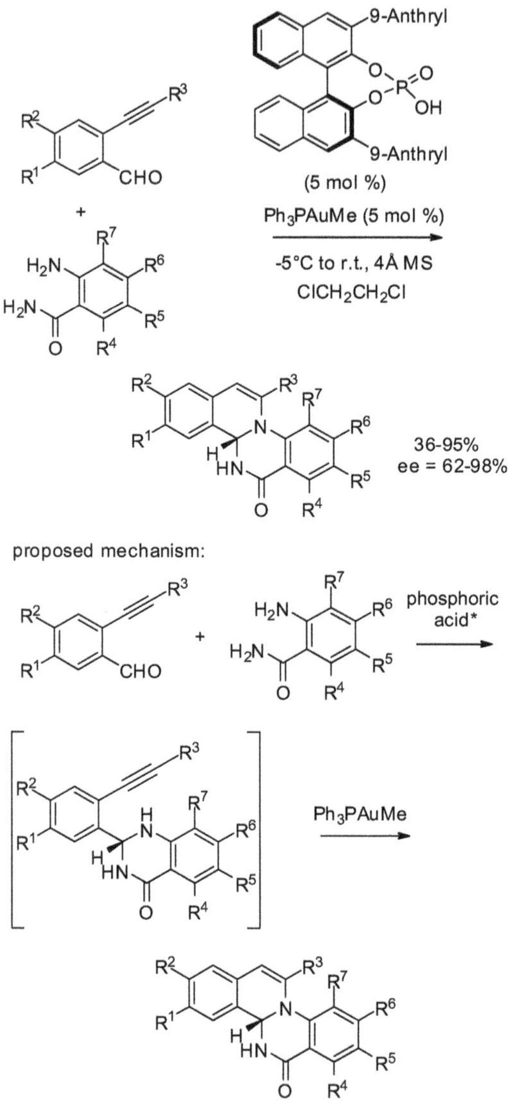

Scheme 7.38 Domino aminalisation–hydroamination reaction catalysed by chiral phosphoric acid catalysis and gold catalysis.

racemisation of relatively labile chiral aminal intermediates. After screening various gold(ı) complexes, the authors found that Ph₃PAuMe was suitable for the domino reaction. As shown in Scheme 7.38, the combination of this gold catalyst with a chiral phosphoric acid allowed chiral fused 1,2-dihydroisoquinoline products to be achieved in good to high yields and enantioselectivities of up to 98% ee.

In 2013, Zhang and Qian developed an enantioselective domino redox-pinacol–Mannich reaction, allowing an easy entry to chiral spirocyclic diketones.[50] This process employed a combination of an achiral gold complex, such as the electron-rich and bulky JohnPhos-derived cationic gold(I) complex with SbF_6^- as the counterion, with a chiral phosphoric acid, allowing the domino products to be achieved in high enantioselectivities of up to 98% ee, as shown in Scheme 7.39. The substrate scope was widely

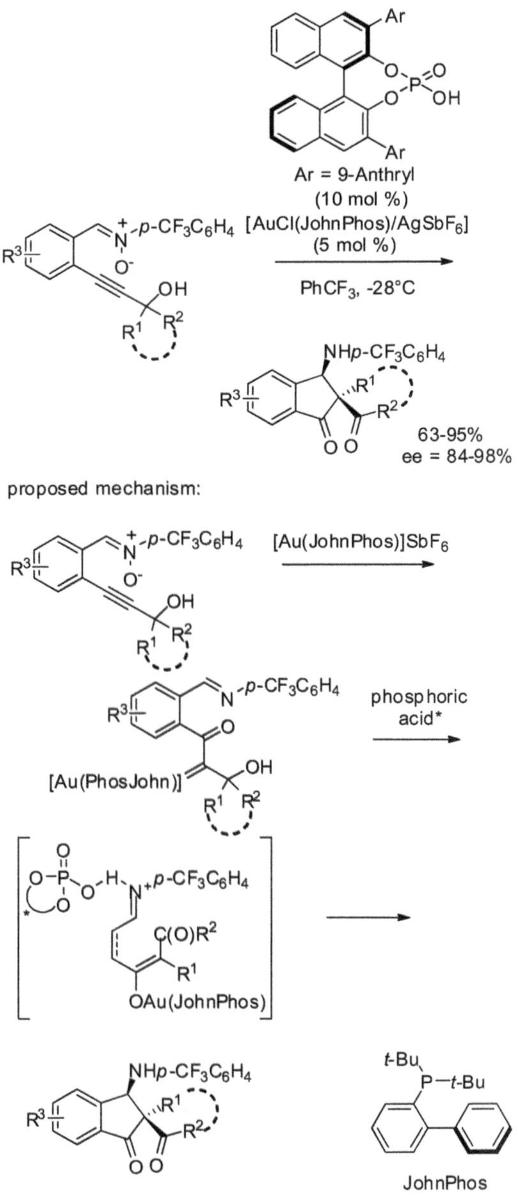

Scheme 7.39 Domino redox–pinacol–Mannich reaction catalysed by chiral phosphoric acid catalysis and gold catalysis.

extended so that the formation of chiral spirocyclic diketones including five-to nine-membered ring structures could be well constructed. The authors have proposed the mechanism depicted in Scheme 7.39 to explain the results, in which an imine intermediate generated *in situ* was activated by the chiral phosphoric acid

In 2013, Gong and co-workers disclosed an asymmetric relay catalytic three-component reaction by using an achiral gold(I) complex in combination with a chiral phosphoric acid, which was able to assemble readily available salicylaldehydes, anilines, and alkynols into chiral aromatic spiroacetals with high optical purities of up to 95% ee.[51] As shown in Scheme 7.40 the process began with the gold-catalysed cyclisation of an alkynol to afford the corresponding aromatic enol ether. This intermediate could further participate in a formal [4 + 2] cyclisation reaction, consisting of an asymmetric Mannich-type reaction with the salicylaldehydimine

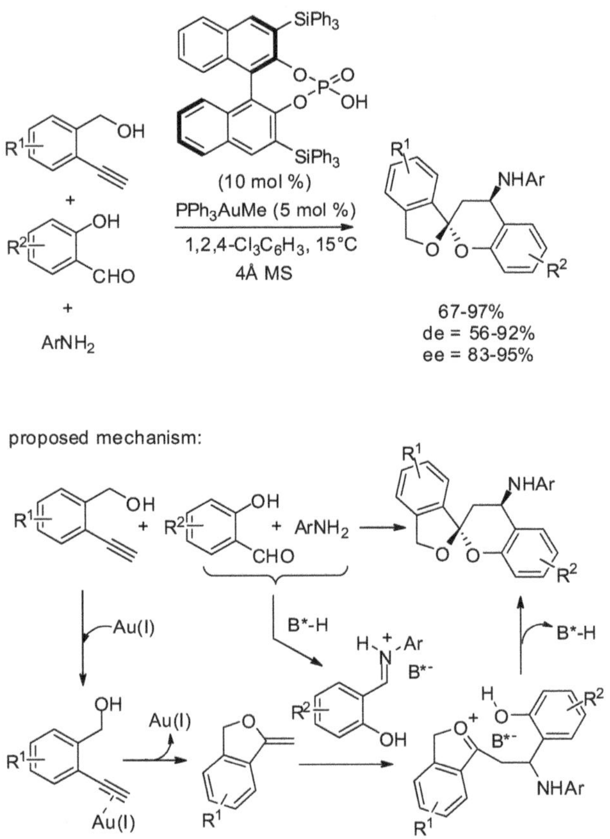

Scheme 7.40 Three-component domino cyclisation–Mannich-type–acetalisation reaction catalysed by chiral phosphoric acid catalysis and gold catalysis.

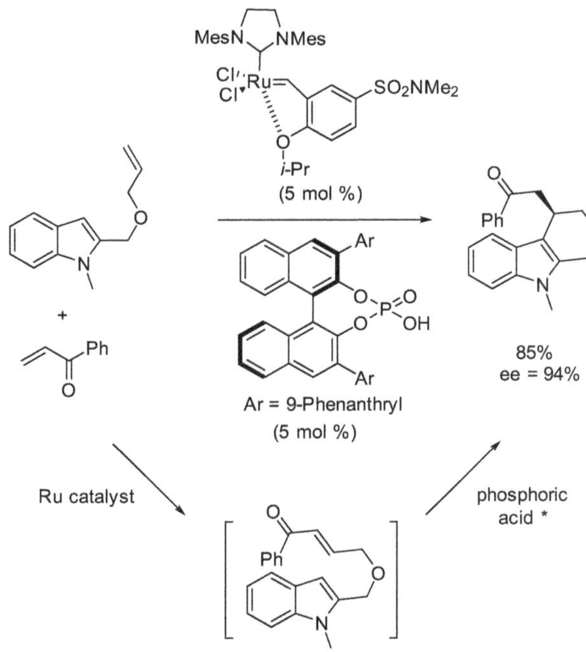

Scheme 7.41 Domino cross-metathesis–Friedel–Crafts reaction catalysed by chiral phosphoric acid catalysis and ruthenium catalysis.

generated *in situ* from the condensation between the salicylaldehyde and the aniline under the catalysis of the chiral phosphoric acid, and a subsequent acetalisation to result in the generation of the final domino product.

On the other hand, several groups have also recently developed asymmetric domino reactions through relay catalysis with combinations of organocatalysts with ruthenium catalysts. For example, You *et al.* demonstrated in 2009 that ruthenium catalyst could be compatible with Brønsted acid catalyst.[52] They reported a practical and economical synthesis of chiral tetrahydropyrano[3,4-b]indols and tetrahydro-β-carbolines by the combination of ruthenium-catalysed olefin cross-metathesis and a chiral phosphoric acid-catalysed Friedel–Crafts alkylation reaction, as shown in Scheme 7.41. This domino reaction allowed the use of readily available materials to highly enantioselectively construct synthetically valuable polycyclic indole frameworks in enantioselectivity of up to 94% ee.

Later, You *et al.* investigated relay catalysis consisting of a combination of the same ruthenium catalyst and a closely related chiral BINOL-derived phosphoric acid.[53] As shown in Scheme 7.42, the use of this catalyst system in an asymmetric domino intramolecular Friedel–Crafts-type–aza-Michael reaction allowed a range of chiral fused indoles to be achieved in high yields and moderate to high enantioselectivities from the corresponding enones and indolyl olefins.

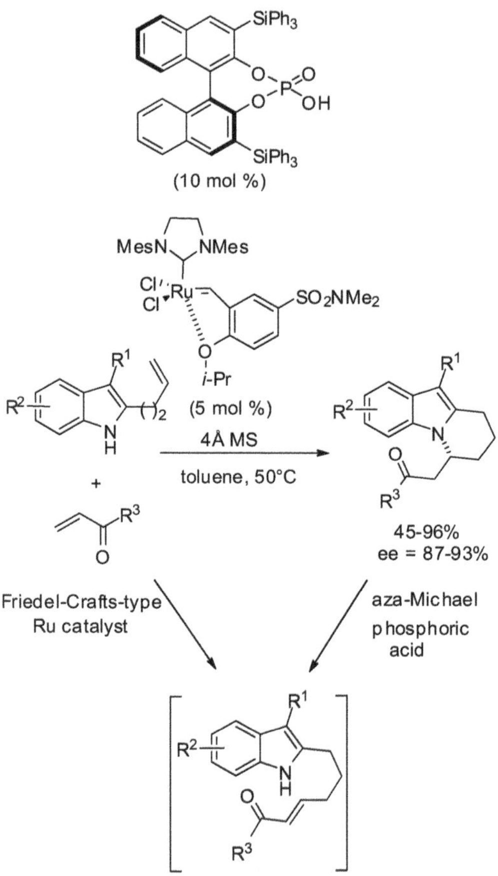

Scheme 7.42 Domino intramolecular Friedel–Crafts-type–aza-Michael reaction catalysed by chiral phosphoric acid catalysis and ruthenium catalysis.

The same authors have used the same chiral phosphoric acid in combination with the closely related Hoveyda–Grubbs II catalyst, enabling a highly efficient synthesis of chiral tetrahydro-β-carbolines in both excellent yields and enantioselectivities of up to 98% and 99% ee, respectively, through a domino ring-closing metathesis–isomerisation–Pictet–Spengler; reaction *via* relay catalysis.[54] These remarkable results, shown in Scheme 7.43, were obtained using readily available tryptamine derivatives as starting materials.

A remarkable multicatalytic relay system consisting of tetra-propylammonium perruthenate/*N*-methylmorpholine *N*-oxide (TPAP/NMO) as oxidant, and chiral diarylprolinol trimethylsilyl ethers as organocatalysts, has recently been developed by Rueping *et al.* and applied in the efficient construction of various valuable chiral molecules through domino reactions.[55] The latter were all based on the *in situ* generation of α,β-unsaturated

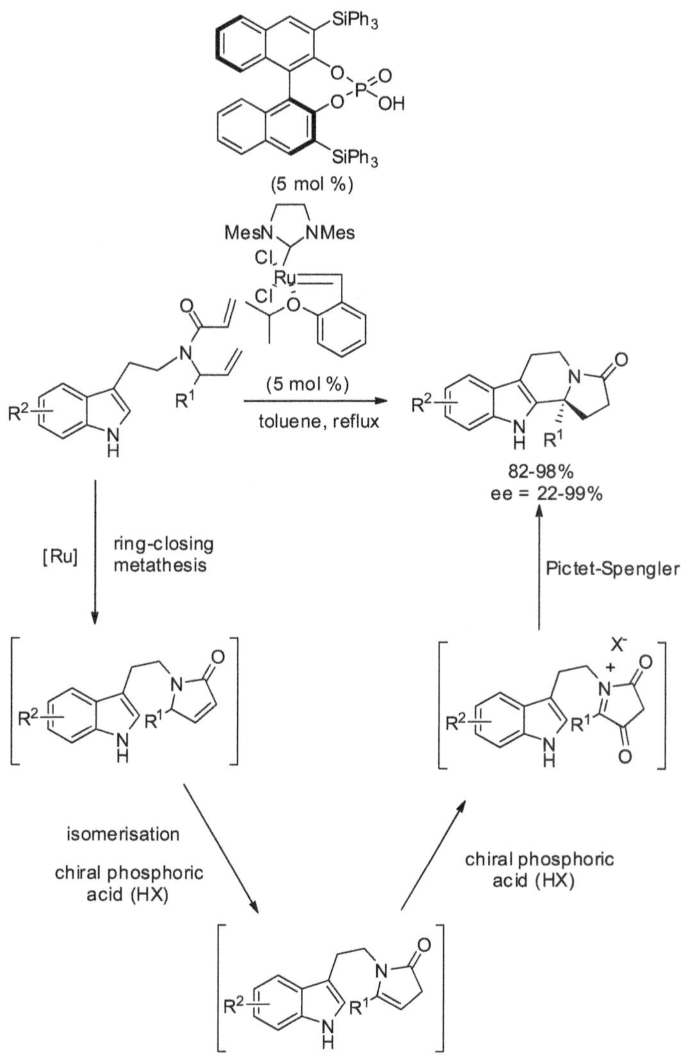

Scheme 7.43 Domino ring-closing metathesis–isomerisation–Pictet–Spengler reaction catalysed by chiral phosphoric acid catalysis and ruthenium catalysis.

aldehydes from the corresponding allylic alcohols and their subsequent use in various asymmetric domino transformations, such as domino Michael–intramolecular alkylation, domino Michael addition–hemiacetalisation, or domino oxa-Michael–Michael reaction. It was shown that TPAP as a substrate-selective redox catalyst was well tolerated by the amine catalysts and the following enantioselective domino reactions proceeded in good yields and high enantioselectivities of up to 95, 96, and 99% ee, respectively, as shown in Scheme 7.44.

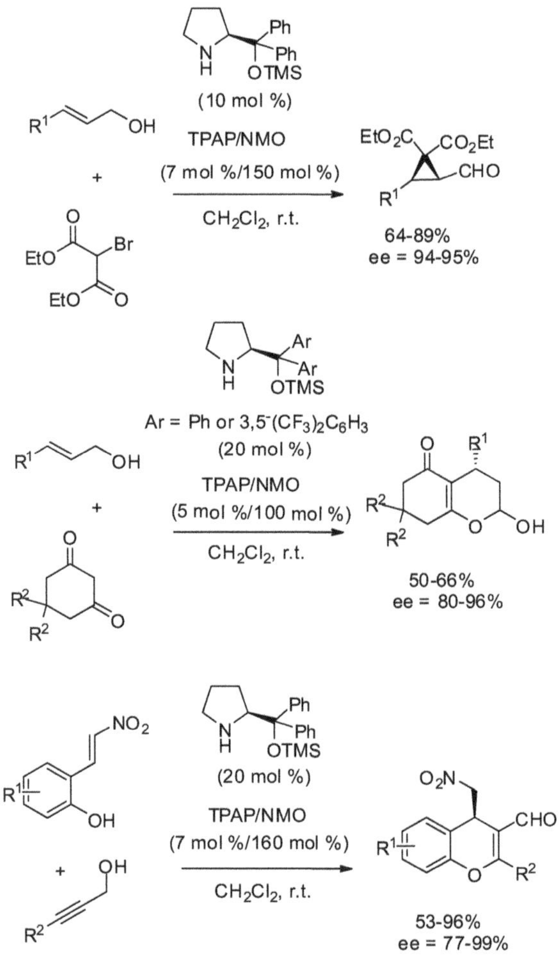

Scheme 7.44 Domino oxidation–Michael–intramolecular alkylation reaction, domino oxidation–Michael–hemiacetalisation reaction, and domino oxidation–oxa-Michael–Michael reaction catalysed by chiral amine catalysis and ruthenium catalysis.

Another ruthenium-based relay catalysis system was described by Zhou *et al.* in 2011.[56] It consisted of the combination of [Ru(*p*-cymene)I$_2$]$_2$ with a chiral phosphoric acid, which was applied to the highly enantioselective hydrogenation of quinoxalines through convergent asymmetric disproportionation of dihydroquinoxalines, employing hydrogen gas as the reductant. As shown in Scheme 7.45, the corresponding chiral tetrahydroquinoxalines were generated in high yields and enantioselectivities of up to 94% ee. Concerning the mechanism, the authors have proposed that the hydrogenation of quinoxalines first generated the corresponding dihydroquinoxalines through catalysis with the ruthenium complex.

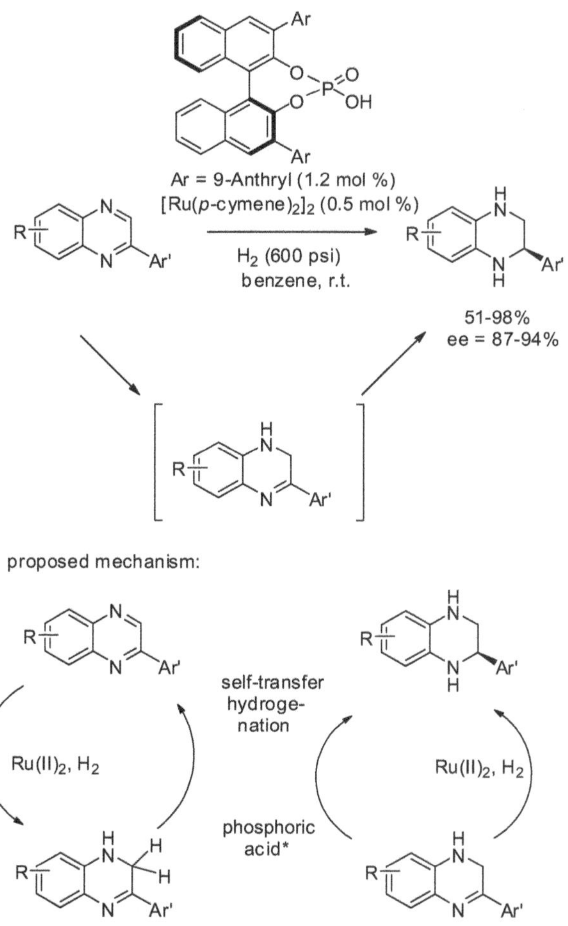

Scheme 7.45 Domino double hydrogenation reaction catalysed by chiral phosphoric acid catalysis and ruthenium catalysis.

Subsequently, these intermediates underwent a self-hydrogenation to deliver the primary starting quinoxalines and the final chiral tetrahydroquinoxalines in the presence of the chiral phosphoric acid (Scheme 7.45).

In 2013, You *et al.* combined a chiral phosphoric acid with Hoveyda–Grubbs II catalyst to promote an enantioselective domino isomerisation–isomerisation–Pictet–Spengler; reaction of *N*-allyltryptamines and *N*-crotyltryptamines.[57] The reaction afforded the corresponding chiral 1,2,3,4-tetrahydro-β-carbolines in moderate to high enantioselectivities of up to 87% ee, as shown in Scheme 7.46. A wide range of substrates bearing either electron-donating or electron-withdrawing substituents on the indole group have been investigated. In general, substrates bearing an electron-donating group gave better yields than those bearing an

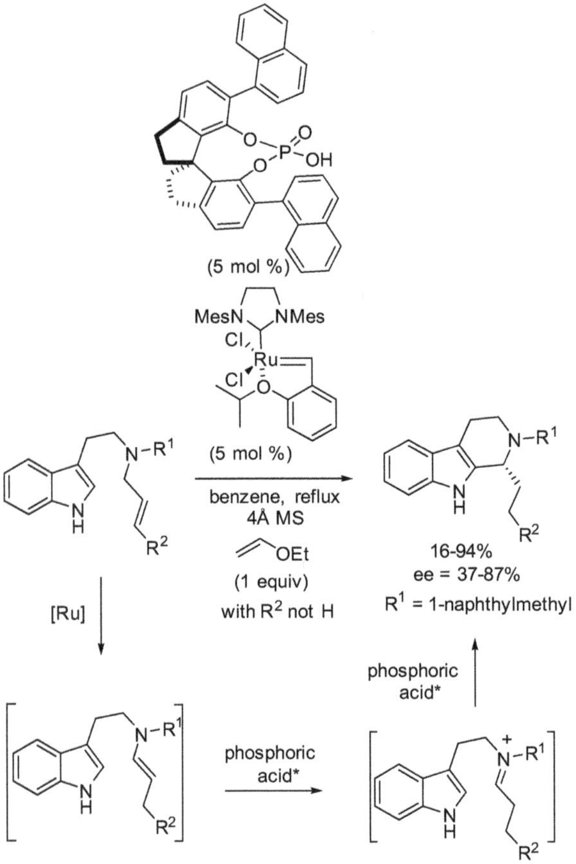

Scheme 7.46 Domino   isomerisation–isomerisation–Pictet–Spengler;   reaction catalysed by chiral phosphoric acid catalysis and ruthenium catalysis.

electron-withdrawing group. In the case of *N*-crotyltryptamines as substrates, high yields were reached when the reaction was carried out in the presence of an additive such as vinyl ethyl ether.

In 2013, Terada and Toda reported a relay catalysis for a ternary reaction sequence composed of double bond isomerisation, protonation of the double bond, and enantioselective Pictet–Spengler-type cyclisation, which was accomplished using a binary catalytic system consisting of a ruthenium hydride complex and a chiral phosphoric acid.[58] As shown in Scheme 7.47, the intramolecular reaction of allylamides led to the corresponding chiral tetrahydroisoquinoline derivatives in moderate to good yields and insufficient enantioselectivities of 18 to 53% ee.

In order to develop a more simple and economic procedure, Rueping *et al.* have also investigated cheap, non-toxic, heterogeneous oxidising reagents,

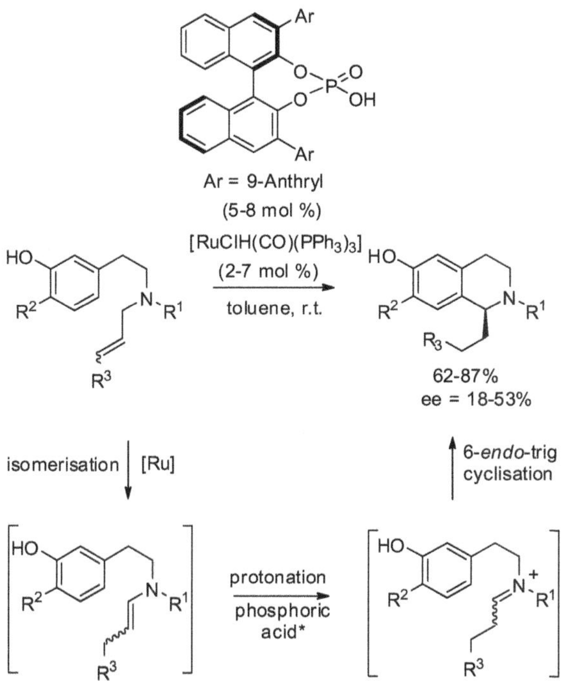

Scheme 7.47 Domino isomerisation–protonation–6-*endo*-trig cyclisation reaction catalysed by chiral phosphoric acid catalysis and ruthenium catalysis.

such as MnO_2, being easily separated by filtration.[59] Thus, they have found that this oxidising agent could be used in combination with chiral secondary amines, such as chiral diarylprolinol trimethylsilyl ethers, in asymmetric domino oxidation–Michael reaction of allylic alcohols and dimethylmalonate. As shown in Scheme 7.48, the relay catalytic system allowed the corresponding chiral functionalised aldehydes to be achieved in good yields and enantioselectivities of up to 91% ee.

Hydroformylation of olefins has been established as an important industrial tool for the production of aldehydes. In recent years, novel asymmetric tandem reactions have included a rhodium-catalysed enantioselective hydroformylation. In this context, in 2007 Abillard and Breit[60] and Chercheja and Eilbracht[61] independently reported a novel domino hydroformylation–aldol reaction catalysed by an achiral rhodium catalyst and L-proline catalyst (Scheme 7.49). Possibly owing to the fact that proline is hard but the rhodium catalyst is soft, the proline can be compatible with the rhodium catalyst to allow this domino reaction to be achieved. By fine adjustment of the hydroformylation rate to that of the L-proline-catalysed aldol addition, the undesired homodimerisation of the aldehyde could be avoided. As a result, by *in situ* hydroformylation reaction, the donor aldehyde of a

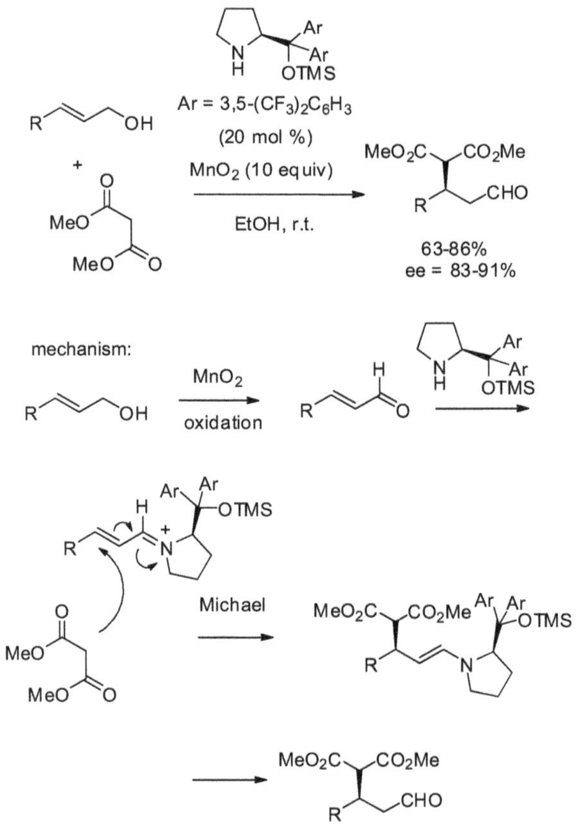

Scheme 7.48 Domino oxidation–Michael reaction catalysed by chiral amine catalysis and manganese catalysis.

cross-aldol reaction could be generated and kept in low concentration, which allowed its homo-aldolate formation. As shown in Scheme 7.49, the domino hydroformylation–aldol reaction product was reached in 81% yield, 86% de and 99% ee.

Later, Eilbracht *et al.* developed a related enantioselective domino hydroformylation–aldol reaction catalysed by a rhodium complex in combination with a chiral amine derived from L-proline. In this process, the reaction of acetone with vinylbenzene afforded the corresponding chiral β-hydroxy ketone in 69% yield, a moderate diastereoselectivity of 74% de, and a high enantioselectivity of 93% ee, as shown in Scheme 7.50.[62]

Earlier, the same authors employed the same strategy to the synthesis of optically active amine compounds from alkenes through a three-component domino hydroformylation–Mannich reaction.[63] The combination of rhodium catalyst hydroformylation with L-proline-catalysed asymmetric Mannich reaction worked well to afford the domino product in moderate yield and good enantioselectivity of 74% ee, as shown in Scheme 7.51.

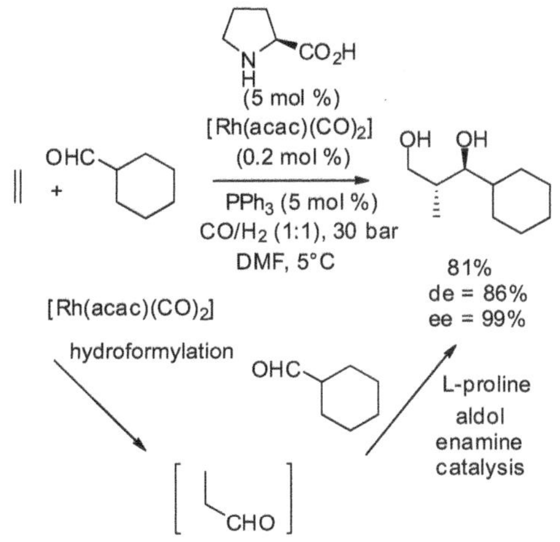

Scheme 7.49 Domino hydroformylation–aldol reaction catalysed by chiral amine catalysis and rhodium catalysis.

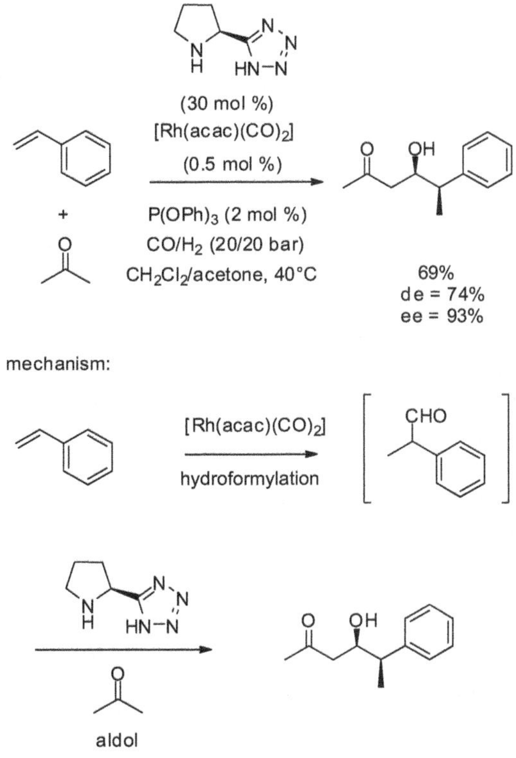

Scheme 7.50 Domino hydroformylation–aldol reaction catalysed by chiral amine catalysis and rhodium catalysis.

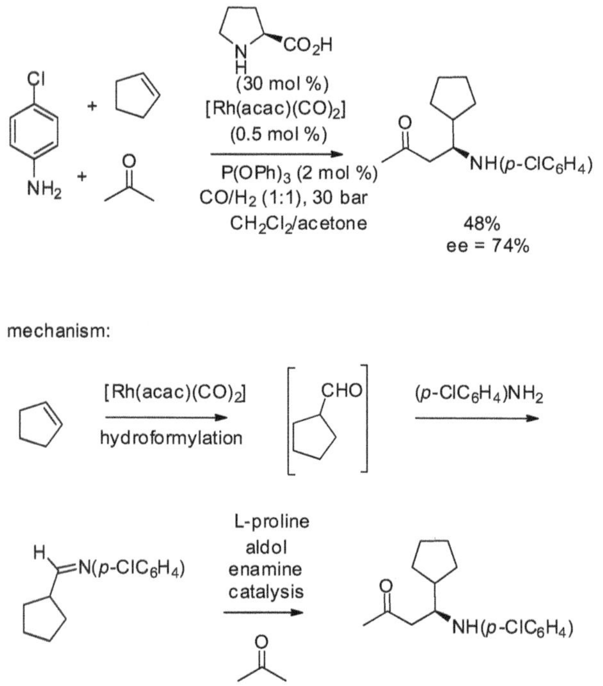

Scheme 7.51   Three-component domino hydroformylation–Mannich reaction catalysed by chiral amine catalysis and rhodium catalysis.

In 2012, Terada and Toda reported another type of relay catalysis based on the combination of a rhodium complex, such as [Rh$_2$(OAc)$_4$], and a chiral phosphoric acid.[64] This catalyst system was employed to induce the domino carbonyl ylide formation–enantioselective reduction reaction of α-diazocarbonyl compounds by methyl Hantzsch ester into the corresponding chiral intermediate isochromanone derivatives, which were subsequently entrapped by a benzoyl group through treatment with benzoyl chloride to give the corresponding final benzoyloxy isochromene derivatives in high yields and enantioselectivities of up to 92% ee, as shown in Scheme 7.52. The proposed relay catalysis involved a four-step transformation including: (a) decomposition of the α-diazocarbonyl compound by the rhodium catalyst to generate the corresponding rhodium carbene complex **47**; (b) subsequent intramolecular cyclisation of this intermediate **47** to afford the carbonyl ylide equivalent **48** or its oxidopyrylium equivalent **48′** through tautomerisation; (c) protonation of this transient species by the chiral phosphoric acid catalyst to afford ions pairs of the stable isobenzopyrylium ion **49** and the conjugate base of the chiral phosphoric acid; (d) termination through a reduction of the cationic intermediate **49** using methyl Hantzsch ester under the influence of the chiral anion of phosphoric acid to afford the chiral isochroman-4-one derivatives.

In 2013, Gong *et al.* reported a highly enantioselective three-component reaction of 3-diazooxindoles and anilines with glyoxylates cooperatively

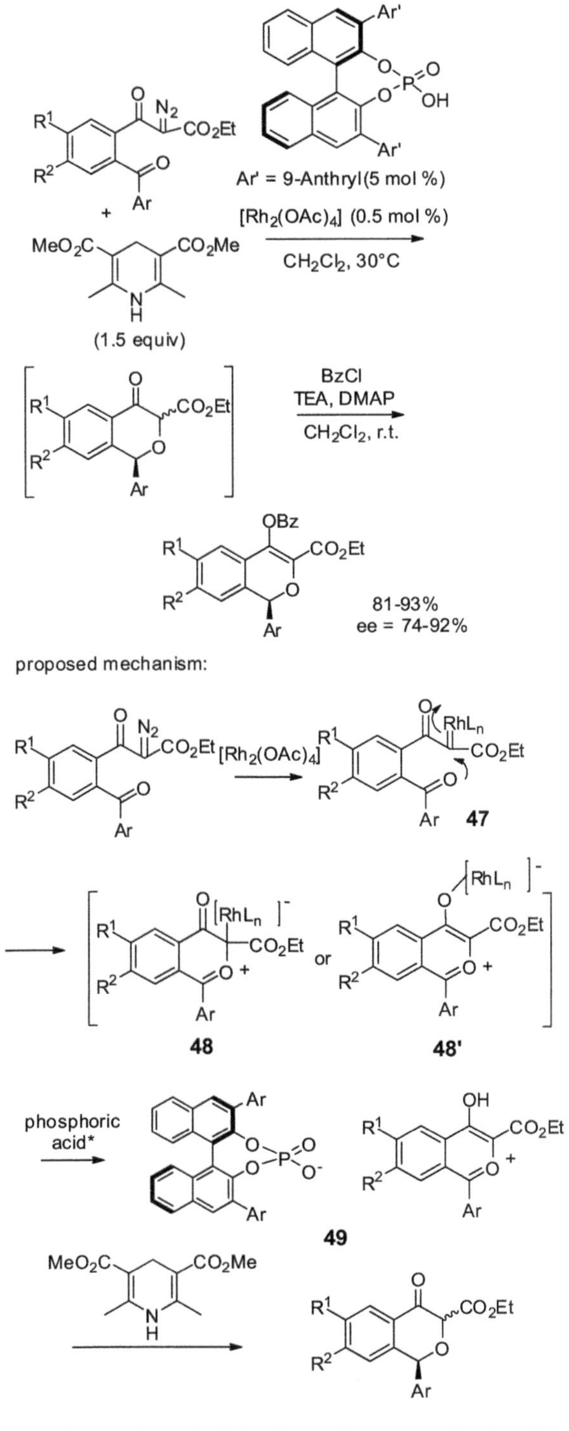

Scheme 7.52 Domino carbonyl ylide formation–enantioselective reduction reaction catalysed by chiral phosphoric acid catalysis and rhodium catalysis followed by benzoylation.

catalysed by an achiral rhodium complex and a chiral phosphoric acid, which afforded highly functionalised 3-amino oxindoles with moderate to excellent diastereo- and enantioselectivities of up to 94% de, and 99% ee, respectively.[65] The process basically proceeded *via* a rhodium-catalysed generation of ammonium ylides from 3-diazooxindoles and anilines, followed by a chiral phosphoric acid-catalysed enantioselective aldol-type reaction with glyoxylates to give the final domino products. In this reaction, the phosphoric acid presumably activated the formyl group of the glyoxylates through a hydrogen-bonding interaction, and simultaneously the phosphoryl oxygen could be able to function as a Lewis base capable of forming an additional hydrogen bond with ammonium ylides to stabilise the transition state (Scheme 7.53).

On the other hand, platinum complexes have also been demonstrated by Aleman *et al.* to be compatible with chiral diphenylprolinol trimethylsilyl ether and DABCO, allowing a domino Michael–lactone-opening–aldol–dehydration reaction to be achieved through relay catalysis.[66] As shown in Scheme 7.54, the reaction began with the Michael addition of a lactone to an α,β-unsaturated aldehyde catalysed by the chiral amine catalyst, providing the corresponding lactone aldehyde intermediate through iminium catalysis. This intermediate was further opened by a water molecule through catalysis by the platinum complex to give a novel aldehyde intermediate, which was submitted to intramolecular amine catalysed-aldol reaction followed by dehydration to give the corresponding carboxylic acids, which were subsequently converted into dimethyl esters by addition of N_2CH_2TMS to the reaction media (Scheme 7.54). These functionalised cyclopentenes were achieved in moderate yields and enantioselectivities of up to 84% ee, as shown in Scheme 7.54.

Relay nickel–organocatalysis has also been recently applied to develop highly efficient asymmetric multicomponent reactions. As an example, McQuade and co-workers have developed an original one-pot tandem reaction catalysed by a microencapsulated amine catalyst and a chiral nickel complex (Scheme 7.55).[67] Although the enantioselectivity of this process was not so high (72% ee), the site-isolation of two otherwise incompatible catalysts provided by microencapsulation brought new insight into the development of amine–Lewis acid tandem reactions. The encapsulation of the amine catalyst was the key for the success of the reaction for the following reasons: (1) the use of soluble amine catalyst led to catalyst deactivation by complexation with nickel catalyst; (2) silica MCM-41 or polystyrene-supported amine catalyst failed to catalyse the nitroalkene formation at room temperature, but the encapsulated poly(ethyleneimine) could; (3) the microencapsulated amine swollen in methanol retained their catalytic potency when in toluene, which allowed the one-pot reaction to be run in a mixture of two different solvents, and the microencapsulated amine and nickel catalyst could work under their respective ideal solvents of methanol and toluene.

In 2012, Gong and co-workers reported an enantioselective three-component reaction based on an asymmetric relay catalytic domino Friedländer condensation–transfer hydrogenation reaction of 2-aminophenyl ketones,

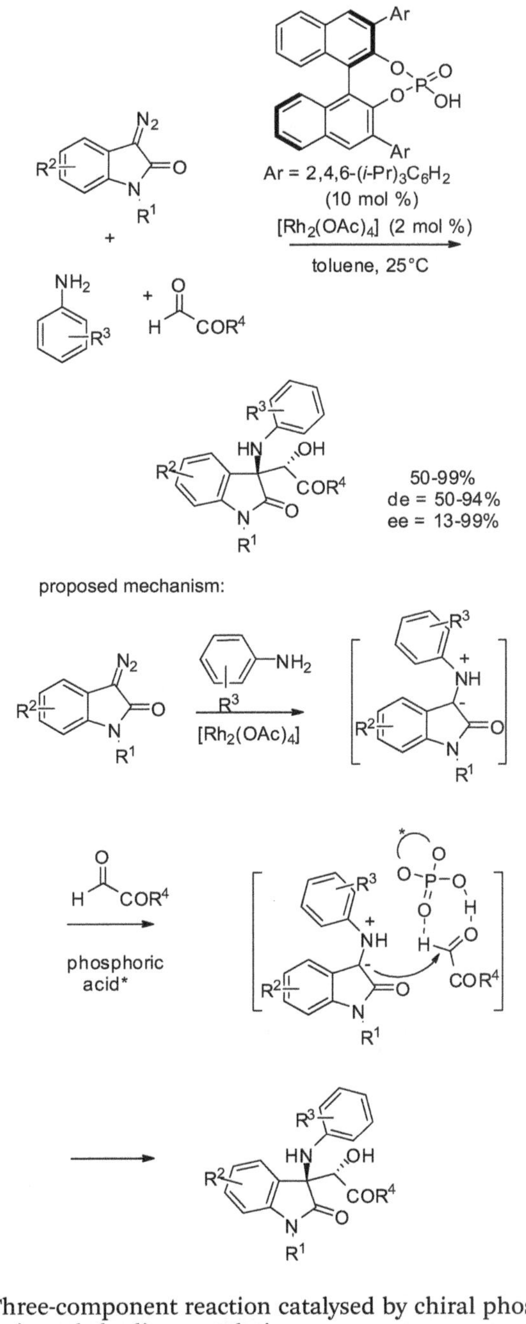

proposed mechanism:

Scheme 7.53 Three-component reaction catalysed by chiral phosphoric acid catalysis and rhodium catalysis.

methyl Hantzsch ester and ethyl acetoacetate, providing almost enantiopure tetrahydroquinolines in high yields, diastereoselectivity of >90% de in all cases of substrates studied, combined with high to excellent

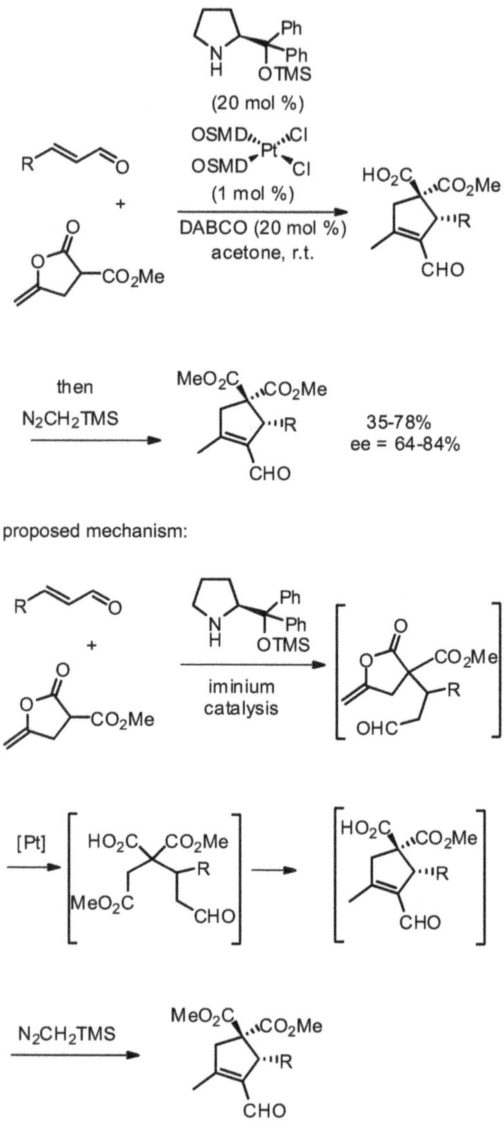

Scheme 7.54 Domino Michael–lactone-opening–aldol–dehydration reaction catalysed by chiral amine catalysis and platinum catalysis followed by
esterification.

enantioselectivities of up to 98% ee.[68] In this case of a multicomponent
reaction, the catalyst system constituted a magnesium catalyst such as
Mg(OTf)$_2$ and a chiral phosphoric acid, depicted in Scheme 7.56. The
authors assumed that the process could evolve through a Friedländer condensation catalysed by either the chiral phosphoric acid or the Lewis acid,
while the following asymmetric transfer hydrogenation was promoted solely
by the chiral Brønsted acid (Scheme 7.56).

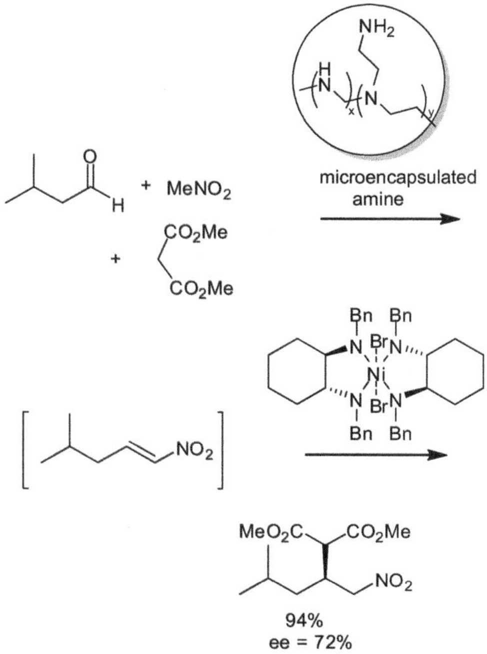

Scheme 7.55 Three-component domino Henry–Michael reaction catalysed by microencapsulated amine catalysis and chiral nickel catalysis.

7.4 Sequential Catalysis

Early in 2003, Choudary *et al.* studied the catalytic activity of a unique tri-functional heterogeneous catalyst system consisting of palladium, osmium, and tungsten species for tandem Heck olefination followed by asymmetric dihydroxylation reaction induced by cinchona alkaloid (DHQD)$_2$PHAL in the presence of a tertiary amine such as *N*-methylmorpholine (NMM) in one pot.[69] The trimetal catalyst system of Pd–Os–W was embedded into hexagonal layered double-hydroxides (LDHs). As shown in Scheme 7.57, the corresponding almost enantiopure diol was achieved in high yield. This remarkable result was not clearly understood by the authors.

In 2009, Terada and Toda employed a combination of nickel(II) hydride complexes with a chiral phosphoric acid in an enantioselective sequential tandem isomerisation–aza-Petasis–Ferrier rearrangement reaction.[70] The nickel hydride complex, formed *in situ* from the corresponding NiI$_2$ complexes, isomerised the starting allylic alcohol to the corresponding Z-configured vinyl ether with high levels of diastereocontrol. The enantiospecific acid-catalysed rearrangement of this vinyl ether with subsequent reduction of the formed aldehyde with NaBH$_4$ furnished valuable chiral β-amino primary alcohols in good yields and enantioselectivities of up to >99% ee, as shown in Scheme 7.58.

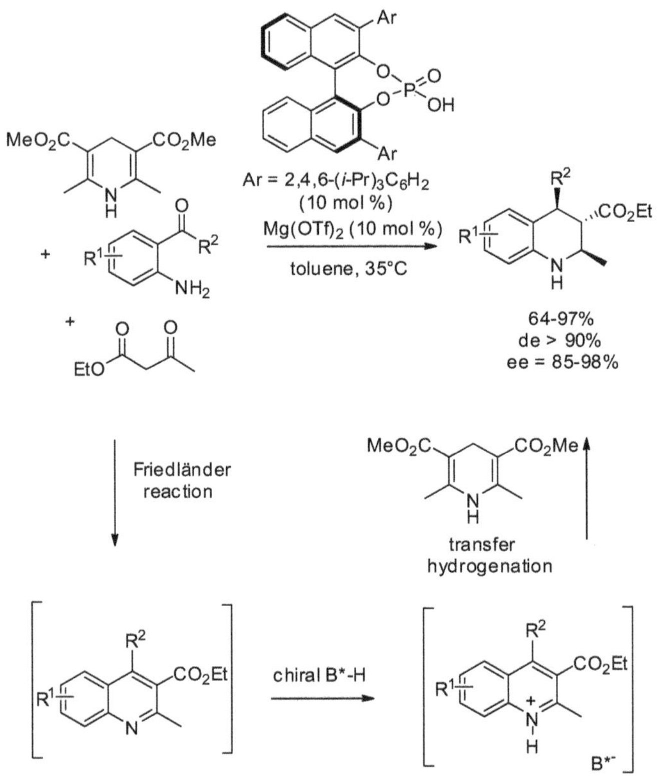

Scheme 7.56   Three-component domino Friedländer–transfer hydrogenation reaction catalysed by chiral phosphoric acid catalysis and magnesium catalysis.

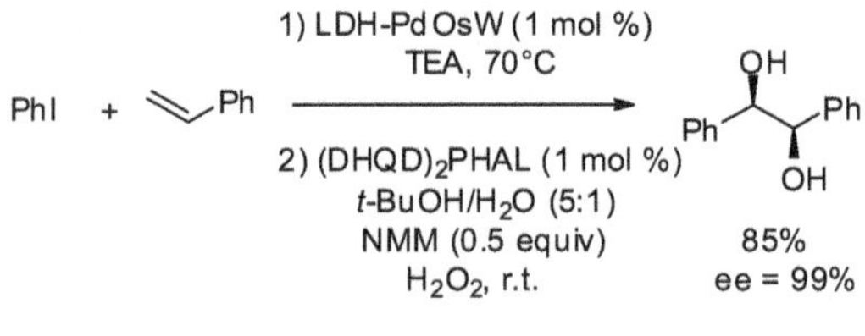

Scheme 7.57   Tandem Heck–dihydroxylation reaction catalysed by palladium, osmium, and tungsten catalysis and chiral cinchona alkaloid catalysis.

In 2009, MacMillan *et al.* reported an example of a cycle-specific cascade reaction blending Grubbs catalyst with chiral iminium and enamine catalysts.[71] In this case, the sequential addition of Grubbs catalyst, a chiral iminium catalyst and L-proline together with the respective addition of 5-hexene-2-one, crotonaldehyde, and trimethylsilyloxyfuran afforded the desired diastereomer of the product in 64% yield, 95% ee and with 66% de, as shown in Scheme 7.59. It must be noted that it is really striking that from simple starting materials, the complex skeleton bearing four stereogenic

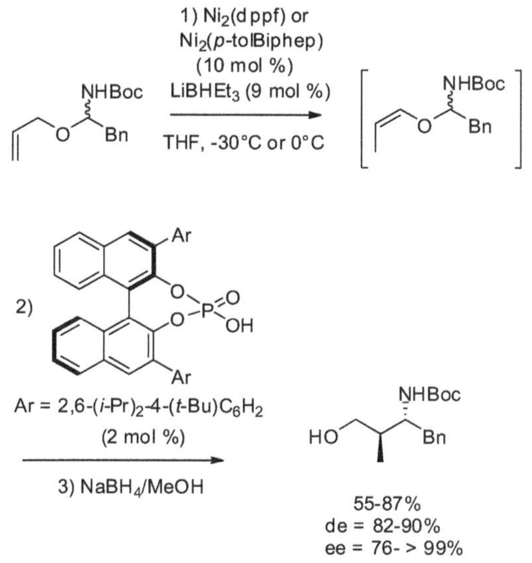

Scheme 7.58 Tandem isomerisation–aza-Petasis–Ferrier rearrangement reaction catalysed by nickel catalysis and chiral phosphoric acid catalysis followed by reduction.

centres formed could be constructed through a one-pot multicatalyst-promoted tandem reaction in a very simple operation. This process was applied to the total synthesis of (−)-aromadendranediol in 8 steps.

In 2009, Dixon *et al.* described an example of enantioselective gold(I)- and chiral phosphoric acid-catalysed sequential reaction to give enantioenriched indole containing tetracyclic products.[72] In this process, a first gold-catalysed cycloisomerisation of an alkynoic acid took place to generate the corresponding enol lactone which subsequently underwent a further reaction after addition of tryptamines and the chiral phosphoric acid catalyst to achieve the final products in good yields and enantioselectivities of 83 to 95% ee, as shown in Scheme 7.60.

On the other hand, Alexakis and co-workers reported an enantioselective tandem Michael–acetalisation–cyclisation reaction using an L-proline-derived chiral amine and a gold complex as catalytic system.[73] This process, involving alkyne-tethered nitroalkenes and aldehydes as substrates, provided the corresponding chiral nitrosubstituted tetrahydrofuranyl ethers in good yields and high diastereo- and enantioselectivities of up to 94% de and >99% ee, respectively, as shown in Scheme 7.61. The key to this methodology was the design of the bifunctional alkyne-tethered nitroalkene substrates. In this work, the Au(I) complex was added sequentially to the reaction flask to achieve the acetalisation–cyclisation sequence after completion of the first amine-catalysed asymmetric Michael reaction. It must be noted that *p*-TsOH was crucial in ensuring that the Au(I) catalyst was not deactivated by the secondary amine catalyst.

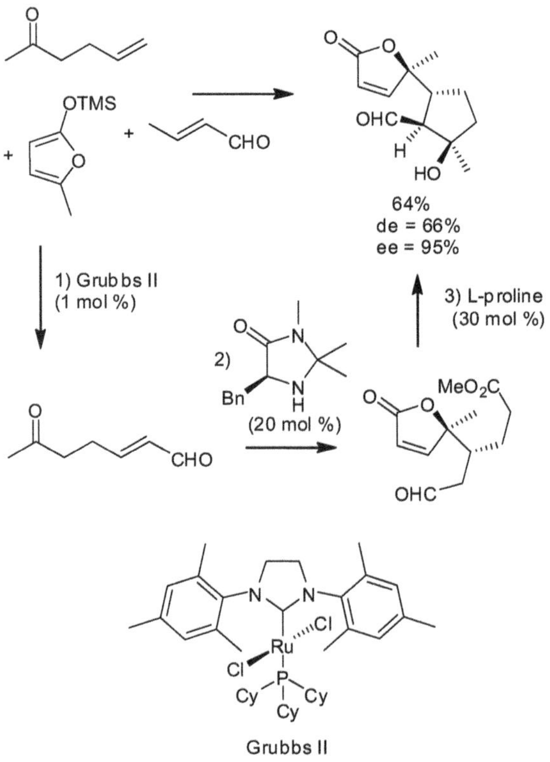

Scheme 7.59   Three-component reaction of 5-hexene-2-one, crotonaldehyde, and trimethylsilyloxyfuran catalysed by ruthenium catalysis and chiral amine catalysis.

In 2010, Jørgensen *et al.* developed an enantioselective tandem reaction of propargylated malononitriles with cyclic enones sequentially catalysed by a cinchona alkaloid-derived primary amine catalyst in the presence of (*R*)-mandelic acid as an additive for the first Michael step, and a gold catalyst for the second tandem *exo-dig* cyclisation–isomerisation reaction.[74] As shown in Scheme 7.62, the corresponding chiral bicyclic enones were achieved in good yields and high enantioselectivities of up to 96% ee, albeit low to moderate diastereoselectivities (34–66% de).

In addition, the same authors reported a novel synthetic approach towards optically active dihydropyrroles on the basis of a sequential catalysis, in 2010.[75] As shown in Scheme 7.63, the sequence involved, as first step, a chiral cinchona thiourea-catalysed Mannich-type reaction of *N*-Boc-pro-tected imines with propargylated malononitriles, which was followed by a gold-catalysed alkyne hydroamination–isomerisation reaction, providing the corresponding chiral 2,3-dihydropyrroles in moderate to high yields and enantioselectivities of up to 88% ee. This work constituted the first example of combining hydrogen-bonding catalysis and transition metal catalysis.

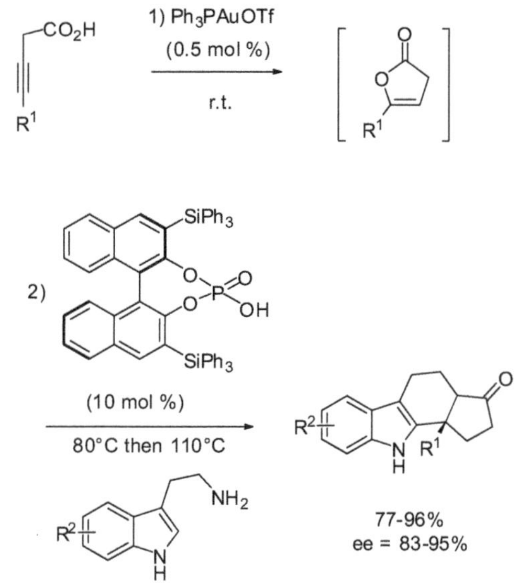

Scheme 7.60 Tandem cycloisomerisation–condensation–cyclisation reaction catalysed by gold catalysis and chiral phosphoric acid catalysis.

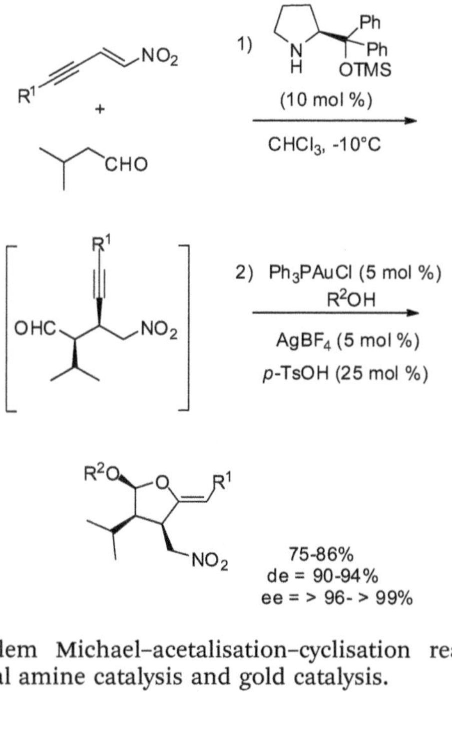

Scheme 7.61 Tandem Michael–acetalisation–cyclisation reaction catalysed by chiral amine catalysis and gold catalysis.

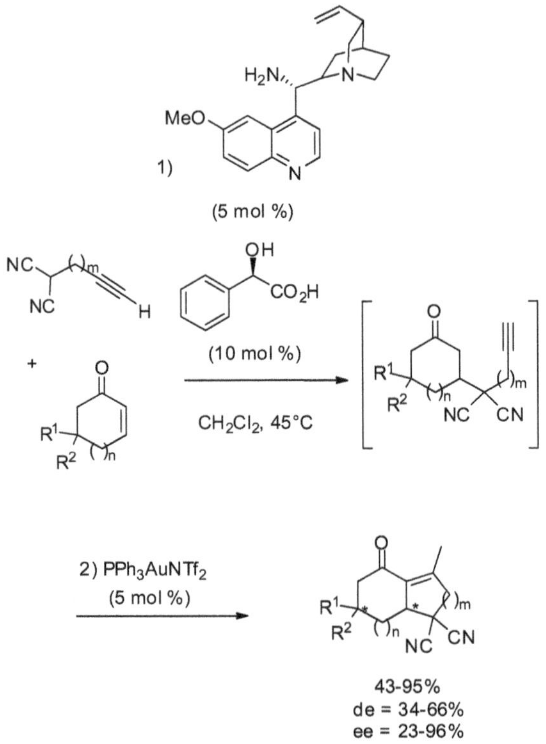

Scheme 7.62 Tandem Michael–cyclisation–isomerisation reaction catalysed by chiral amine catalysis and gold catalysis.

In the same area and more recently, Dixon *et al.* described a one-pot enantioselective tandem nitro-Mannich–hydroamination–isomerisation reaction of nitro alkynes with *N*-protected aldimines, which provided the corresponding chiral tetrahydropyridine derivatives in moderate to good yields, good to high diastereoselectivities of up to >96% de, and high enantioselectivities ranging from 87 to 96% ee.[76] As shown in Scheme 7.64, this process was sequentially catalysed by a chiral urea to give the intermediate nitro-Mannich products, which were subsequently submitted to hydroamination followed by isomerisation by treatment with an achiral gold(I) catalyst which was added to the reaction media to give the final products. In order to quench the basic N-atom of the bifunctional catalyst, a Brønsted acid, such as diphenylphosphate (DPP), was used as an additive prior to the addition of the gold catalyst.

Later, an enantioselective one-pot tandem Mannich–hydroamination reaction was reported by Liu and co-workers on the basis of a sequential organo- and gold catalysis.[77] The process involved propargylated malonitrile and oxindole imine derivatives as substrates and employed a chiral cinchona alkaloid, such as a quinidine phenol derivative, to induce the enantioselective Mannich reaction and a gold catalyst, such as XPhosAuNTf$_2$

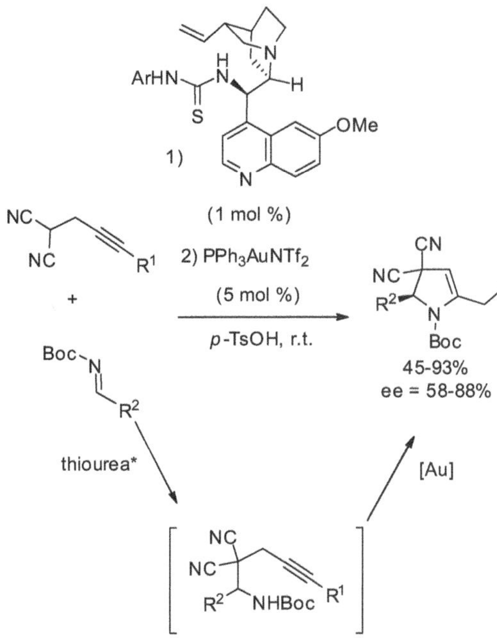

Scheme 7.63 Tandem Mannich-type–hydroamination–isomerisation reaction catalysed by chiral thiourea catalysis and gold catalysis.

(XPhos = 2-(dicyclohexylphosphino)-2′,4′,6′-triisopropylbiphenyl), to promote the following hydroamination reaction. As shown in Scheme 7.65, the corresponding chiral spiro[pyrrolidin-3,2′-oxindole]derivatives were obtained in good yields and good to excellent enantioselectivities of up to 97% ee.

In 2011, remarkable enantioselectivities ranging from 97 to 99% ee were achieved by Enders *et al.* in a novel enantioselective tandem double Friedel–Crafts reaction of indoles with *ortho*-alkyne-substituted nitrostyrenes.[78] This process was sequentially catalysed by a chiral thiourea-based catalyst and achiral gold complex [Au(PPh₃)]NTf₂, providing the corresponding enantiopure seven-membered ring containing tetracyclic indole derivatives in good to excellent yields, as shown in Scheme 7.66. The process was supposed to evolve through a first Friedel–Crafts reaction of indole with *ortho*-alkyne-substituted nitrostyrene to give the corresponding C3-substituted alkyne intermediate (Scheme 7.66), which subsequently underwent a second Friedel–Crafts reaction with gold(ɪ)-activated alkyne to generate a spirocyclic intermediate. The latter rearranged through a 1,2-shift to effect an expansion from a six- to a seven-membered ring. Subsequent rearomatisation and protodeauration led to the final product.

More recently, Che *et al.* reported a tandem asymmetric reaction of aminobenzaldehydes (or aminophenones) with alkynes to give chiral bioactive diversely substituted tetrahydroquinolines.[79] The first step catalysed by a

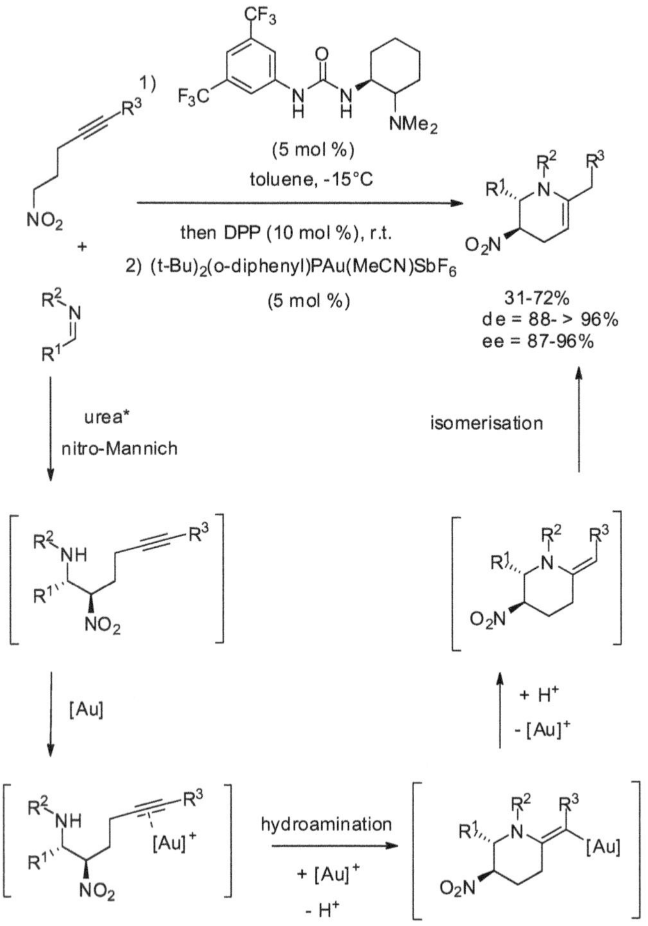

Scheme 7.64 Tandem nitro-Mannich–hydroamination–isomerisation reaction catalysed by chiral urea catalysis and gold catalysis.

gold complex led to the corresponding quinoline intermediates through a hydroamination–hydroarylation sequence, which subsequently underwent an asymmetric transfer hydrogenation catalysed by a chiral phosphoric acid in the presence of ethyl Hantzsch ester. As shown in Scheme 7.67, the products of the cascade were achieved in good yields, moderate to high diastereoselectivities of up to >90% de, and high enantioselectivities ranging from 82 to 97% ee.

In 2010, Alexakis and Quintard reported a tandem reaction beginning with a copper-catalysed 1,4-addition of dialkylzincs to α,β-unsaturated aldehydes which was followed by an organocatalysed Michael reaction with vinyl sulfone.[80] As shown in Scheme 7.68, this simple procedure, using two chiral entities, such as a chiral copper complex of (R)-BINAP as the first catalyst,

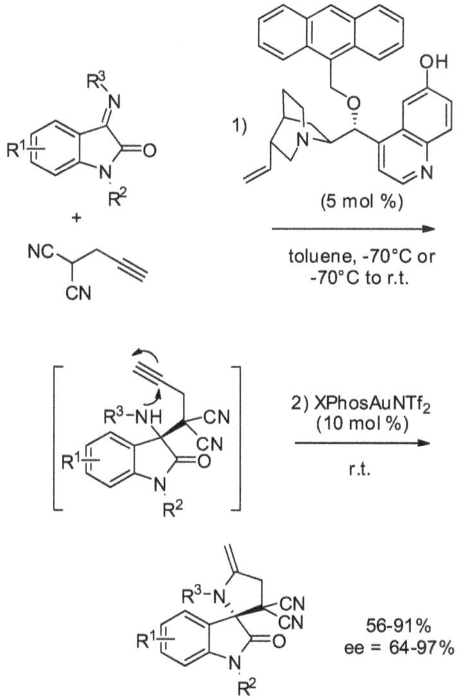

Scheme 7.65 Tandem Mannich–hydroamination reaction catalysed by chiral cinchona alkaloid catalysis and gold catalysis.

and chiral diphenylprolinol trimethylsilyl ether as the second catalyst, afforded the corresponding chiral aldehydes bearing two contiguous stereocentres in good yields, moderate diastereoselectivities of up to 70% de, and a remarkably excellent enantioselectivity of 99% ee in all cases of substrate studied. It must be noted that this work constituted the first example of an asymmetric copper-catalysed step combined with an enantioselective organocatalytic step. This procedure was applied to the synthesis of the (2*S*,3*S*) isomer of valnoctamide. As an extension of the methodology, the authors also developed another tandem asymmetric reaction consisting of an enantioselective copper-catalysed 1,4-addition of dialkylzincs to α,β-unsaturated aldehydes, which was followed by an enantioselective organocatalysed fluorination by treatment with NFSI, which provided the corresponding fluorinated products in good yields (57–74%), moderate diastereoselectivities (60–62% de) and excellent general enantioselectivity of 99% ee.

In 2012, Christmann *et al.* developed an enantioselective tandem oxidation–Diels–Alder reaction of a trienol which constituted the key step in the synthesis of the key decalin subunit of UCS1025A.[81] The first step of this one-pot process was a copper-catalysed oxidation of this trienol into

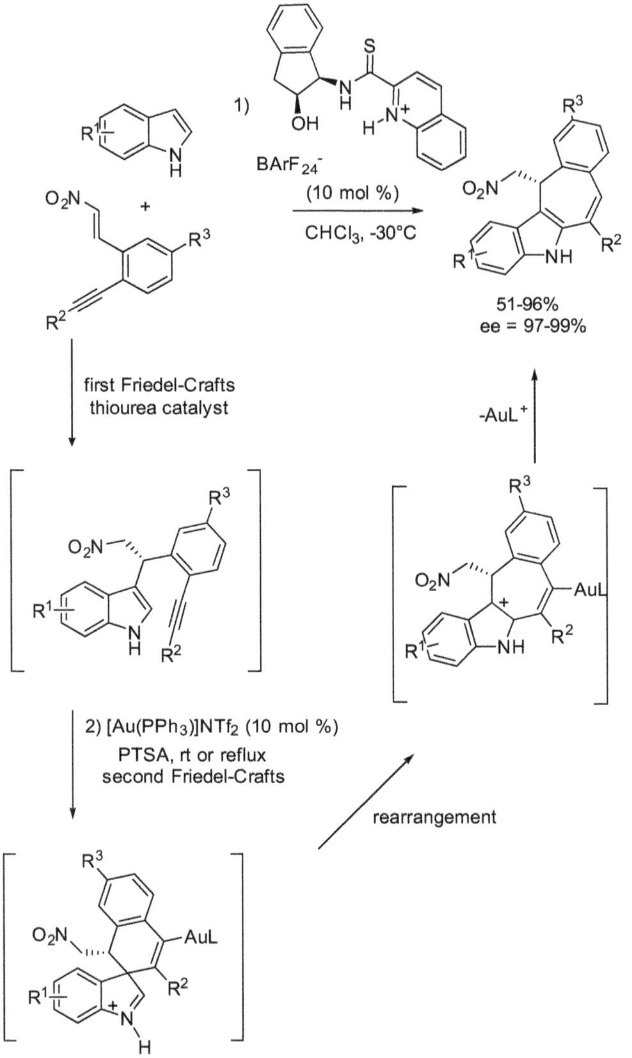

Scheme 7.66 Tandem double Friedel – Crafts reaction catalysed by chiral thiourea catalysis and gold catalysis followed by a rearrangement.

the corresponding aldehyde, which was followed by an intramolecular Diels–Alder cycloaddition induced by a chiral imidazolidinone catalyst subsequently added to the reaction media. The use of these compatible catalysts allowed the corresponding chiral bicyclic product to be achieved in good yield, moderate diastereoselectivity of 58% de and high enantioselectivity of 90% ee, as shown in Scheme 7.69.

In 2013, a remarkable highly enantio- and diastereoselective synthesis of 2,6-*cis*-substituted tetrahydropyrans was achieved by Zhao *et al.* on the basis

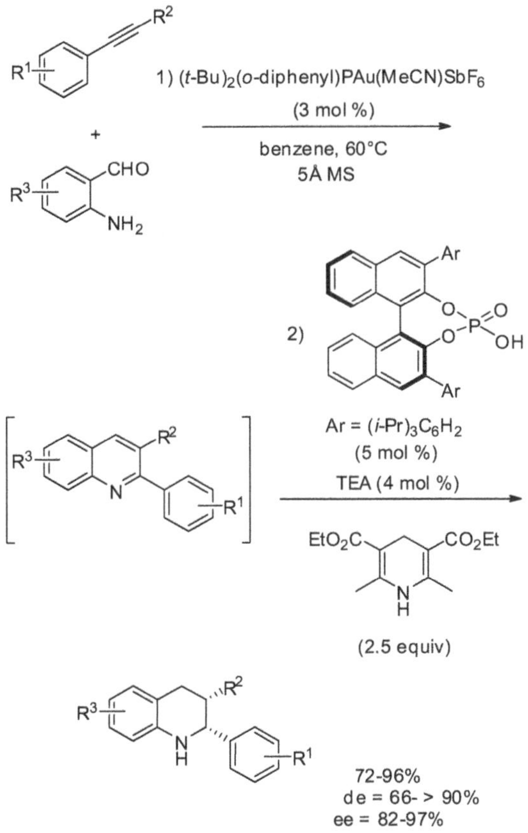

Scheme 7.67 Tandem hydroamination–hydroarylation–transfer hydrogenation reaction catalysed by gold catalysis and chiral phosphoric acid catalysis.

of a one-pot sequential catalysis involving Henry and oxa-Michael reactions.[82] The first Henry reaction of nitromethane with a 7-oxo-hept-5-enal was catalysed by $Cu(OAc)_2$ in the presence of a chiral tetraamine to give the corresponding alcohol, which underwent an intramolecular oxa-Michael reaction by treatment with camphorsulfonic acid (CSA) to finally afforded the corresponding 2,6-*cis*-substituted tetrahydropyran in high yield and exceptional diastereo- and enantioselectivity (Scheme 7.70).

On the other hand, Alexakis *et al.* have developed a sequential reaction exploiting the compatibility between a cationic iridium catalyst and chiral diphenylprolinol trimethylsilyl ether.[83] The role of the iridium complex was proposed to isomerise the starting allylic primary alcohols into the corresponding aldehydes in the presence of H_2, while the chiral amine catalyst promoted the α-functionalisation of these aldehydes through reaction with vinyl sulfone to give the corresponding Michael adducts as a mixture of *syn*- and *anti*-diastereomers obtained in moderate yields, moderate to excellent

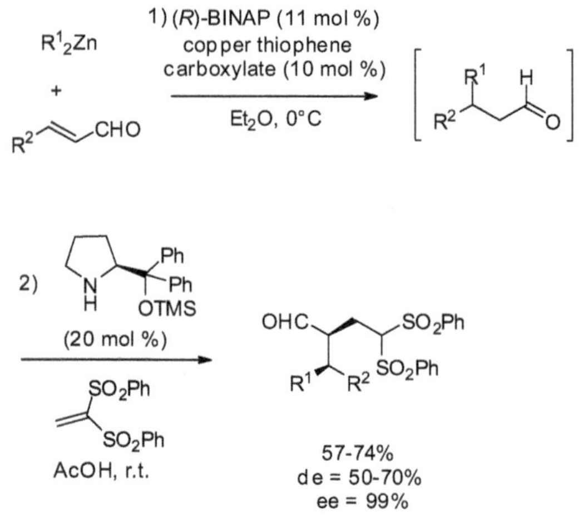

Scheme 7.68 Tandem Michael–Michael reaction catalysed by chiral copper catalysis and chiral amine catalysis.

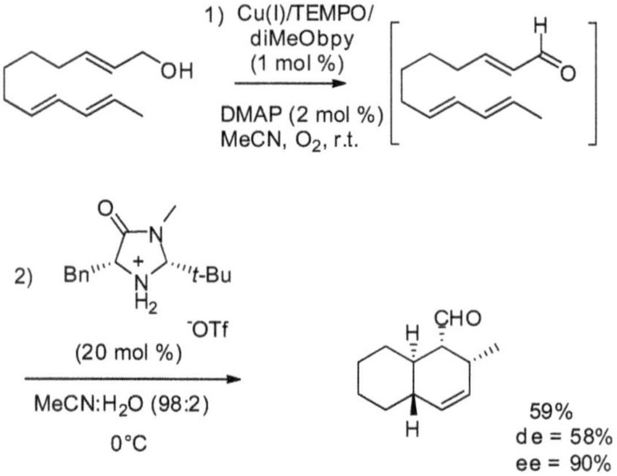

Scheme 7.69 Tandem oxidation–Diels–Alder reaction catalysed by copper catalysis and chiral amine catalysis.

diastereoselectivities of up to 96% de, and excellent enantioselectivities of up to 99% ee, as shown in Scheme 7.71.

Palladium has also been recently employed in combination with organocatalysts in enantioselective sequential tandem reactions. For example, in 2013 Sudalai and co-workers reported a novel synthesis of chiral 3-substituted tetrahydroquinoline derivatives based on an α-aminooxylation or -amination, using nitrobenzene as an oxygen source or diisopropyl azodicarboxylate as an amine source, respectively, followed by reductive cyclisation of o-nitrohydrocinnamaldehydes.[84] The first step was simply catalysed by L-proline and the second one by Pd/C, providing a range of chiral bicyclic products in moderate to good yields and enantioselectivities

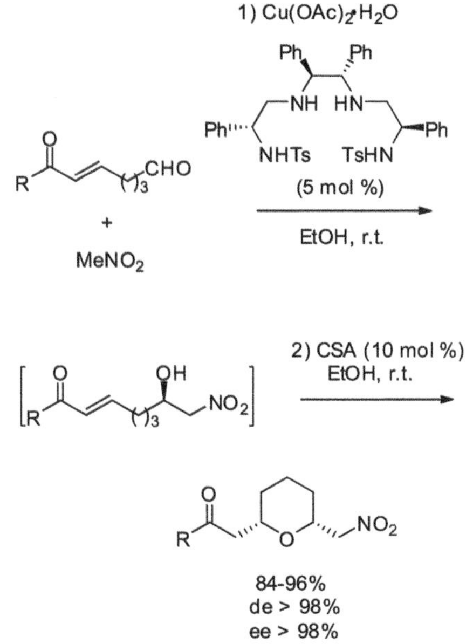

Scheme 7.70 Tandem Henry–oxa-Michael reaction catalysed by chiral copper catalysis and camphorsulfonic acid catalysis.

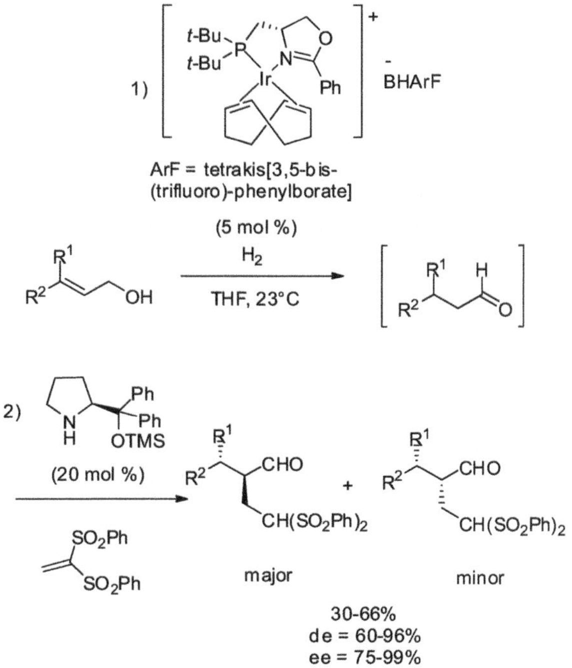

Scheme 7.71 Tandem isomerisation–Michael reaction catalysed by iridium catalysis and chiral amine catalysis.

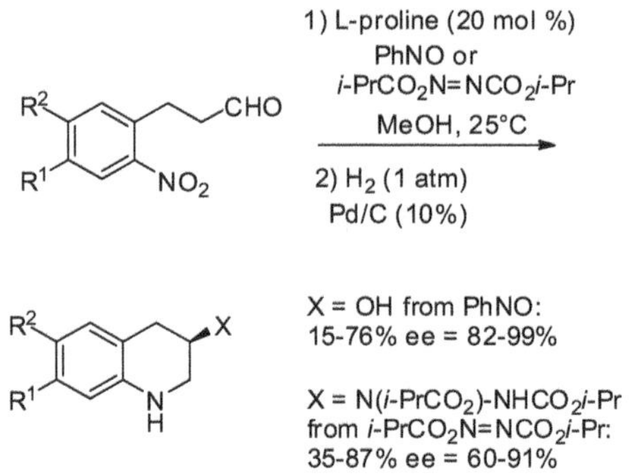

Scheme 7.72 Tandem α-aminooxylation or -amination–reductive cyclisation reaction catalysed by chiral amine catalysis and palladium catalysis.

of up to 99% ee, as shown in Scheme 7.72. The utility of this novel methodology was demonstrated by the synthesis of two important bioactive
molecules, such as (−)-sumanirole and 1-[(S)-3-(dimethylamino)-3,4-dihydro-
6,7-dimethoxy-quinolin-1(2H)-yl]-propanone.

References

1. Y. Ito, M. Sawamura and T. Hayashi, *J. Am. Chem. Soc.*, 1986, **108**, 6405–6406.
2. S. D. Pastor and A. Togni, *J. Am. Chem. Soc.*, 1989, **111**, 2333–2334.
3. B. G. Jellerichs, J.-R. Kong and M. J. Krische, *J. Am. Chem. Soc.*, 2003, **125**, 7758–7759.
4. (a) M. Klussmann, *Angew. Chem., Int. Ed.*, 2009, **48**, 7124–7125; (b) Z. Shao and H. Zhang, *Chem. Soc. Rev.*, 2009, **38**, 2745–2755.
5. C. C. J. Loh and D. Enders, *Chem. – Eur. J.*, 2012, **18**, 10212–10225.
6. T. Yue, M.-X. Wang, D.-X. Wang, G. Masson and J. Zhu, *J. Org. Chem.*, 2009, **74**, 8396–8399.
7. S. Murkherjee and B. List, *J. Am. Chem. Soc.*, 2007, **129**, 11336–11337.
8. J. Erb, D. H. Paull, T. Dudding, L. Belding and T. Lectka, *J. Am. Chem. Soc.*, 2011, **133**, 7536–7546.
9. S. Lin, G.-L. Zhao, L. Deiana, J. Sun, Q. Zhang, H. Leijonmarck and A. Cordova, *Chem. – Eur. J.*, 2010, **16**, 13930–13934.
10. G.-L. Zhao, F. Ullah, L. Deiana, S. Lin, Q. Zhang, J. Sun, I. Ibrahem, P. Dziedzic and A. Cordova, *Chem. – Eur. J.*, 2010, **16**, 1585–1591.
11. C. Yu, Y. Zhang, S. Zhang, J. He and W. Wang, *Tetrahedron Lett.*, 2010, **51**, 1742–1744.
12. W. Sun, G. Zhu, L. Hong and R. Wang, *Chem. – Eur. J.*, 2011, **17**, 13958–13962.
13. I. Ibrahem, P. Breistein and A. Cordova, *Angew. Chem., Int. Ed.*, 2011, **50**, 12036–12041.

14. J.-H. Kim, E.-J. Park, H.-J. Lee, X.-H. Ho, H.-S. Yoon, P. Kim, H. Yun and H.-Y. Jang, *Eur. J. Org. Chem.*, 2013, 4337–4344.
15. L. Stegbauer, F. Sladojevich and D. J. Dixon, *Chem. Sci.*, 2012, **3** 942–958.
16. C. Arroniz, A. Gil-Gonzalez, V. Semak, C. Escolano, J. Bosch and M. Amat, *Eur. J. Org. Chem.*, 2011, 3755–3760.
17. M.-X. Zhao, D.-K. Wei, F.-H. Ji, X.-L. Zhao and M. Shi, *Chem. – Asian J.*, 2012, 7, 2777–2781.
18. F. Sladojevich, A. Trabocchi, A. Guarna and D. J. Dixon, *J. Am. Chem. Soc.*, 2011, **133**, 1710–1713.
19. H. Y. Kim and K. Oh, *Org. Lett.*, 2011, **13**, 1306–1309.
20. (a) N. T. Patil, *Angew. Chem., Int. Ed.*, 2011, **50**, 1759–1761; (b) D. T. Cohen and K. A. Scheidt, *Chem. Sci.*, 2012, **3**, 53–57.
21. D. E. A. Raup, B. Cardinal-David, D. Holte and K. A. Scheidt, *Nature Chem.*, 2010, **2**, 766–771.
22. B. Cardinal-David, D. E. A. Raup and K. A. Scheidt, *J. Am. Chem. Soc.*, 2010, **132**, 5345–5347.
23. (a) D. T. Cohen, B. Cardinal-David and K. A. Scheidt, *Angew. Chem., Int. Ed.*, 2011, **50**, 1678–1682; (b) D. T. Cohen, B. Cardinal-David and K. A. Scheidt, *Angew. Chem., Int. Ed.*, 2011, 7, 1687–1692.
24. D. T. Cohen, B. Cardinal-David, J. M. Roberts, A. A. Sarjeant and K. A. Scheidt, *Org. Lett.*, 2011, **13**, 1068–1071.
25. (a) E. Wingstrand and C. Moberg, *Synlett*, 2010, **3**, 355–367; (b) L. Fransson and C. Moberg, *ChemCatChem*, 2010, **2**, 1523–1532; (c) R. Hertzberg, K. Widyan, B. Heid and C. Moberg, *Tetrahedron*, 2012, **68**, 7680–7684.
26. F. Li, K. Widyan, E. Wingstrand and C. Moberg, *Eur. J. Org. Chem.*, 2009, 3917–3922.
27. S. Ogawa, N. Iida, E. Tokunaga, M. Shiro and N. Shibata, *Chem. – Eur. J.*, 2010, **16**, 7090–7095.
28. C. Li, B. Villa-Marcos and J. Xiao, *J. Am. Chem. Soc.*, 2009, **131**, 6967–6969.
29. M. Rueping and R. M. Koenigs, *Chem. Commun.*, 2011, **47**, 304–306.
30. N. T. Patil, V. S. Raut and R. B. Tella, *Chem. Commun.*, 2013, **49**, 570–572.
31. B. Xu, S.-F. Zhu, X.-L. Xie, J.-J. Shen and Q.-L. Zhou, *Angew. Chem., Int. Ed.*, 2011, **50**, 11483–11486.
32. (a) X. Xu, J. Zhou, L. Yang and W. Hu, *Chem. Commun.*, 2008, 6564–6566; (b) W. Hu, X. Xu, J. Zhou, L. W.-J. Liu, H. Huang, J. Hu, L. Yang and L.-Z. Gong, *J. Am. Chem. Soc.*, 2008, **130**, 7782–7783.
33. J. Jiang, H.-D. Xu, J.-B. Xi, B.-Y. Ren, F.-P. Lv, X. Guoi, L.-Q. Jiang, Z.-Y. Zhang and W.-H. Hu, *J. Am. Chem. Soc.*, 2011, **133**, 8428–8431.
34. H. Qiu, M. Li, L.-Q. Jiang, F.-P. Lv, L. Zan, C.-W. Zhai, M. P. Doyle and W.-H. Hu, *Nat. Chem.*, 2012, **4**, 733–738.
35. X. Xu, Y. Qian, L. Yang and W. Hu, *Chem. Commun.*, 2011, **47**, 797–799.
36. Y. Qian, X. Xu, L. Jiang, D. Prajapati and W. Hu, *J. Org. Chem.*, 2010, **75**, 7483–7486.

37. Y. Qian, C. Jing, T. Shi, J. Ji, M. Tang, J. Zhou, C. Zhai and W. Hu, *ChemCatChem*, 2011, **3**, 653–656.
38. P. Biginelli, *Gazz. Chim. Ital.*, 1893, **23**, 360–413.
39. Y.-F. Cai, H.-M. Yang, L. Li, K.-Z. Jiang, G.-Q. Lai, J.-X. Jiang and L.-W. Xu, *Eur. J. Org. Chem.*, 2010, 4986–4990.
40. A. S. K. Hashmi and C. Hubbert, *Angew. Chem., Int. Ed.*, 2010, **49**, 1010–1012.
41. Z.-Y. Han, H. Xiao, X.-H. Chen and L.-Z. Gong, *J. Am. Chem. Soc.*, 2009, **131**, 9182–9183.
42. X.-Y. Liu and C.-M. Che, *Org. Lett.*, 2009, **11**, 4204–4207.
43. Y.-P. He, H. Wu, D.-F. Chen, J. Yu and L.-Z. Gong, *Chem. – Eur. J.*, 2013, **19**, 5232–5237.
44. K. L. Jensen, P. T. Franke, C. Arróniz, S. Kobbelgaard and K. A. Jørgensen, *Chem. – Eur. J.*, 2010, **16**, 1750–1753.
45. Z.-Y. Han, R. Guo, P.-S. Wang, D.-F. Chen, H. Xiao and L.-Z. Gong, *Tetrahedron Lett.*, 2011, **52**, 5963–5967.
46. C. Wang, Z.-Y. Han, H.-W. Luo and L.-Z. Gong, *Org. Lett.*, 2010, **12**, 2266–2269.
47. Z.-Y. Han, D.-F. Chen, Y.-Y. Wang, R. Guo, P.-S. Wang and C. Wang, *J. Am. Chem. Soc.*, 2012, **134**, 6532–6535.
48. P.-S. Wang, K.-N. Li, X.-L. Zhou, X. Wu, Z.-Y. Han, R. Guo and L.-Z. Gong, *Chem. – Eur. J.*, 2013, **19**, 6234–6238.
49. N. T. Patil, A. K. Mutyala, A. Konala and R. B. Tella, *Chem. Commun.*, 2012, **48**, 3094–3096.
50. D. Qian and J. Zhang, *Chem. – Eur. J.*, 2013, **19**, 6984–6988.
51. H. Wu, Y.-P. He and L.-Z. Gong, *Org. Lett.*, 2013, **15**, 460–463.
52. Q. Cai, Z.-A. Zhao and S.-L. You, *Angew. Chem., Int. Ed.*, 2009, **48**, 7428–7431.
53. Q. Cai, C. Zheng and S.-L. You, *Angew. Chem., Int. Ed.*, 2010, **49**, 8666–8669.
54. Q. Cai, X.-W. Liang, S.-G. Wang, J.-W. Zhang, X. Zhang and S.-L. You, *Org. Lett.*, 2012, **14**, 5022–5025.
55. (a) M. Rueping, H. Sunden and E. Sugiono, *Chem. – Eur. J*, 2012, **18**, 3649–3653; (b) M. Rueping, J. Dufour and M. S. Maji, *Chem. Commun.*, 2012, **48**, 3406–3408.
56. Q.-A. Chen, D.-S. Wang, Y.-G. Zhou, Y. Duan, H.-J. Fan, Y. Yang and Z. Zhang, *J. Am. Chem. Soc.*, 2011, **133**, 6126–6129.
57. Q. Cai, X.-W. Liang, S.-G. Wang and S.-L. You, *Org. Biomol. Chem.*, 2013, **11**, 1602–1605.
58. Y. Toda and M. Terada, *Synlett*, 2013, **24**, 752–756.
59. M. Rueping, H. Sunden, L. Hubener and E. Sugiono, *Chem. Commun.*, 2012, **18**, 2201–2203.
60. O. Abillard and B. Breit, *Adv. Synth. Catal.*, 2007, **349**, 1891–1895.
61. S. Chercheja and P. Eilbracht, *Adv. Synth. Catal.*, 2007, **349**, 1897–1905.
62. S. Chercheja, S. K. Nadakudity and P. Eilbracht, *Adv. Synth. Catal.*, 2010, **352**, 637–643.

63. S. Checheja, T. Rothenbücher and P. Eilbracht, *Adv. Synth. Catal.*, 2009, **351**, 339–344.
64. M. Terada and Y. Toda, *Angew. Chem., Int. Ed.*, 2012, **51**, 2093–2097.
65. L. Ren, X.-L. Lian and L.-Z. Gong, *Chem. – Eur. J.*, 2013, **19**, 3315–3318.
66. J. Aleman, V. del Solar, C. Martin-Santos, C. N. Cubo and L. Ranninger, *J. Org. Chem.*, 2011, **76**, 7287–7293.
67. (a) S. L. Poe, M. Kobaslija and D. T. McQuade, *J. Am. Chem. Soc.*, 2006, **128**, 15586–15587; (b) S. L. Poe, M. Kobaslija and D. T. McQuade, *J. Am. Chem. Soc.*, 2007, **129**, 9216–9221.
68. L. Ren, T. Lei, J.-X. Ye and L.-Z. Gong, *Angew. Chem., Int. Ed.*, 2012, **51**, 771–774.
69. B. M. Choudary, N. S. Chowdari, S. Madhi and M. L. Kantam, *J. Org. Chem.*, 2003, **68**, 1736–1746.
70. M. Terada and Y. Toda, *J. Am. Chem. Soc.*, 2009, **131**, 6354–6355.
71. B. Simmons, A. M. WAlji and D. W. C. MacMillan, *Angew. Chem., Int. Ed.*, 2009, **48**, 4349–4353.
72. M. E. Muratore, C. A. Holloway, A. W. Pilling, R. I. Storer, G. Trevitt and D. J. Dixon, *J. Am. Chem. Soc.*, 2009, **131**, 10796–10797.
73. (a) S. Belot, K. A. Vogt, C. Besnard, N. Krause and A. Alexakis, *Angew. Chem., Int. Ed.*, 2009, **48**, 8923–8926; (b) S. Belot, A. Quintard, N. Krause and A. Alexakis, *Adv. Synth. Catal.*, 2010, **352**, 667–695.
74. T. Zweifel, D. Hollmann, B. Prüger, M. Nielsen and K. A. Jørgensen, *Tetrahedron: Asymmetry*, 2010, **21**, 1624–1629.
75. D. Monge, K. L. Jensen, P. T. Franke, L. Lykke and K. A. Jørgensen, *Chem. – Eur. J.*, 2010, **16**, 9478–9484.
76. D. M. Barber, H. J. Sanganee and D. J. Dixon, *Org. Lett.*, 2012, **14**, 5290–5293.
77. X. Chen, H. Chen, X. Ji, H. Jiang, Z.-J. Yao and H. Liu, *Org. Lett.*, 2013, **15**, 1846–1849.
78. C. C. J. Loh, J. Badorrek, G. Raabe and D. Enders, *Chem. – Eur. J.*, 2011, **17**, 13409–13414.
79. X.-Y. Liu, Y.-P. Xiao, F.-M. Siu, L.-C. Ni, Y. Chen, L. Wang and C.-M. Che, *Org. Biomol. Chem.*, 2012, **10**, 7208–7219.
80. A. Quintard and A. Alexakis, *Adv. Synth. Catal.*, 2010, **352**, 1856–1860.
81. D. Könning, W. Hiller and M. Christmann, *Org. Lett.*, 2012, **14**, 5258–5261.
82. Q. Dai, N. K. Rana and J. C.-G. Zhao, *Org. Lett.*, 2013, **15**, 2922–2925.
83. A. Quintard, A. Alexakis and C. Mazet, *Angew. Chem., Int. Ed.*, 2011, **50**, 2354–2358.
84. V. Rawat, B. S. Kumar and A. Sudalai, *Org. Biomol. Chem.*, 2013, **11**, 3608–3611.

Reactions Catalysed by a Combination of Metals and Enzymes

8.1 Chemoenzymatic Reactions Not Based on DKR

The combination of an enzyme with either organometallic or organic molecules is still a relatively unexplored field. Since the prototypes of tandem processes are the sequential transformations catalysed in nature by biocatalysts, the incorporation of enzymatic transformations in a series of sequential nonenzymatic reactions opens up new and promising opportunities for organic synthesis.[1] In contrast with the traditional asymmetric chemocatalysed tandem reactions, only a few examples of asymmetric reactions have been reported in which the initiation of the reaction cascade consisted of a biotransformation. In the case of the sequence of events being triggered by a biocatalyst, the cascade may proceed in a highly asymmetric fashion to furnish products in a nonracemic form. In the first step, the enzyme modifies an enzyme-labile trigger group within the starting material, *e.g. via* reduction, oxidation, hydrolysis of an ester or epoxide, transesterification of an alcohol, *etc.*, giving access to a reactive intermediate, which can immediately undergo a subsequent reaction, which may consist of a cyclisation, fragmentation, or rearrangement, *etc.* It must be remembered that the first successful combination of enzymatic with nonenzymatic transformations in a nonasymmetric domino reaction sequence was reported by Waldmann *et al.*, in 1996.[2] A nice recent example of enzyme-triggered one-pot tandem reaction was reported by Moberg *et al.* with the addition of α-ketonitriles to aldehydes mediated by a chiral titanium catalyst and lipase CALB as enzyme in the presence of DBU or DMAP.[3] The salen metal complex promoted the

RSC Catalysis Series No. 20
Enantioselective Multicatalysed Tandem Reactions
By Hélène Pellissier
© Hélène Pellissier 2014
Published by the Royal Society of Chemistry, www.rsc.org

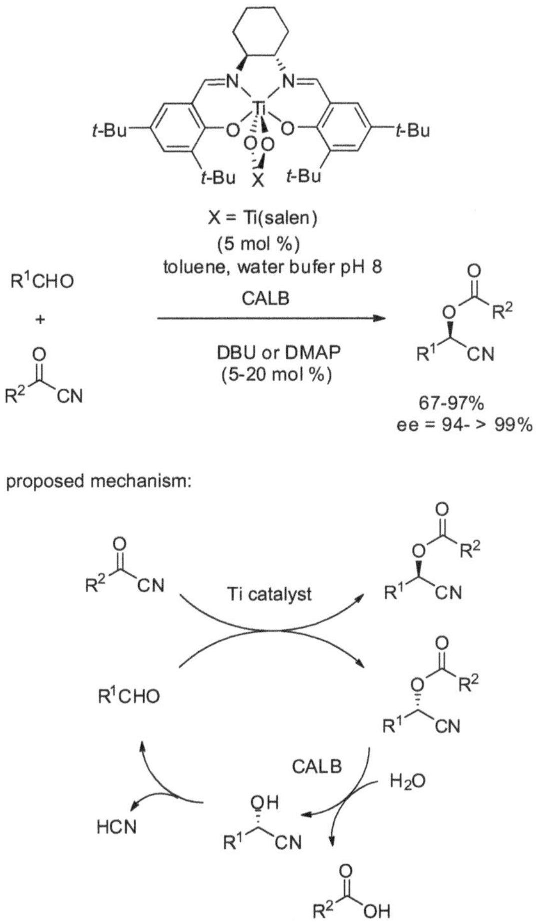

Scheme 8.1 Domino nucleophilic addition–kinetic resolution reaction catalysed by biocatalysis and chiral titanium catalysis.

addition of the α-ketonitrile to the aldehyde, and the enzyme selectively hydrolysed the minor enantiomer of the product by removing the carboxylic acid. In this way, the undesired enantiomer was converted back into the aldehyde, which could enter the catalytic cycle again. This allowed chiral esters to be achieved in good to high yields and excellent enantioselectivities of up to >99% ee, as shown in Scheme 8.1.

In 2010, a beautiful highly enantioselective tandem reaction was developed by Schmitzer *et al.*, beginning with a palladium-catalysed Suzuki coupling reaction of bromo- or iodobenzene with *p*-tolylboronic acids to give the corresponding diaryl intermediates, which were subsequently reduced by treatment with *E. coli*/ADH-A cells added to the Suzuki reaction media.[4] As shown in Scheme 8.2, the corresponding enantiopure biaryl alcohols were achieved in good to high yields, and complete enantioselectivities in all

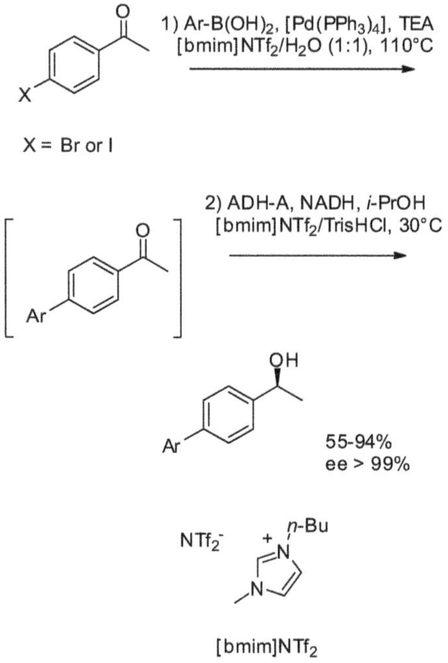

Scheme 8.2 Tandem Suzuki coupling–reduction reaction catalysed by palladium catalysis and biocatalysis.

cases of substrates studied. Gröger *et al.* have also investigated this process by using a water-soluble palladium catalyst prepared from palladium chloride and TPPTS (tris(3-sulfonatophenyl)phosphine hydrate, sodium salt), and performing the reaction in aqueous medium.[5] Indeed, when this palladium catalyst was combined with alcohol dehydrogenase in a mixture of isopropanol and water as the solvent at room temperature and in the presence of sodium carbonate as a base, the enantioselective tandem Suzuki coupling–reduction reaction led to the corresponding biaryl alcohols in higher yields (86–97%) and the same complete enantioselectivity of >99% ee.

In 2010, another example of an enzyme-triggered one-pot asymmetric tandem reaction was described by Janssen and co-workers, dealing with a one-pot four-step preparation of enantiopure β-hydroxytriazoles.[6] As shown in Scheme 8.3, the process began with a biocatalytic cascade incorporating four processes which transformed prochiral α-halo ketones into the corresponding enantiopure β-hydroxy azides. In the first reaction, the ketone was stereoselectively reduced to the corresponding β-halo alcohol by an alcohol dehydrogenase (ADH). In this process, NADPH was used as a cofactor. The recycling of NADPH was provided by the same enzyme in a reaction which converted isopropyl alcohol into acetone. The β-halo alcohol served as a substrate for halohydrin dehalogenase (Hhe), which catalysed the ring-closure towards the corresponding epoxide, and subsequent ring-opening of

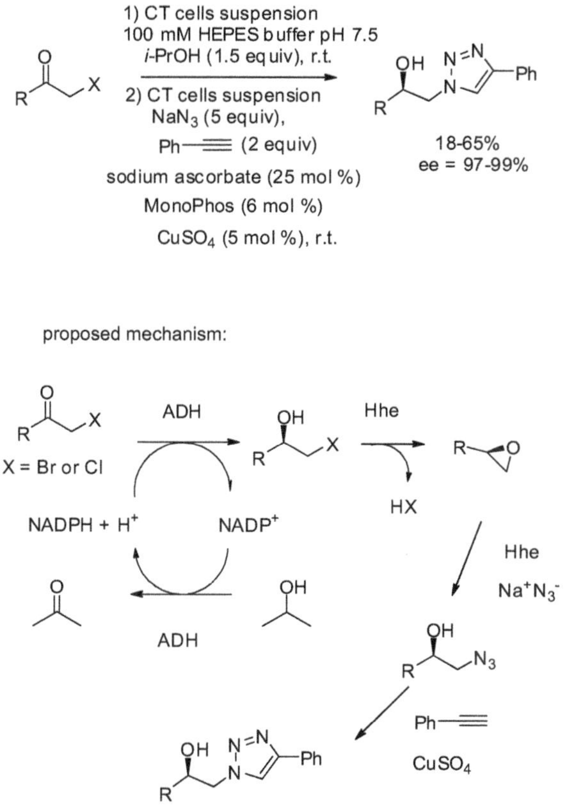

1) CT cells suspension
100 mM HEPES buffer pH 7.5
i-PrOH (1.5 equiv), r.t.

2) CT cells suspension
NaN₃ (5 equiv),

Ph———≡ (2 equiv)

sodium ascorbate (25 mol %)

MonoPhos (6 mol %)

CuSO₄ (5 mol %), r.t.

18-65%
ee = 97-99%

proposed mechanism:

Scheme 8.3 Tandem reduction–ring-closure–ring-opening–1,3-dipolar cycloaddition reaction catalysed by whole-cell catalysis and copper catalysis.

this epoxide by azide anion, forming an azido alcohol, which was the final product of the biocatalytic cascade. Then, this azido alcohol was, in the same reaction vessel, converted into the corresponding chiral β-hydroxytriazole through a copper(I)-catalysed 1,3-dipolar cycloaddition. Moderate to good yields combined with a general enantioselectivity of 99% ee were obtained for this one-pot whole-cell biocatalytic and copper-catalysed tandem reaction.

In 2013, Omori and co-workers reported a useful and environmentally friendly method for the chemoenzymatic synthesis of enantiopure disubstituted 1,2,3-triazoles.[7] This procedure consisted of using cheap, versatile and readily available reagents, such as water as solvent, carrot bits as biocatalysts, and CuSO₄ as metal catalyst. As shown in Scheme 8.4, the process began with the enantioselective reduction of commercially available *meta*- and *para*-substituted aminoacetophenone azides by treatment with *Daucus carota*, providing the corresponding (S)-alcohols, which subsequently underwent a copper-catalysed 1,3-dipolar cycloaddition reaction with an alkyne to give the final enantiopure disubstituted 1,2,3-triazoles in moderate to good yields.

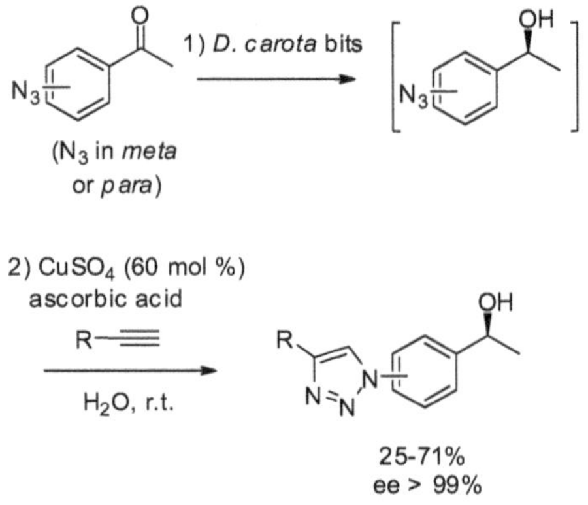

Scheme 8.4 Tandem reduction–1,3-dipolar cycloaddition reaction catalysed by biocatalysis and copper catalysis.

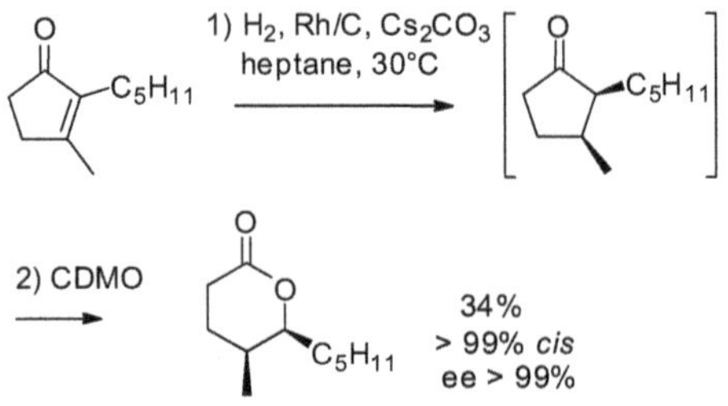

Scheme 8.5 Tandem hydrogenation–Baeyer–Villiger oxidation reaction catalysed by rhodium catalysis and biocatalysis.

In 2013, Mihovilovic and co-workers reported an enantioselective synthesis of *Aerangis* lactones on the basis of a chemoenzymatic tandem hydrogenation–Baeyer–Villiger oxidation reaction.[8] For example, (5*S*,6*S*)-5-methyl-6-pentyltetrahydro-2*H*-pyran-2-one was diastereo- and enantio-selectively produced from dihydrojasmone through a single operation consisting of a rhodium-catalysed hydrogenation in continuous flow providing the corresponding non-isolated *cis*-cyclopentanone, which was further submitted to Baeyer–Villiger oxidation by treatment with Baeyer–Villiger monooxygenases, such as cyclododecanone monooxygenase (CDMO), in a subsequent batch reactor to give the final diastereo- and enantiopure product in 34% yield (Scheme 8.5).

8.2   Chemoenzymatic Reactions Based on DKR

The versatility of the combination of enzymes with metal catalysts is also well demonstrated by chemoenzymatic dynamic kinetic resolutions (DKRs). Indeed, to overcome the major drawback of kinetic resolution for which the maximum yield is limited to 50%,[9] the combination of a metal-catalysed racemisation of the slow-reacting enantiomer with an enzyme-catalysed

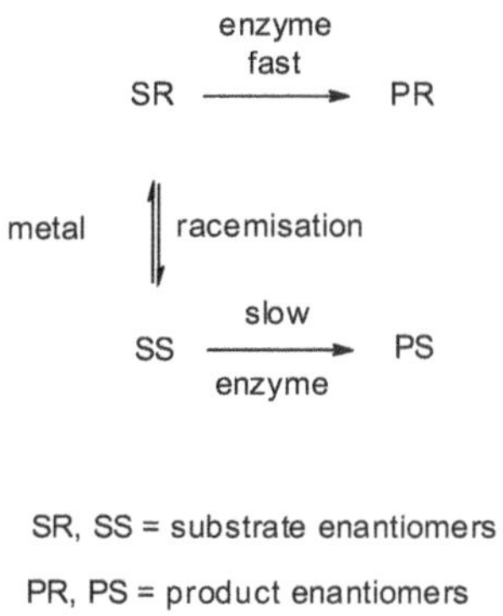

$$SR \xrightarrow[\text{fast}]{\text{enzyme}} PR$$

$$\text{metal} \Updownarrow \text{racemisation}$$

$$SS \xrightarrow[\text{enzyme}]{\text{slow}} PS$$

SR, SS = substrate enantiomers

PR, PS = product enantiomers

Scheme 8.6 Concept of dynamic kinetic resolution.

resolution of a racemate turned out to be a powerful strategy to achieve DKR processes.[10] In such processes, if the enantioselectivity is sufficient, then isolation of a highly enriched non-racemic product is possible with a theoretical yield of 100% based on the racemic substrate. In this way, all of the substrate can be converted into a product as a single enantiomer with a 100% theoretical yield (Scheme 8.6).

Since the demonstration of the compatibility of enzymes with metal complexes in one pot,[11] this powerful concept has attracted much attention[10i,l,n–s,12] Indeed, the use of transition metal–enzyme combinations, independently highlighted by Reetz and Schimossek,[11c] Stürmer,[13] and Bäckvall and co-workers,[14] to effect tandem *in situ* racemisation and resolution has widely extended the scope of DKRs.[10c–e,g,15] In this powerful approach, the enzyme acts as an enantioselective resolving catalyst and the metal serves as a racemising catalyst for efficient DKR. In order to use these reactions, a few conditions must be met. A selective enzyme is crucial and the organometallic catalyst must facilitate a fast racemisation of the substrate. Last, but not least, the catalyst should not influence the enzyme in terms of selectivity and reactivity. In the ideal case, the enzyme transforms one enantiomer of the substrate, giving rise to the corresponding product, which is not susceptible to metal-catalysed racemisation. Three major types of enzyme–metal combinations, lipase–ruthenium, subtilisin–ruthenium, and lipase combined with a metal other than ruthenium, have been developed primarily as the catalysts for the DKRs of various secondary alcohols but also for diols, amines, and esters. Meanwhile, the lipase–ruthenium combination has been the most used method up to the present time.

8.2.1 Ruthenium and Enzyme-Catalysed DKRs

8.2.1.1 DKRs of Alcohols

In 1996, Williams *et al.* described the first examples of DKRs based on the use of a combination of enzymes and metal catalysts, which involved a lipase–palladium combination for the DKR of allyl acetates,[11b] and a lipase–rhodium combination for the DKR of secondary alcohols.[11a] In the same year, Reetz and Schimossek reported the DKR of 1-phenylethylamine by a lipase–palladium combination[11c] These examples will be detailed in Section 8.2.2, dealing with DKRs based on the use of metals other than ruthenium. In 1997, Bäckvall *et al.* reported substantial improvements in the chemoenzymatic DKR processes, by employing a dinuclear ruthenium complex, called Shvo's catalyst,[16] as a racemisation catalyst in combination with an enzyme, such as the immobilised lipase from *Candida antarctica* (Novozym 435).[14,17] Therefore, a range of chiral 1-phenylethanol derivatives could be synthesised from the corresponding racemates in excellent yields and enantioselectivities of >99% ee by using this powerful binuclear ruthenium catalyst combined with an immobilised lipase and a specifically-designed acyl donor (*p*-chlorophenyl acetate). A variety of benzylic- and phenoxy-substituted *sec*-alcohols were resolved under such conditions (Scheme 8.7). These authors, and earlier also Cuccia *et al.*, demonstrated that the treatment of *meso/dl* mixtures of symmetrical diols under comparable conditions resulted in the formation of the corresponding enantio-merically-pure (*R,R*)-diacetates (Scheme 8.7).[18] The extension of this methodology to the DKR of various functionalised alcohols, such as α-hydroxycarboxylic acid esters, was also successfully achieved.[19] Moreover, a number of ring-substituted (*S*)-*O*-acetylmandelic acid esters were obtained in 69–80% yields combined with enantioselectivities of 94–98% ee (Scheme 8.7). In the same area, the DKR of β-azidoalcohols led to the corresponding enantiomerically pure-β-azidoacetates, providing an efficient route to the important class of chiral aziridines (Scheme 8.7).[20] In 2002, the scope of the same methodology was extended to the DKR of a series of β-hydroxynitriles, providing the corresponding acetates in good yields (72–98%) and high enantioselectivities (74–99% ee) (Scheme 8.7).[21]

The scope of this methodology was also applied by the same authors to the DKR of β-hydroxyesters.[22] The reaction was carried out in tandem with an aldol reaction and the β-hydroxyester formed, after neutralisation, underwent DKR using a combination of Novozym 435 and Shvo's ruthenium catalyst, as shown in Scheme 8.8. Excellent enantioselectivities of up to 99% ee were obtained, albeit associated with moderate to good yields for the expected acetates.

As another application of this methodology, an efficient route to chiral epoxides was reported by Pamies and Bäckvall based on the DKR of β-haloalcohols, in 2002 (Scheme 8.9).[23] In this case, Shvo's catalyst was associated with *Pseudomonas* species (LPS) lipase.

In 2002, the same authors demonstrated that a combination of *Pseudomonas cepacia* (PS-C) lipase and Shvo's catalyst allowed the DKR of δ- or γ-hydroxyacid derivatives to be achieved, as shown in Scheme 8.10.[24,25]

Earlier in 1999, Park *et al.* showed that the indenyl ruthenium complex, depicted in Scheme 8.11, could be applied to the DKR of 1-phenylethanol with *Pseudomonas cepacia* lipase (PCL) in the presence of molecular oxygen and TEA, without the presence of hydrogen mediators.[26] Both excellent yield

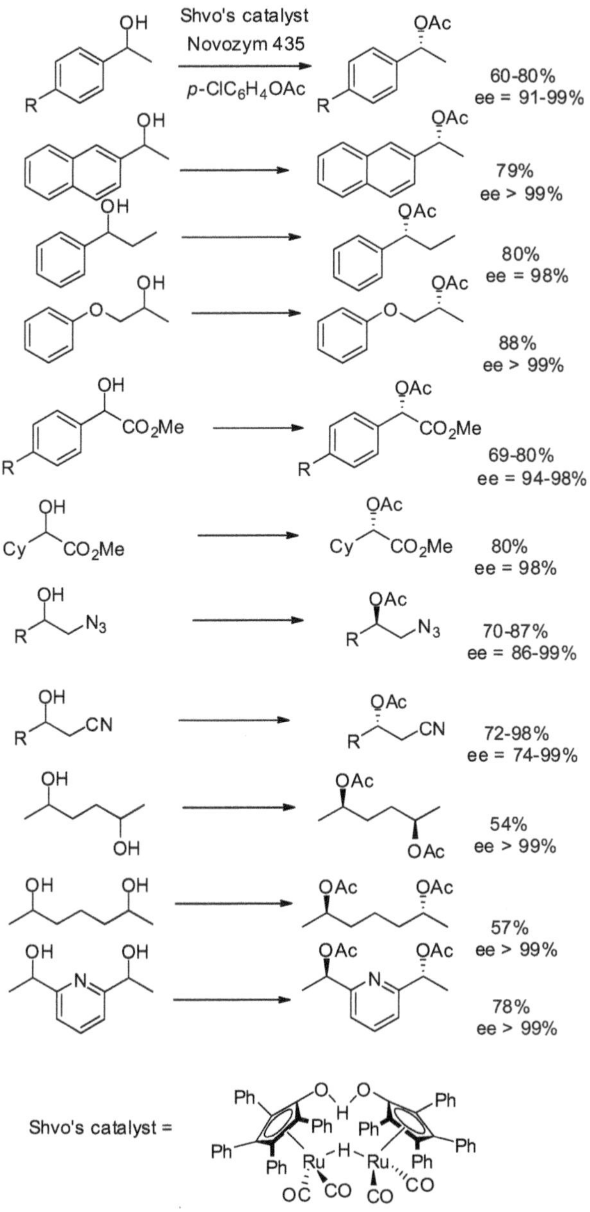

Scheme 8.7 Enzymatic DKRs of various alcohols with Shvo's catalyst.

and enantioselectivity were obtained by using *p*-chlorophenyl acetate as the acyl donor, as shown in Scheme 8.11.

The same authors also found that cymene ruthenium complexes, depicted in Scheme 8.12, were also active for the racemisation of alcohols in the presence of TEA.[27] As shown in Scheme 8.12, a noticeable feature of these

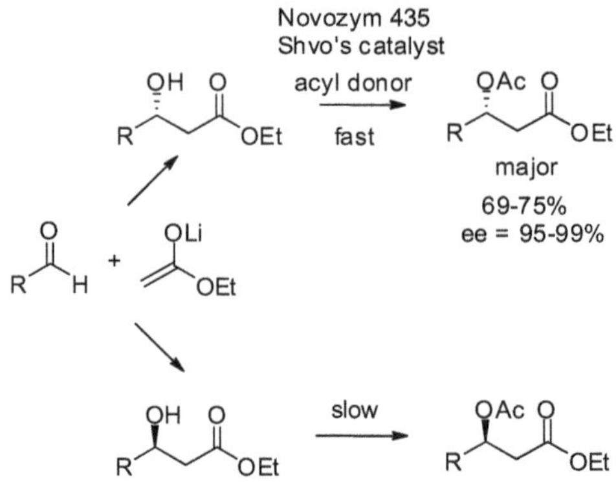

Scheme 8.8 Enzymatic DKR of β-hydroxyesters with Shvo's catalyst.

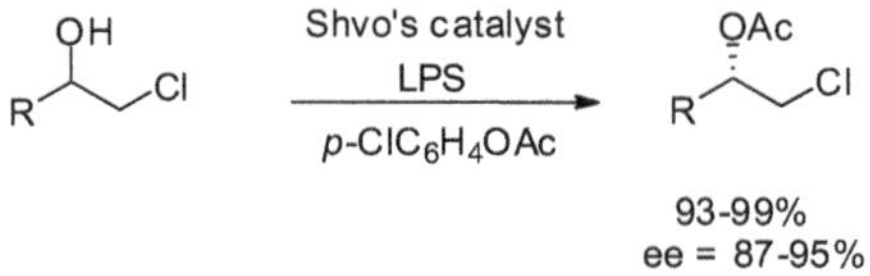

Scheme 8.9 Enzymatic DKR of β-haloalcohols with Shvo's catalyst.

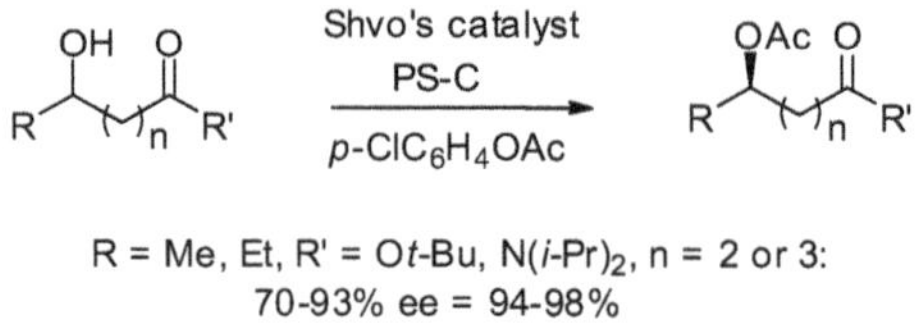

R = Me, Et, R' = O*t*-Bu, N(*i*-Pr)$_2$, n = 2 or 3:
70–93% ee = 94–98%

Scheme 8.10 Enzymatic DKR of δ- and γ-hydroxyacid derivatives with Shvo's catalyst.

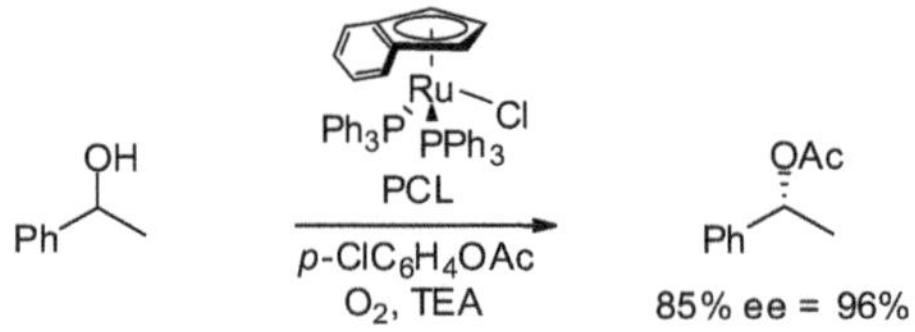

Scheme 8.11 Enzymatic DKR of 1-phenylethanol with an indenyl ruthenium catalyst.

catalysts was their high activity towards allylic alcohols, since their DKR was possible even at room temperature by using PCL lipase and *p*-chlorophenyl acetate as the acyl donor.

One of the useful strategies for enhancing the enzyme enantioselectivity is the use of structurally-modified substrates. In this way, various substrates, such as β-hydroxyacids, diols and hydroxyaldehydes were protected by Park

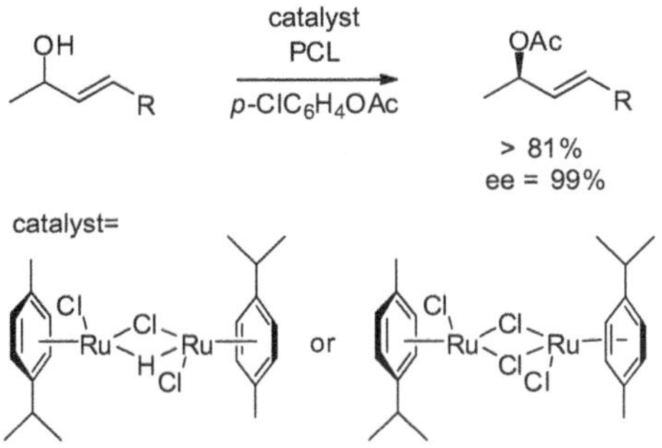

Scheme 8.12 Enzymatic DKR of allylic alcohols with cymene ruthenium catalysts.

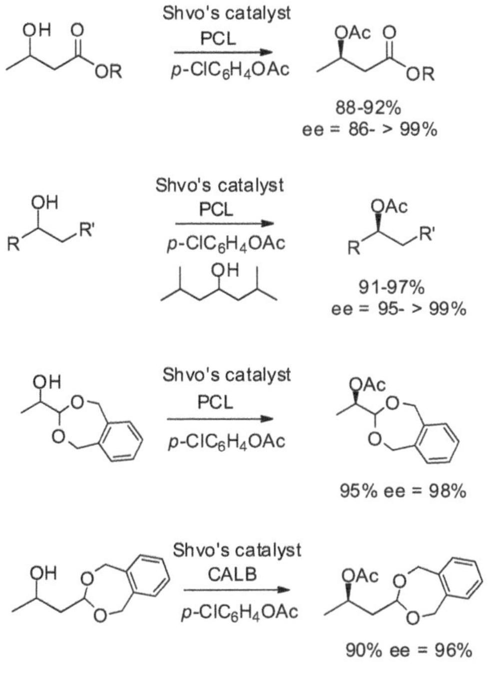

Scheme 8.13 Enzymatic DKRs of alcohols bearing bulky groups with Shvo's catalyst.

et al. with a bulky group and then submitted to lipase–ruthenium-catalysed DKR.[28] In all cases, the reactions provided the products of *R* configuration in good yields and high enantioselectivities by using Shvo's catalyst in combination with PCL or CALB lipase, as shown in Scheme 8.13.

In 2000, an original concerted catalytic reaction for the conversion of ketones or enol acetates to chiral acetates was developed by Park and co-workers.[29] This conversion proceeded through a five-step process, as shown in Scheme 8.14, which comprised: the deacetylation of the enol acetate to give the corresponding enol and the acetylated lipase, the keto–enol

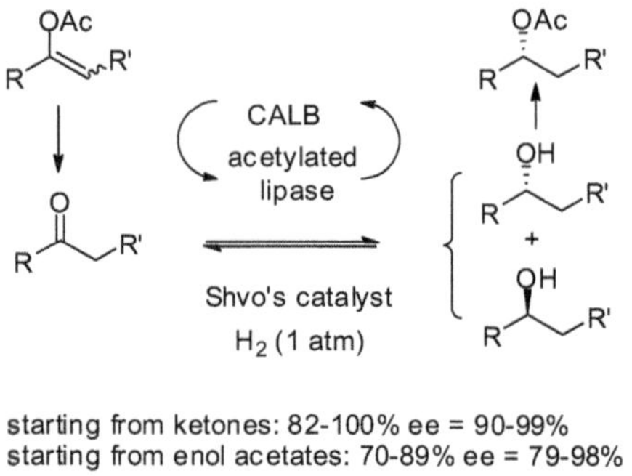

starting from ketones: 82-100% ee = 90-99%
starting from enol acetates: 70-89% ee = 79-98%

Scheme 8.14 Concerted catalytic reactions for the conversion of enol acetates or ketones to acetates.

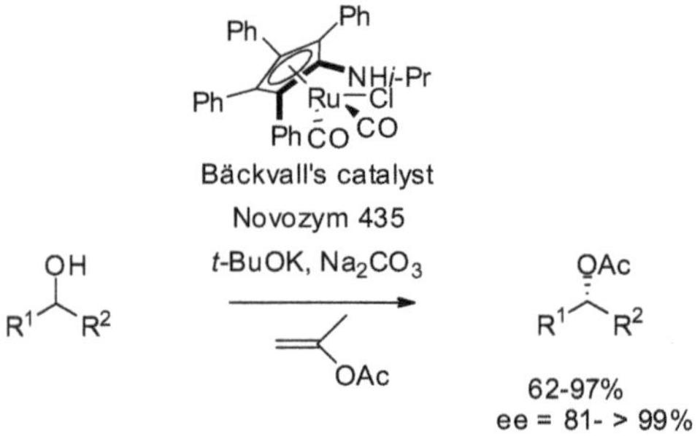

Scheme 8.15 Enzymatic DKR of alcohols with a pentaphenylcyclopentadienyl ruthenium chloride catalyst.

tautomerisation for the formation of ketone, the reduction of the ketone through hydrogenation to the racemic mixture of alcohols, the enantiose-lective acetylation of the (R)-alcohol with the acetylated lipase to produce the chiral acetate, and the reversible transformation between the two enantiomeric alcohols.

Later, a novel aminocyclopentadienyl ruthenium chloride complex was introduced by the same authors, involving a new mode of catalytic racemi-sation allowing the use of more reactive isopropenyl acetate as an acyl donor and much less lipase.[30] This catalytic system was particularly efficient for the DKR of various aliphatic or aromatic alcohols as shown in Scheme 8.15. Not only simple alcohols, but also functionalised alcohols, such as allylic alcohols, alkynyl alcohols, diols, hydroxyl esters, and chlorhydrins, were successfully transformed into the corresponding chiral acetates.[31]

Another novel ruthenium-based catalytic system, such as [TosN(CH$_2$)$_2$-NH$_2$]RuCl(p-cymene)–TEMPO, capable of promoting *in situ* racemisation during enzymatic resolution, was reported by Sheldon *et al.* and employed for the DKR of 1-phenylethanol, providing the corresponding acetate in 76% yield and enantioselectivity of >99% ee by using Novozym 435.[32] The only side product observed was acetophenone, formed by oxidation of the

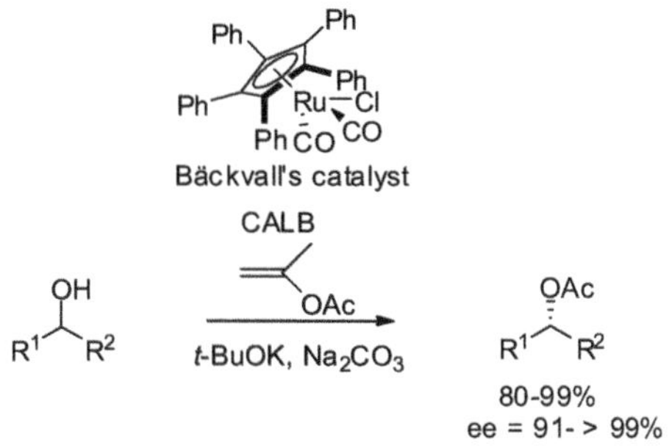

Scheme 8.16 Enzymatic DKR of alcohols with Bäckvall's catalyst.

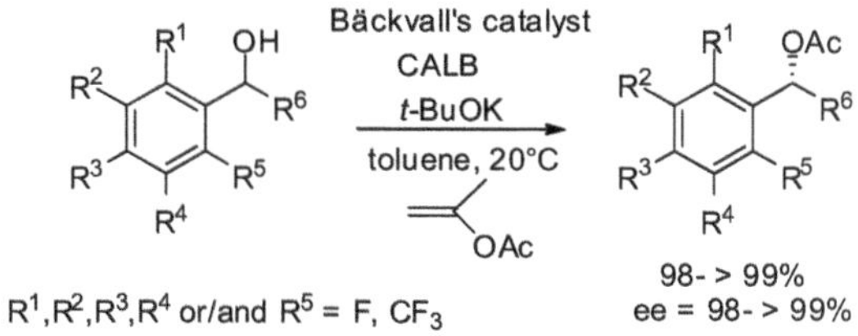

Scheme 8.17 Enzymatic DKR of fluorinated aryl alcohols with Bäckvall's catalyst.

substrate by the TEMPO cocatalyst. Soon after, Bäckvall *et al.* investigated a highly performing pentaphenylcyclopentadienylruthenium chloride complex, called Bäckvall's catalyst, which did not bear an amino group, with lipases in tandem for the DKR of a wide variety of functionalised secondary alcohols including heteroaromatic alcohols, leading, for many of the latter, to the corresponding enantiopure acetates prepared for the first time *via* DKR (Scheme 8.16).[33] The reaction took place in very short times, and isoproprenyl acetate was employed as the acyl donor, which made the purification of the products very easy. A study of the racemisation of (*S*)-1-phenylethanol indicated that the racemisation took place within the coordination sphere of the ruthenium catalyst.

The asymmetric synthesis of chiral fluoroorganic compounds plays an important role in the development of medicines, agrochemicals, and materials due to the influence of fluorine's unique properties.[34] In this context, Bogar and Bäckvall investigated the DKR of various fluorinated aryl alcohols, in 2007.[35] As shown in Scheme 8.17, a series of chiral fluorinated acetates were produced in remarkably high yields and enantioselectivities by using a combination of CALB with Bäckvall's catalyst in the presence of isopropenyl acetate as the acyl donor at room temperature.

In 2007, Park *et al.* showed that one of the two carbonyl ligands of Bäckvall's catalyst could easily be replaced with triphenylphosphane with the aid of trimethylamine *N*-oxide to give a novel analogue of Bäckvall's catalyst, depicted in Scheme 8.18.[36] Interestingly, this catalyst promoted the racemisation of alcohols at room temperature in the presence of a catalytic

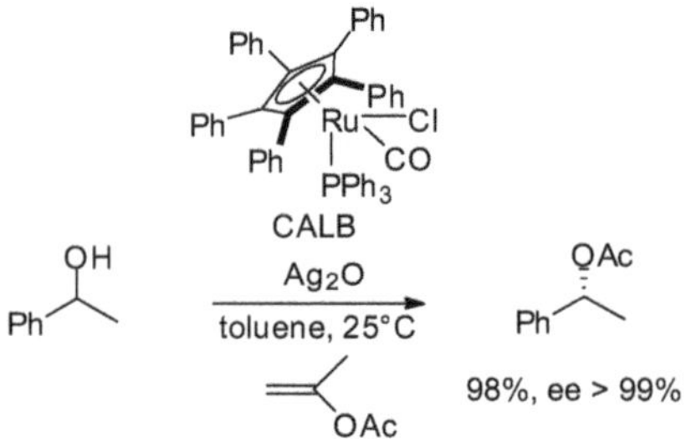

Scheme 8.18 Enzymatic DKR of phenylethanol with a Bäckvall's catalyst analogue.

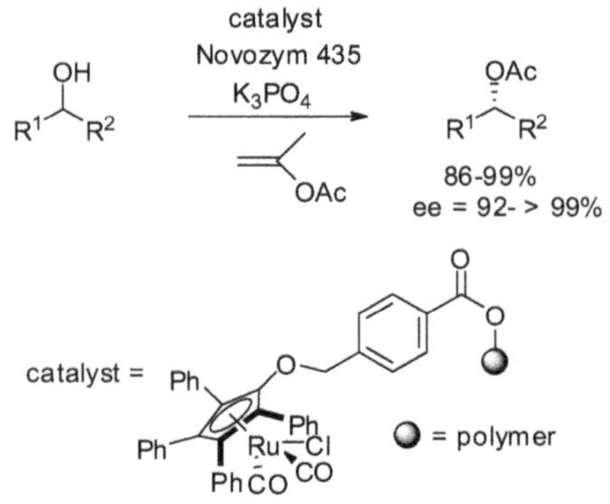

Scheme 8.19 Enzymatic DKR of alcohols with a polymer-bound ruthenium catalyst.

amount of silver oxide. The catalytic species were stable in air and reusable at least ten times for the DKR of alcohols, although stoichiometric amounts of Ag$_2$O were required. The DKR of phenylethanol under these conditions is described in Scheme 8.18.

Earlier in 2005, another racemisation ruthenium catalyst, bearing a benzyloxy function, was synthesised by the same group and was successfully applied to similar reactions, providing the DKR of a wide range of functionalised alcohols in excellent yields and enantioselectivities (>99% ee).[37] The corresponding polymer-supported derivative was also synthesised and tested as a recyclable catalyst for the aerobic DKR of alcohols (Scheme 8.19), and its catalytic activity was found to be practically the same as that of the non-polymeric catalyst.

In the same year, Trauthwein *et al.* reported the synthesis of an easy-to-handle and stable racemisation catalyst for secondary alcohols by an *in situ* mixture of readily available [Ru(*p*-cymene)Cl$_2$]$_2$ with chelating aliphatic amines.[38] The optimisation of the reaction revealed that *N,N,N',N'*-tetramethyl-1,3-propanediamine as the ligand racemised aromatic alcohols completely within 5 h. The combination of this catalyst with lipase CALB showed a good performance for the DKR of various alcohols in the presence of *p*-chlorophenyl acetate as the acyl donor, as shown in Scheme 8.20.

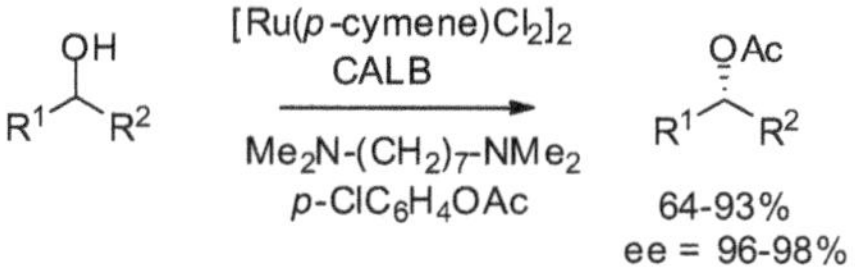

[Ru(*p*-cymene)Cl₂]₂
CALB

$$\xrightarrow{\text{Me}_2\text{N-(CH}_2)_7\text{-NMe}_2}$$
p-ClC₆H₄OAc

64–93%
ee = 96–98%

Scheme 8.20 Enzymatic DKR of alcohols with an *in situ* generated ruthenium catalyst.

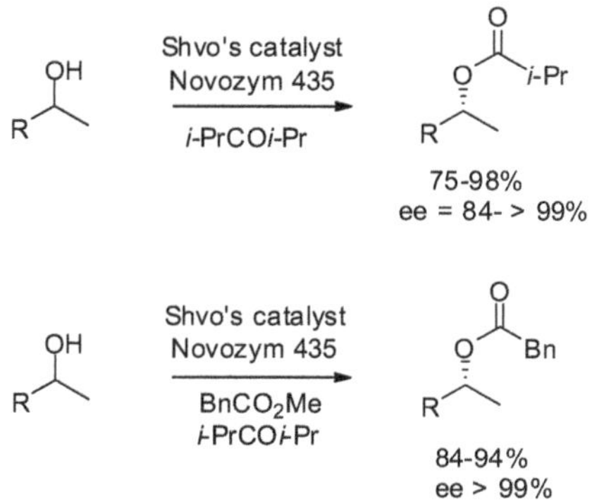

Shvo's catalyst
Novozym 435

i-PrCO*i*-Pr

75–98%
ee = 84– > 99%

Shvo's catalyst
Novozym 435

BnCO₂Me
i-PrCO*i*-Pr

84–94%
ee > 99%

Scheme 8.21 Enzymatic DKRs of alcohols with Shvo's catalyst.

In 2006, Livingston *et al.* studied the efficiency of ruthenium *p*-cymene catalyst combined with Novozym 435, in the presence of a number of different bases, for the DKR of allylic alcohols.[39] Using this method, the DKR of methyl styryl carbinol, performed in the presence of a base, such as TEA or trioctylamine, and vinyl acetate as the acyl donor, led to the corresponding chiral acetate in yields above 68% and enantioselectivities of 80–97% ee. In 2006, Wolfson *et al.* reported the DKR of 1-phenylethanol by hydrated ruthenium chloride in an aqueous medium using Novozym 435 as the lipase.[40] This novel process, involving phenyl acetate as the acyl donor, led to the formation of the corresponding chiral acetate in 82% yield and 98% ee. Besides its low price and ideal environmental impact, performing the reaction in an aqueous medium allowed an easy separation of the product. In 2005, Verzijl *et al.* demonstrated that the continuous removal of the acyl donor residue during the reaction allowed the use of simple alkyl esters as the acyl donors for the DKR of various secondary alcohols in the presence of Novozym 435 and Shvo's catalyst, as depicted in Scheme 8.21.[41] The addition of a ketone, such as 2,4-dimethyl-3-pentanone, sped up the racemisation process and allowed the amounts of enzyme and ruthenium catalyst to be reduced. Hence, various benzylic and aliphatic alcohols were reacted using isopropyl butyrate or methyl phenyl acetate as the acyl donor and, in most cases, the ester was isolated in >95% yield and 99% ee.

In 2004, the activated hydride form of a cymene ruthenium complex was shown to be effective as a racemising catalyst in ionic liquids, such as [EMIm]BF₄ and [BMIm]PF₆ ([EMIm] = 1-ethyl-3-methylimidazolium and

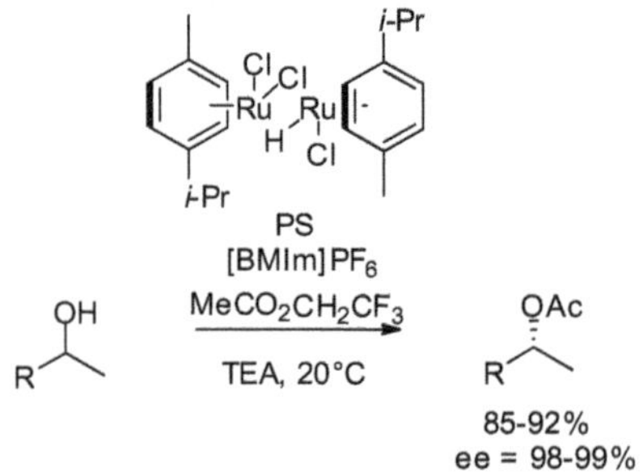

i-Pr

Cl Cl
Rú   Ru
  H   Cl

i-Pr

PS
[BMIm]PF$_6$
MeCO$_2$CH$_2$CF$_3$

OH
R

TEA, 20°C

QAc
R

85-92%
ee = 98-99%

Scheme 8.22   Enzymatic DKR of alcohols with a diruthenium catalyst in ionic liquids.

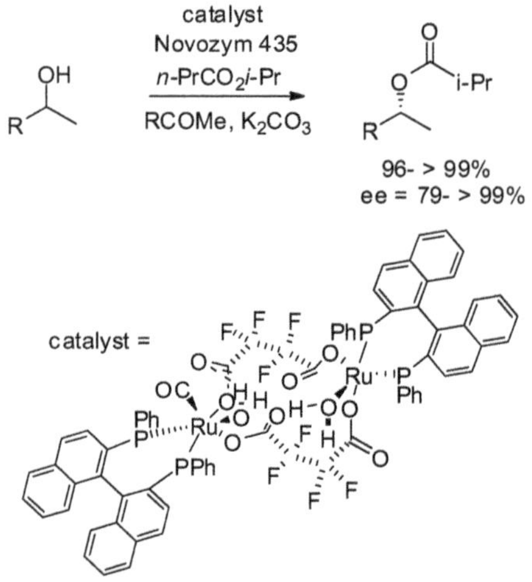

catalyst
Novozym 435
n-PrCO$_2$i-Pr

OH
R

RCOMe, K$_2$CO$_3$

O
O   i-Pr
R

96- > 99%
ee = 79- > 99%

catalyst =

Scheme 8.23   Enzymatic DKR of alcohols with a BINAP-based diruthenium catalyst.

[BMIm] = 1-butyl-3-methylimidazolium).[42] In these conditions, the DKR of various secondary alcohols occurred in the ionic liquids at room temperature, allowing the catalyst and enzyme in the ionic liquid layer to be reusable after extracting the products with ether (Scheme 8.22).

In 2006, Hulshof *et al.* reported the synthesis of a novel dinuclear ruthenium catalyst, bearing tetrafluorosuccinate and racemic BINAP ligands.[43] This catalyst was applied to the DKR of various secondary alcohols in the presence of isopropyl butyrate as the acyl donor and Novozym 435 as the enzyme (Scheme 8.23). The activation of the ruthenium catalyst with K$_2$CO$_3$ was necessary. When the reaction was performed in the presence of the ketone corresponding to the substrate, it was complete within 10 hours with an excellent enantioselectivity, whereas, without this ketone, the complete reaction was achieved in 23 hours, also giving an excellent enantioselectivity.

The combination of a lipase and Shvo's catalyst has also been applied to the DKR of α- and β-hydroxyphosphonates.[44] Hydroxyphosphonates are an

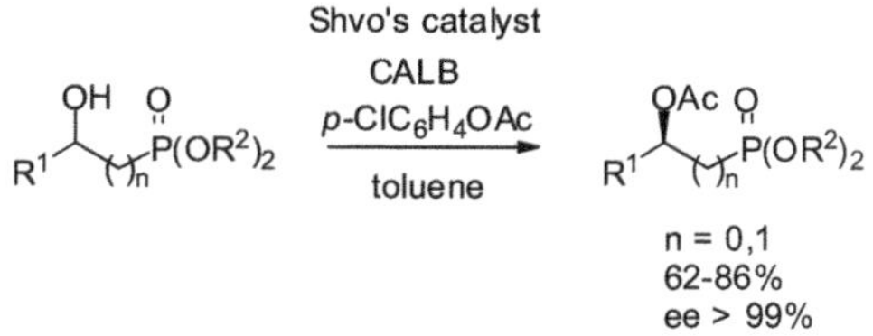

Scheme 8.24 Enzymatic DKR of dimethyl- and diethyl-α- and β-hydroxyphosphonates with Shvo's catalyst.

important class of substrates, with applications in medicinal chemistry (haptens of catalytic antibodies, phosphonic acid-based antibiotics), biochemistry (enzyme inhibitors), and organic synthesis. Under typical conditions, Pamies and Bäckvall have shown that the DKR of several dimethyl- and diethyl-α-hydroxyphosphonates proceeded with excellent enantioselectivities and moderate to good yields (Scheme 8.24). This was attributed to the coordination of the phosphonate moiety to the ruthenium catalyst at a low alcohol concentration. This DKR procedure was also applied to the deracemisation of diethyl β-hydroxyphosphonates. However, in contrast to the DKR results on the α-hydroxyphosphonates, the formation of large amounts of the corresponding ketones was observed. To increase the efficiency of the process by reducing the amount of ketone, the authors completely suppressed the ketone formation by adding 2,4-dimethyl-3-pentanol as a hydrogen source after 24 hours. Under these conditions, the DKR of β-hydroxyphosphonates proceeded with excellent enantioselectivities and moderate yields and without ketone formation. On the other hand, these authors showed that the DKR of β-hydroxynitriles by a combination of CALB with Shvo's catalyst required a high temperature (100 °C), which had the consequence of lowering the yields (≤85%) because hydrogenation occurred as a side reaction. The optical purities of the products were, however, satisfactory in all cases (94–99% ee).[45]

Later, a similar methodology was applied to the DKR of functionalised γ-hydroxy amides by using Shvo's ruthenium catalyst in combination with *Pseudomonas cepacia* lipase (lipase PS).[46] This enzyme tolerated both variations in the chain length and different functionalities, giving fair to high enantioselectivities, as shown in Scheme 8.25. The synthetic utility of this procedure was illustrated by the practical synthesis of the versatile intermediate γ-lactone, (*R*)-5-methyltetrahydrofuran-2-one.

Other functionalised alcohols have recently been submitted to DKR under closely related conditions, such as aryl β-hydroxyalkyl sulfones, which have been successfully transformed into the corresponding optically active *O*-acetyl derivatives in high yields and enantioselectivities by using CALB combined with Shvo's ruthenium catalyst (Scheme 8.26).[47]

In 2006, Alcantara *et al.* reported the first DKR of different benzoins by using *Pseudomonas stutzeri* lipase (lipase TL) and Shvo's catalyst in organic solvents, obtaining the corresponding (*S*)-acylated products with yields of up

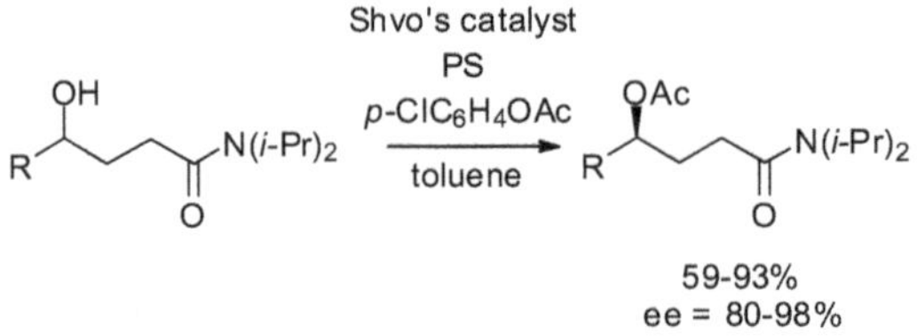

59-93%
ee = 80-98%

Scheme 8.25   Enzymatic DKR of functionalised γ-hydroxy amides with Shvo's catalyst.

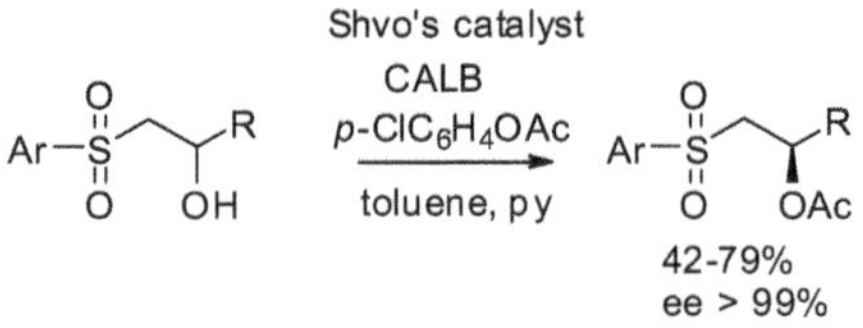

42-79%
ee > 99%

Scheme 8.26   Enzymatic DKR of β-hydroxyalkyl sulfones with Shvo's catalyst.

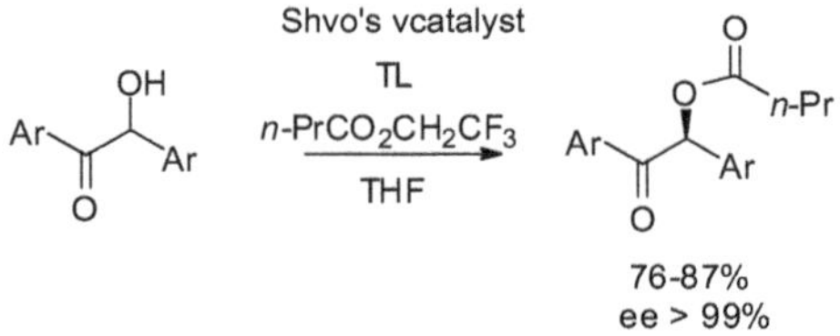

76-87%
ee > 99%

Scheme 8.27   Enzymatic DKR of benzoins with Shvo's catalyst.

to 87% and enantioselectivities of >99% ee (Scheme 8.27).[48] In all cases, the particular stereobias of the lipase towards the racemic substrates allowed the production of the opposite enantiomer to that prepared through a different enzymatic methodology.

On the other hand, Kita *et al.* reported a combination of the domino re-action concept and the DKR protocol,[49] comprising the first lipase-catalysed domino process that combined the DKR of alcohols by using 1-ethoxyvinyl esters and the Diels–Alder reaction of the intermediates. The finding that ruthenium catalysts produced a rapid racemisation of the slow-reacting (*S*)-enantiomers was the key to the success of this process, which provided useful chiral intermediates for natural products, such as compactin and forskolin (Scheme 8.28).

The lipase–ruthenium-catalysed DKR of other functionalised alcohols, such as diols, has been widely studied by Bäckvall *et al.* in recent years.[50] These authors have developed a highly efficient synthesis of enantiopure diacetates of the symmetric diols, 2,4-pentanediol and 2,5-hexanediol, by combining Bäckvall's catalyst with lipase CALB, in the presence of vinyl acetate or isoproprenyl acetate as the acyl donor, respectively. Excellent yields, and diastereo- and enantioselectivities were obtained in both cases,

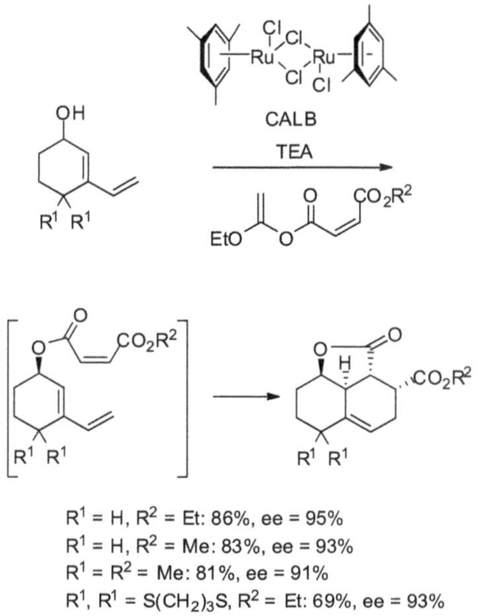

R^1 = H, R^2 = Et: 86%, ee = 95%
R^1 = H, R^2 = Me: 83%, ee = 93%
R^1 = R^2 = Me: 81%, ee = 91%
R^1, R^1 = S(CH$_2$)$_3$S, R^2 = Et: 69%, ee = 93%

Scheme 8.28 Enzymatic domino DKR–Diels–Alder reaction with a diruthenium catalyst.

as shown in Scheme 8.29. The scope of the methodology was extended to 1,3-cyclohexanediol, providing the corresponding diacetate with high *syn*-diastereoselectivity and enantioselectivity (Scheme 8.29).[51] Moreover, in 2006, the DKR of a series of 1,2-diols was achieved in similar conditions, affording the corresponding enantioenriched *syn*-diacetates as the main diastereomers (Scheme 8.29).[52] This procedure provided a useful alternative to the Sharpless asymmetric dihydroxylation, since the costs of the ruthenium catalyst and the CALB lipase are not very high. In addition, a similar methodology was applied to the synthesis of a *cis*-3,5-piperidine diacetate with excellent yield and diastereo- and enantioselectivities (Scheme 8.29).[53] This product was further converted into various interesting 3,5-disubstituted piperidines.

In 2003, Bäckvall *et al.* employed Shvo's catalyst for the lipase-catalysed acylation of unsymmetrical alkanediols.[54] Hence, enantiomerically pure *syn*-1,3-diacetates, containing one large and one small group, could be prepared, starting from the corresponding 1,3-diols (Scheme 8.30). Surprisingly, when a similar methodology was extended to unsymmetrical 1,4-diols, the authors observed the formation of the corresponding enantiomerically enriched γ-acetoxy ketones (Scheme 8.30).[55] The least hindered alcohol was acetylated, whereas oxidation of the second hydroxyl group took place under the reaction conditions. This procedure constituted a new method for the synthesis of chiral γ-hydroxy ketones, which are precursors of versatile building blocks, such as tetrahydrofurans and dihydrofurans.

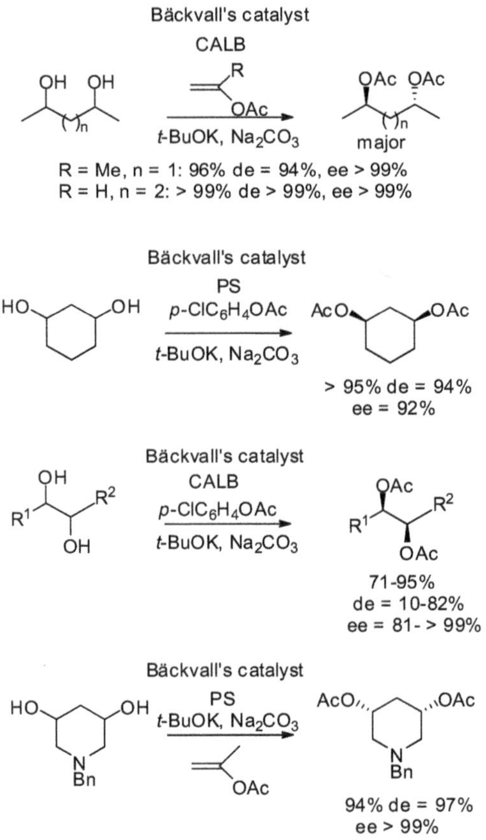

Bäckvall's catalyst
CALB

R = Me, n = 1: 96% de = 94%, ee > 99%
R = H, n = 2: > 99% de > 99%, ee > 99%

Bäckvall's catalyst
PS
p-ClC$_6$H$_4$OAc
t-BuOK, Na$_2$CO$_3$

> 95% de = 94%
ee = 92%

Bäckvall's catalyst
CALB
p-ClC$_6$H$_4$OAc
t-BuOK, Na$_2$CO$_3$

71-95%
de = 10-82%
ee = 81- > 99%

Bäckvall's catalyst
PS
t-BuOK, Na$_2$CO$_3$

94% de = 97%
ee > 99%

Scheme 8.29 Enzymatic DKRs of diols with Bäckvall's catalyst.

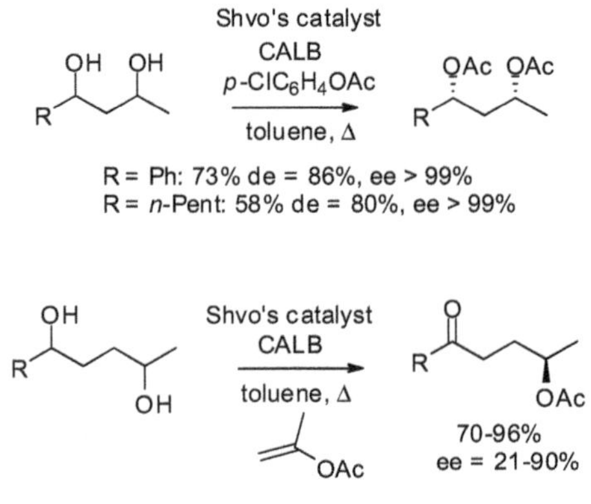

Shvo's catalyst
CALB
p-ClC$_6$H$_4$OAc
toluene, Δ

R = Ph: 73% de = 86%, ee > 99%
R = n-Pent: 58% de = 80%, ee > 99%

Shvo's catalyst
CALB
toluene, Δ

70-96%
ee = 21-90%

Scheme 8.30 Enzymatic DKRs of unsymmetrical diols with Shvo's catalyst.

In 2006, Heise *et al.* reported a novel concept for the synthesis of chiral polyesters based on a lipase-catalysed DKR polymerisation of racemic diols.[56] As shown in Scheme 8.31, a mixture of stereoisomers of a secondary

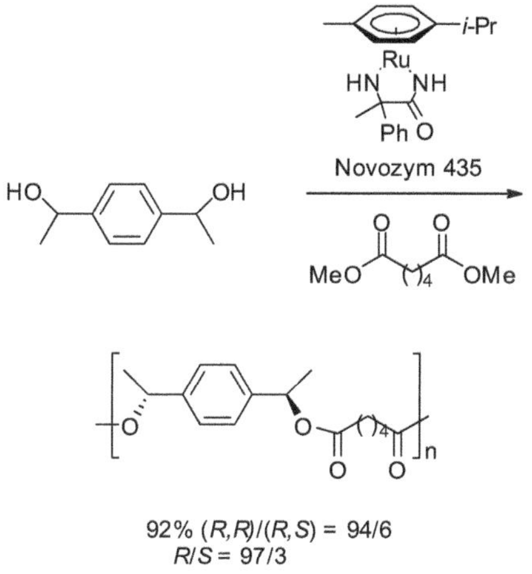

Scheme 8.31 Synthesis of chiral polyesters *via* enzymatic DKR polymerisation with a ruthenium catalyst.

diol was enzymatically polymerised with a difunctional acyl donor (dicarboxylic acid derivative) and, because of its enantioselectivity, the lipase converted only the hydroxyl groups at the *R*-configured centres. *In situ* racemisation of the hydroxyl-substituted stereocentres from the *S* to the *R* configuration allowed the polymerisation to proceed with high conversion. This combination of a DKR with a polymerisation was performed in the presence of lipase Novozym 435, a Noyori-type ruthenium catalyst, and dimethyl adipate as the acyl donor, providing chiral polyesters from non-natural monomers. Later, Meijer *et al.* applied a related methodology to 1,1′-(1,3-phenylene)diethanol and diisopropyl adipate as substrates, performing the reaction with the same catalysts albeit in the presence of 2,4-dimethyl-3-pentanol as hydrogen donor to suppress dehydrogenation of end groups resulting in inactive ketone chain ends.[57] In this case, the corresponding chiral polyester was obtained in 99% conversion and excellent enantioselectivity of 99% ee.

In 2007, a cycloruthenated amine complex, depicted in Scheme 8.32, was successfully applied by Arends *et al.* to promote the DKR of phenylethanol.[58] The reaction was performed with CALB in the presence of isopropyl acetate and *t*-BuOK, providing the corresponding enantiopure acetate in good yield.

In order to obtain (*S*)-selective DKR of secondary alcohols, an enzyme with a complementary (*S*)-stereoselectivity was needed, since the lipase-catalysed DKR provides only (*R*)-products. In this context, Park *et al.* reported, in 2003, the use of subtilisin instead of lipase, but the commercial form of subtilisin was not applicable to DKR, due to its low enzymatic activity and instability in non-aqueous medium.[59] However, these authors succeeded in enhancing its activity and stability by treating it with a surfactant before use. In these conditions, the combination of subtilisin with an analogue of Bäckvall's catalyst and trifluoroethyl butanoate as the acylating agent

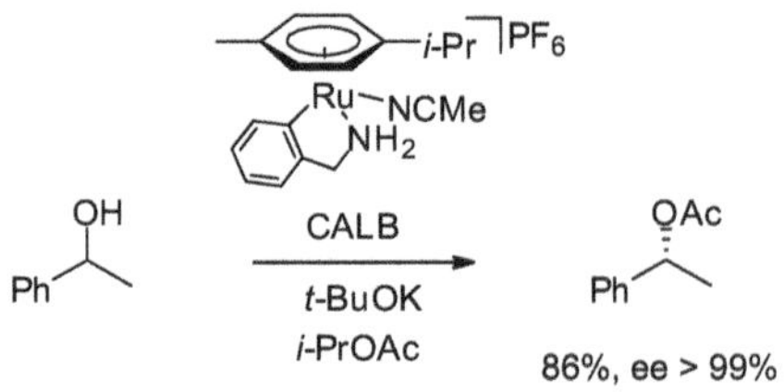

Scheme 8.32   Enzymatic DKR of phenylethanol with a ruthenium catalyst.

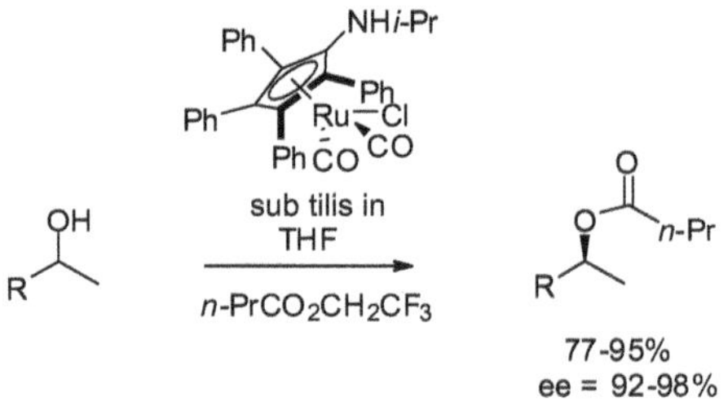

Scheme 8.33   DKR of alcohols with a subtilisin–ruthenium combination.

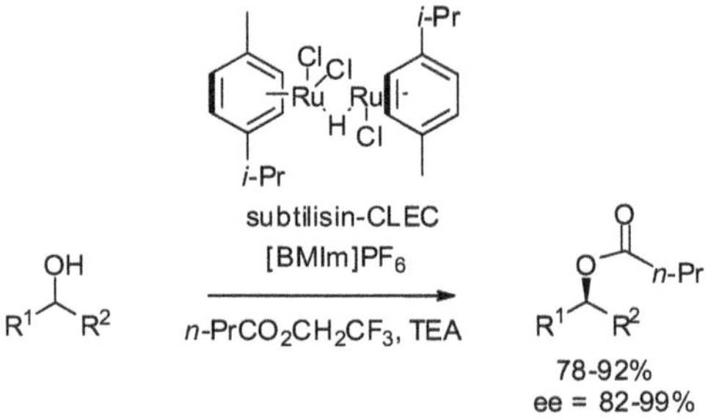

Scheme 8.34   DKR of alcohols with a subtilisin–ruthenium combination in ionic liquids.

allowed the (S)-products to be obtained in good yields with high optical purities (Scheme 8.33). More recently, Bäckvall *et al.* have optimised the DKR of 1-phenylethanol by using a specially treated subtilisin, subtilisin Carlsberg, which was activated by two surfactants, octyl β-D-glycopyranoside, and Brij 56.[60] The latter enzyme was about 4-to-5-fold faster than the previously reported DKR, providing the corresponding chiral butanoate in 96% yield and 95% ee.

In addition, the (S)-selective DKR of alcohols with subtilisin-CLEC (crosslinked crystal) was also possible in ionic liquids at room temperature.[42] In this case, a cymene ruthenium complex was used as the catalyst, and the optical purities of the final (S)-esters were lower than those of the (R)-esters obtained by using lipases, as shown in Scheme 8.34.

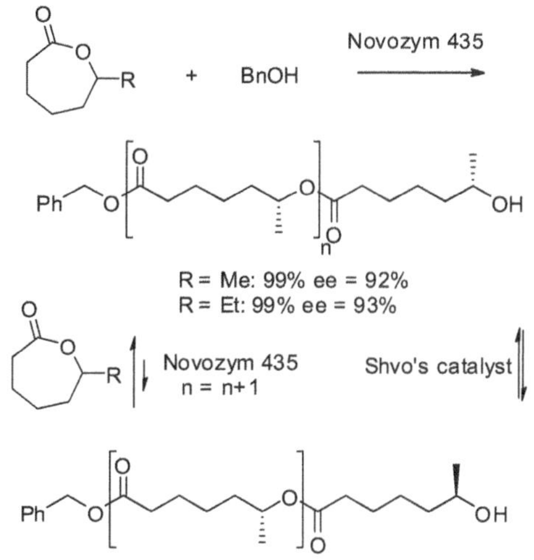

Scheme 8.35 Synthesis of chiral polymers.

The scope of the chemoenzymatic DKR methodology has been extended to the asymmetric synthesis of polymers. Therefore, Meijer *et al.* have developed a particularly elegant example of DKR applied to the synthesis of chiral polymers by iterative tandem catalysis.[61] Hence, these authors have shown that racemic ω-substituted caprolactones could be completely converted into chiral polyesters of remarkably high MW and high enantioselectivity by combining lipase-catalysed ring-opening polymerisation with Ru-catalysed racemisation. Actually, Novozym 435 catalysed the ring-opening of the ω-substituted caprolactone *S*-selectively, yielding an *S*-secondary alcohol, which was the slower-reacting enantiomer in lipase-catalysed reactions. The *in situ* Ru-catalysed racemisation of the terminal secondary alcohol was, therefore, required for propagation, and iterative operation of these two reactions enabled polymerisation. Both 6-methyl- and 6-ethyl-ε-caprolactones were successfully converted into the chiral polymer by using this methodology, as shown in Scheme 8.35.

In 2008, Bäckvall *et al.* employed Bäckvall's catalyst and lipases in tandem for the DKR of a wide variety of alcohols including aromatic chlorohydrins.[62] Therefore, the reaction of these substrates with *Pseudomonas cepacia* lipase (PS-C) combined with this catalyst in the presence of isopropenyl acetate as the acyl donor and a base such as *t*-BuOK in toluene provided the corresponding chiral β-chloroacetates in almost quantitative yields at room temperature for most of the cases. As can be seen in Scheme 8.36, the DKR worked very well for all the substrates, both with activating and deactivating groups, providing the corresponding acetates in enantioselectivities often exceeding 99% ee. For chlorohydrins bearing highly electron-withdrawing groups on the aromatic ring, an elevated temperature (60–80 °C) was required to make the racemisation faster, allowing the selectivity of the enzyme to still be high, since the corresponding products were produced in

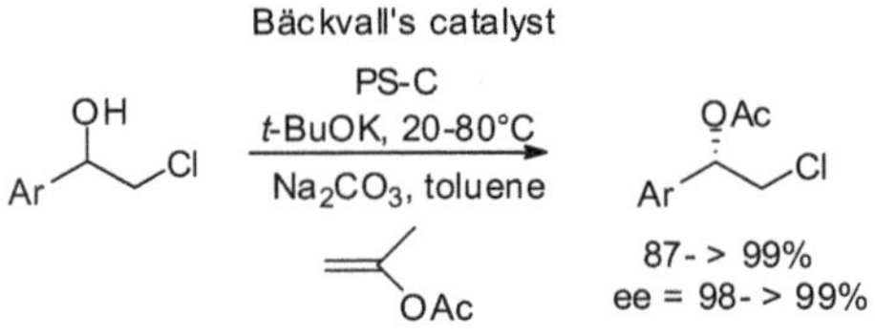

Scheme 8.36 Enzymatic DKR of aromatic chlorohydrins with Bäckvall's catalyst.

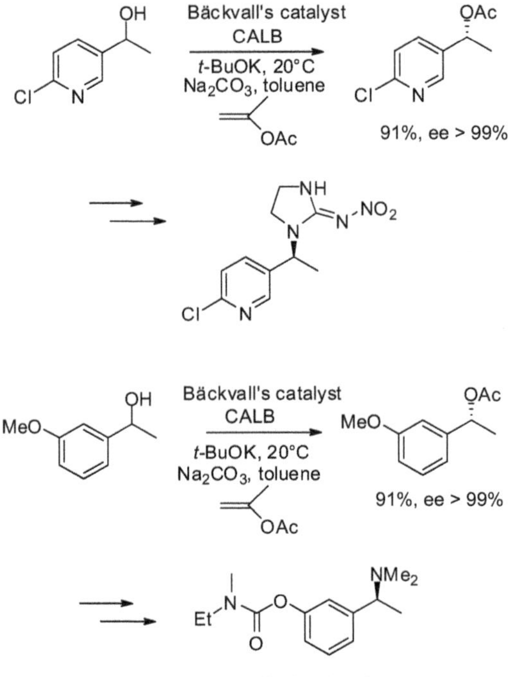

Scheme 8.37 Syntheses of a neonicotinoide pesticide derivative and (*S*)-rivastigmine.

enantioselectivity of 98% ee. These chiral acetates constituted useful synthetic intermediates since they can be transformed to a range of important chiral compounds, such as chiral epoxides.

In 2009, Bäckvall's catalyst was also employed at a low catalyst loading (0.5 mol%) by these authors in combination with CALB to perform the DKR of 1-(6-chloropyridin-3-yl)ethanol by using similar conditions.[63] The corresponding acetate was achieved in 91% yield and enantioselectivity of >99% ee, as shown in Scheme 8.37. This reaction constituted the key step of a synthesis of a neonicotinoide pesticide derivative depicted in Scheme 8.37. These conditions were also applied by Gotor *et al.* to the total synthesis of (*S*)-rivastigmine, employed as a drug for the treatment of dementia of the Alzheimer's type.[64] As shown in Scheme 8.37, the key step of this synthesis was based on the DKR of 1-(3-methoxyphenyl)ethanol, providing the corresponding (*R*)-acetate in excellent yield and enantioselectivity of >99% ee, which was further converted into the expected (*S*)-rivastigmine.

In addition, Bäckvall's catalyst was applied as a combination with *Candida antarctica* lipase B (CALB) and *Pseudomonas cepacia* lipase II (PS-C II), in the presence of *t*-BuOK as the base and isopropenyl acetate as the acyl donor, to the highly efficient DKR of a series of 1,5-diols, resulting in the formation of the corresponding enantiomerically pure 1,5-diacetates in high yields and *anti* selectivity, as shown in Scheme 8.38.[65] These compounds constituted useful building blocks for the enantioselective synthesis of important 2,6-disubstituted and 3,5-disubstituted six-membered heterocycles. As an example, one of these diacetates was employed as an intermediate in the total synthesis of the natural product (+)-solenopsin A (Scheme 8.38).[66] As an extension of this methodology, a series of 1,4-diols could be converted into the corresponding 1,4-diacetates with excellent yields, high *syn* selectivity,

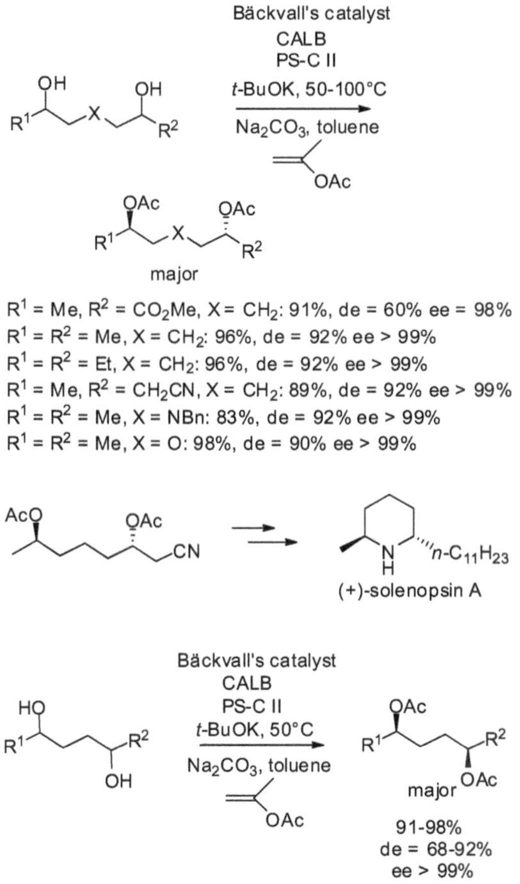
Bäckvall's catalyst
CALB
PS-C II
t-BuOK, 50-100°C
Na2CO3, toluene
OAc
OH
OH
R1
X
R2
OAc
OAc
major
R1 = Me, R2 = CO2Me, X= CH2: 91%, de = 60% ee = 98%
R1 = R2 = Me, X = CH2: 96%, de = 92% ee > 99%
R1 = R2 = Et, X = CH2: 96%, de = 92% ee > 99%
R1 = Me, R2 = CH2CN, X = CH2: 89%, de = 92% ee > 99%
R1 = R2 = Me, X = NBn: 83%, de = 92% ee > 99%
R1 = R2 = Me, X = O: 98%, de = 90% ee > 99%
AcO
CN
N
H
n-C11H23
(+)-solenopsin A
t-BuOK, 50°C
HO
91-98%
de = 68-92%
ee > 99%

Scheme 8.38 Enzymatic DKRs of 1,5-diols with Bäckvall's catalyst.

and enantioselectivity of >99% ee in all cases of substrates, as shown in Scheme 8.38.[67] The usefulness of these enantiopure 1,4-diacetates was demonstrated by the synthesis of enantiopure 2,5-disubstituted pyrrolidines.

Generally, the most frequently employed enzymes, such as CALB, for the DKR of secondary alcohols accepted a limited range of substrates bearing a small (up to three carbon units) and one significantly larger substituent at the hydroxymethine centre. Accordingly, this type of enzyme was inapplicable to the DKR of 1,2-diarylethanols with two bulky substituents at the hydroxymethine centre. In order to overcome this limitation, Park *et al.* found that *Pseudomonas stutzeri* lipase (PSL) was highly enantioselective towards 1,2-diarylethanols.[68] These enzymes were used in the presence of a closely related ruthenium catalyst to Bäckvall's catalyst, which was demonstrated by these authors to be more practical to synthesise. Therefore, the application of this lipase–ruthenium couple, in the presence of isoproprenyl acetate as the acyl donor and K_2CO_3 in toluene at room temperature, to the DKR of a wide series of 1,2-diarylethanols provided the corresponding acetates in excellent yields and enantioselectivities approaching 100% ee in all cases of substrates studied, as shown in Scheme 8.39.

Another analogue of Bäckvall's catalyst was reported in 2009 by Kanerva *et al.* for the DKR of secondary alcohols, such as 1-phenyl- and 1-(furan-2-yl)ethanols. The reaction was performed with CALB in the presence of isoproprenyl acetate as the acyl donor in toluene.[69] This new catalyst showed a higher stability combined with an improved performance as an alcohol racemisation catalyst in comparison with its well-known analogue, Bäckvall's catalyst. The enhanced stability of this catalyst as compared to Bäckvall's catalyst could be related to the better shielding of the metal centre towards hydrolysis, thus hindering catalyst decomposition. As shown in Scheme 8.40, the corresponding acetates were obtained in high to excellent yields and excellent enantioselectivity of >97% ee.

On the other hand, Nanda *et al.* have developed a total synthesis of the naturally occurring phytotoxic noneolide, stagonolide-C, in which two key intermediates were synthesised through the DKRs of the corresponding alcohols.[70] Both these two highly efficient DKRs were performed by using CALB as the enzyme, isopropenyl acetate as the acyl donor, a mixture of

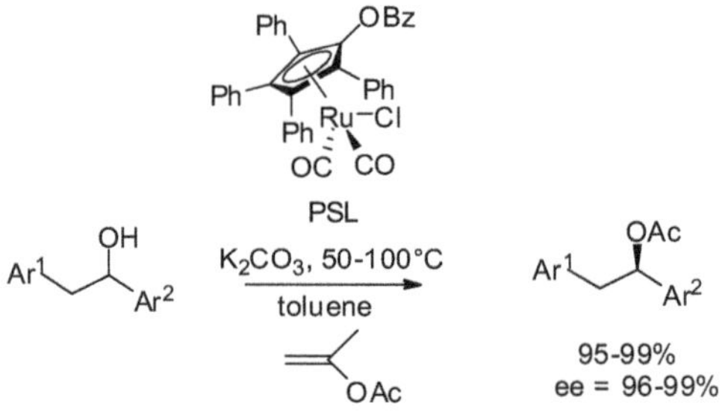

Scheme 8.39   Enzymatic DKR of 1,2-diarylethanols with a Bäckvall's catalyst analogue.

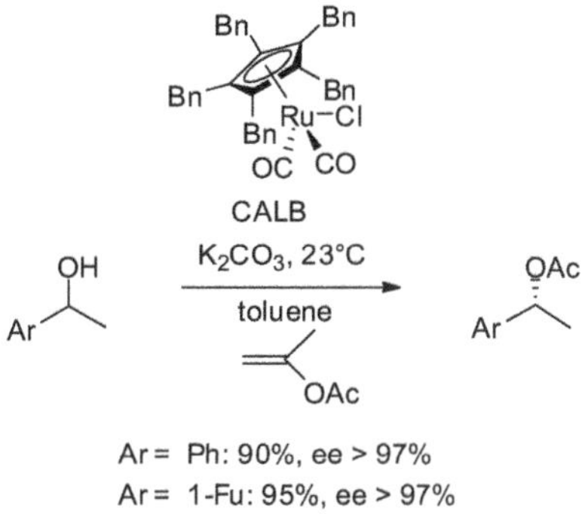

Ar = Ph: 90%, ee > 97%
Ar = 1-Fu: 95%, ee > 97%

Scheme 8.40 Enzymatic DKR of 1-phenyl- and 1-(furan-2-yl)ethanols with a Bäckvall's catalyst analogue.

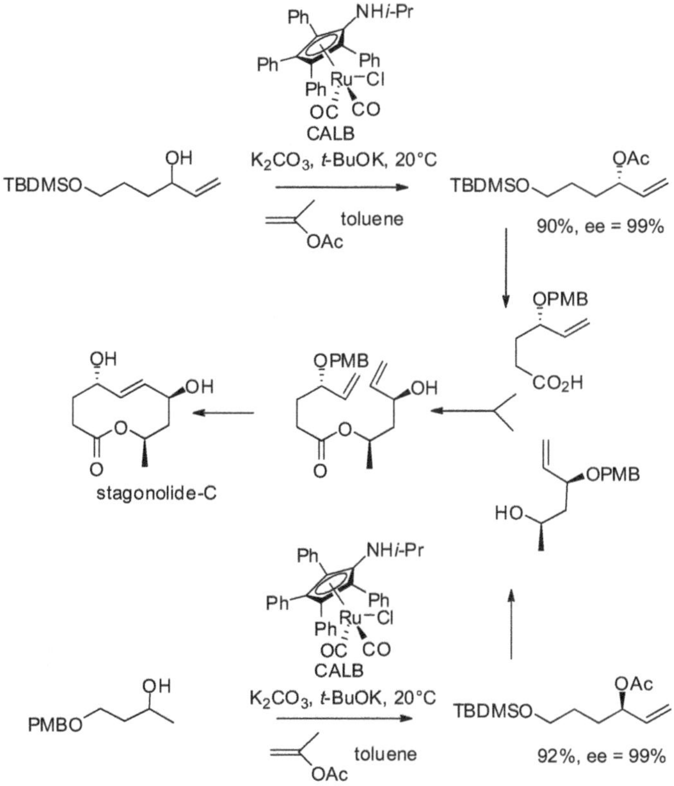

Scheme 8.41 Synthesis of stagonolide-C.

K_2CO_3 and *t*-BuOK as the base, and a novel ruthenium catalyst, analogue to Bäckvall's catalyst. In both reactions, the corresponding key acetates were produced in high yields combined with excellent enantioselectivities, as shown in Scheme 8.41. These acetates were respectively converted into the expected intermediates, which were further coupled to give another alcohol intermediate which was finally cyclised into the expected stagonolide-C.

Bäckvall's catalyst and its analogues usually required a strong base, such as *t*-BuOK, to perform the DKR of secondary alcohols and, moreover, these catalysts could not be reused. Additionally, for secondary alcohols bearing a sulfonyl or phosphonate moiety, which easily coordinates to ruthenium, none of these catalysts could operate smoothly and efficiently. For these reasons, in 2010 Chen and Yuan developed a novel cyclopentadienyl benzoyl ruthenium(II) catalyst, having a unique metallic spiro structure.[71] This novel ruthenium complex was successfully used as a powerful catalyst in combination with CALB in the DKR of a wide range of secondary alcohols under mild conditions. As shown in Scheme 8.42, the reaction provided the corresponding acetates when performed in the presence of K_3PO_4, isopropenyl acetate as the acyl donor in toluene at 25 °C. Both excellent yields and enantioselectivities of up to 99% ee were obtained even with alcohols including sulfonyl and phosphonate functionalities. It must be noted that this catalyst could be recovered with 90% yield and reused with the same activity.

The catalytic activity of Shvo's catalyst is mainly due to the fact that it dissociates into two monomeric ruthenium species in solution under thermal conditions and it can be well combined with various lipases in DKRs. Minidis *et al.* have recently employed a combination of this catalyst with Novozym 435 to achieve the DKR of a series of 1-heteroaryl substituted ethanols, such as oxadiazoles, isoxazoles, 1*H*-pyrazole, or 1*H*-imidazole.[72] In the presence of *p*-chlorophenyl acetate as the acyl donor, the corresponding acetates were produced in moderate to high yields and excellent enantioselectivities, as shown in Scheme 8.43. In order to prepare novel chiral pincer ligands based on the 6-phenyl-2-aminomethylpyridine and 2-aminomethylbenzo[*h*]quinoline scaffolds, Felluga *et al.* have applied similar conditions to the DKR of 2-pyridil and 2-benzoquinolyl ethanols, which provided the corresponding enantiopure acetates in 70% yield and excellent enantioselectivity of >99% ee, as shown in Scheme 8.43.[73] In addition, Alcantara *et al.* demonstrated that the immobilisation of *Pseudomonas stutzeri* lipase in a hydrophobic material by silicon elastomer entrapment

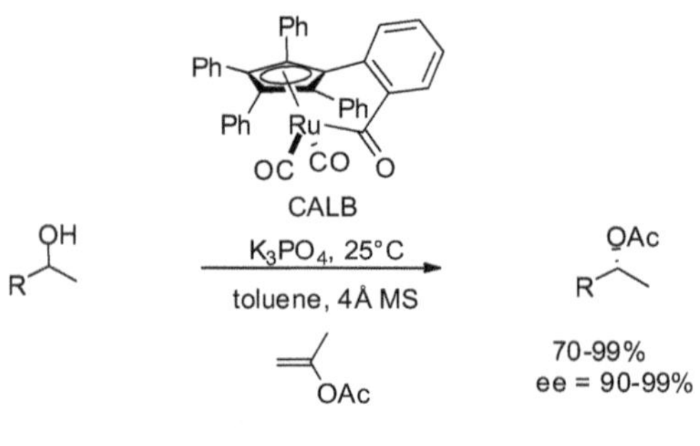

Scheme 8.42   Enzymatic DKR of 1-heteroaryl substituted ethanols, catalysed with a Bäckvall's catalyst analogue.

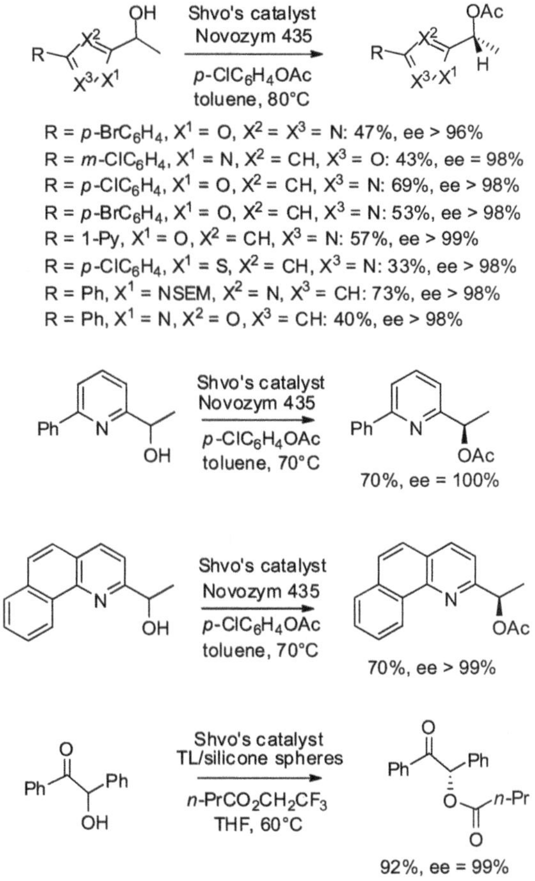
OH
Shvo's catalyst
Novozym 435
p-ClC6H4OAc
toluene, 80°C
OAc
R = p-BrC6H4, X1 = O, X2 = X3 = N: 47%, ee > 96%
R = m-ClC6H4, X1 = N, X2 = CH, X3 = O: 43%, ee = 98%
R = p-ClC6H4, X1 = O, X2 = CH, X3 = N: 69%, ee > 98%
R = p-BrC6H4, X1 = O, X2 = CH, X3 = N: 53%, ee > 98%
R = 1-Py, X1 = O, X2 = CH, X3 = N: 57%, ee > 99%
R = p-ClC6H4, X1 = S, X2 = CH, X3 = N: 33%, ee > 98%
R = Ph, X1 = NSEM, X2 = N, X3 = CH: 73%, ee > 98%
R = Ph, X1 = N, X2 = O, X3 = CH: 40%, ee > 98%
toluene, 70°C
70%, ee = 100%
70%, ee > 99%
Shvo's catalyst
TL/silicone spheres
n-PrCO2CH2CF3
THF, 60°C
92%, ee = 99%

Scheme 8.43 Enzymatic DKRs of 1-heteroaryl ethanols and benzoin with Shvo's catalyst.

resulted in a considerable activation of this enzyme, possibly due to an enhanced mass transfer of hydrophobic compounds like benzoin and the stabilisation of the lipase in its active form, while commercial lipase suffers from a deactivation when incubated at 50 °C.[74] After immobilisation, temperatures of up to 60 °C have been applied to achieve the DKR of benzoin into the corresponding *n*-butyrate in the presence of trifluoroethyl acetate as the acyl donor in THF, as shown in Scheme 8.43. It must be noted that this catalytic system could be reused at least four times without a significant loss of activity.

Carbocyclic nucleosides are structural analogues of natural and synthetic nucleosides, which often exhibit powerful antitumor and antiviral activities. In the course of synthesising chiral members of this family of compounds, Castillon *et al.* have developed the DKR of a 3-hydroxymethyl-cyclopentanol intermediate as key step, providing the corresponding acetate which constituted the key intermediate of this synthesis.[75] This DKR process was

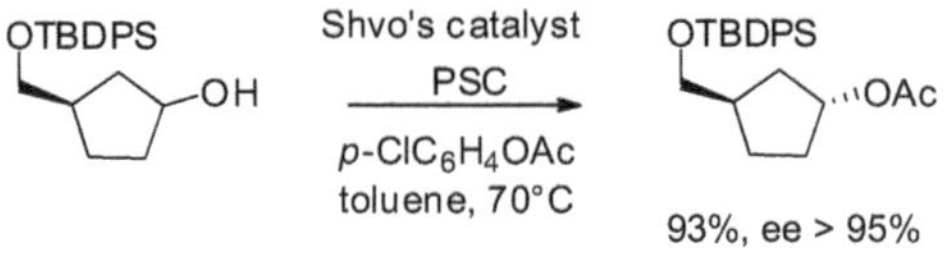

Scheme 8.44   Synthesis of a carbocyclic nucleoside.

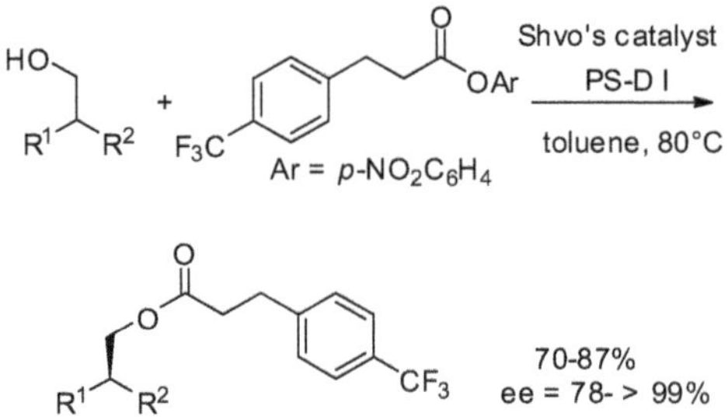

Scheme 8.45   Enzymatic DKR of primary alcohols with Shvo's catalyst.

performed with a combination of Shvo's catalyst and *Pseudomonas cepacia* lipase (PSC) in the presence of *p*-chlorophenyl acetate as the acyl donor in toluene, yielding the corresponding acetate in 93% yield and enantioselectivity of >95% ee, as shown in Scheme 8.44. This intermediate was further converted into the expected carbocyclic-ddA, opening a novel enantioselective approach to carbocyclic nucleosides.

On the other hand, Bäckvall *et al.* demonstrated that the DKR of primary alcohols with an unfunctionalised stereogenic centre at the β-carbon could be achieved by employing a combination of Shvo's catalyst with Amano lipase PS-D I in the presence of vinyl 3-[4-(trifluoromethyl)phenyl]propanoate as the acyl donor.[76] The *in situ* racemisation of the primary alcohol took place through the Ru-catalysed dehydrogenation of the alcohol, followed by the enolisation of the formed aldehyde, and then the Ru-catalysed readdition of hydrogen to the aldehyde. The *in situ* generation of a α-branched aldehyde was of vital importance for the DKR process, since this intermediate could be racemised by means of a straightforward enolisation reaction. As shown in Scheme 8.45, various primary alcohols were converted into the corresponding esters in good to high yields and with a good to high optical purity of up to >99% ee. This process widened the scope of metal- and enzyme-catalysed DKR, which had been limited to α-chiral alcohol and amine derivatives.

In the same area, Atuu and Hossain have developed the DKR of a series of other primary alcohols derived from tropic acid ethyl ester.[77] Indeed, these substrates led to the formation of the corresponding acetates in good yields and moderate to high enantioselectivities when submitted to a combination of lipase PS and Shvo's catalyst in the presence of isopropenyl acetate, as shown in Scheme 8.46. These products constituted precious precursors of

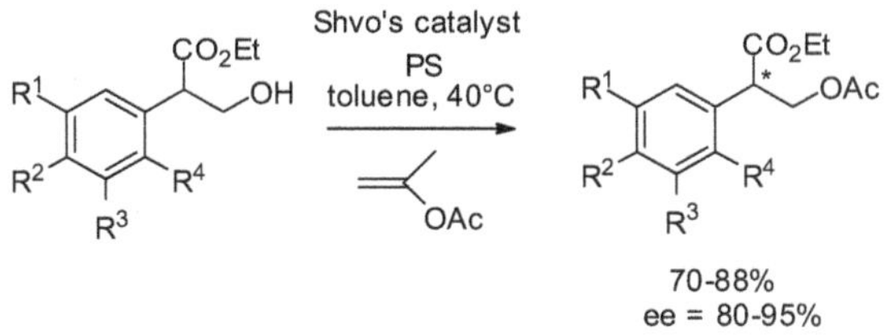

70–88%
ee = 80–95%

Scheme 8.46 Enzymatic DKR of primary alcohols derived from tropic acid ethyl ester with Shvo's catalyst.

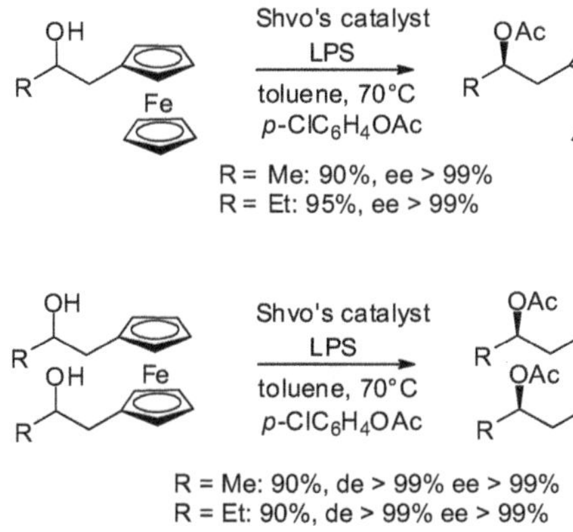

R = Me: 90%, ee > 99%
R = Et: 95%, ee > 99%

R = Me: 90%, de > 99% ee > 99%
R = Et: 90%, de > 99% ee > 99%

Scheme 8.47 Enzymatic DKRs of β-hydroxyalkylferrocene and 1,1′-bis(β-hydroxyalkyl)ferrocene derivatives with Shvo's catalyst.

many chiral 2-aryl propanoic acids, such as ibuprofen and naproxen, which are important non-steroidal anti-inflammatory drugs.

In 2004, Lee and Ahn reported a highly efficient DKR of β-hydroxyalkylferrocene and 1,1′-bis(β-hydroxyalkyl)ferrocene derivatives, which was achieved by using Shvo's catalyst in combination with *Pseudomonas* species (LPS) in the presence of *p*-chlorophenyl acetate as the acyl donor.[78] As shown in Scheme 8.47, excellent yields and stereoselectivities were obtained for the corresponding acetates. These successful DKRs provided an attractive alternative route to overcome the problems of low yields and optical purities in the asymmetric synthesis of the ferrocene derivatives.

Later, Thomas *et al.* investigated the DKR of 1-phenylethanol with *Pseudomonas cepacia* lipase immobilised on ceramic particles (lipase Amano PS CI) in supercritical carbon dioxide, using [Ru(*p*-cymene)Cl₂]₂ as the racemising catalyst.[79] In the presence of acetophenone as a hydrogen acceptor and phenyl acetate as the acyl donor, the corresponding (*R*)-phenylethyl acetate was obtained in 70% yield and enantioselectivity of 96% ee after six days of reaction at 55 °C. It must be noted that this reaction could also be performed in hexane, albeit providing a lower enantioselectivity of 91% ee. In 2011, Kanerva and co-workers also investigated this type of DKR by using *Candida antarctica* lipase B (CALB) in combination with Bäckvall's catalyst.[80] These

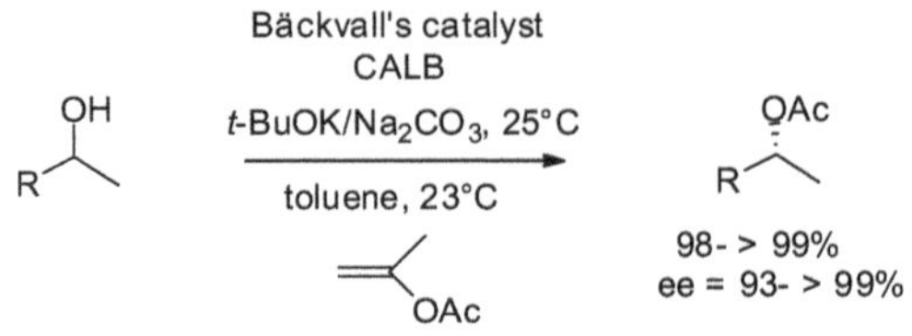

Scheme 8.48   Enzymatic DKR of alcohols with Bäckvall's catalyst.

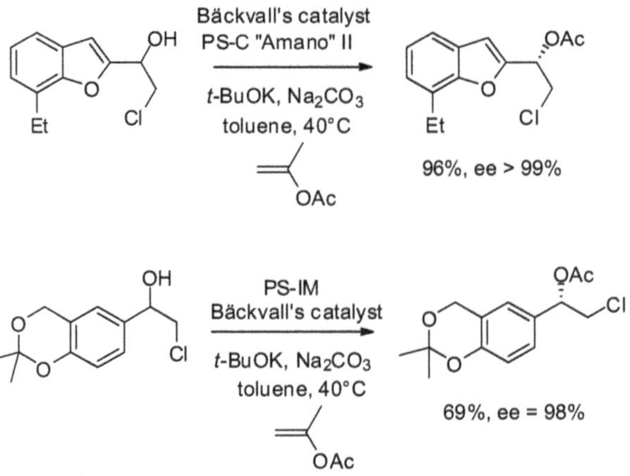

Scheme 8.49   Enzymatic DKRs of chlorohydrins with Bäckvall's catalyst.

two catalysts were shown to function in a highly compatible manner, allowing the conversion of a range of racemic secondary alcohols into the corresponding (R)-acetates in practically theoretical yields and, in most cases of substrates, enantioselectivities exceeding 99% ee, as shown in Scheme 8.48.

An enantioselective synthesis of biologically active (R)-bufuralol was reported by Bäckvall *et al.* on the basis of a DKR of a chlorohydrin as a key step.[81] This reaction was catalysed by Bäckvall's catalyst and enzyme, providing the corresponding enantiopure acetate in 96% yield, as shown in Scheme 8.49. Moreover, the same catalyst was combined by these authors with subtilisin as enzyme in the presence of sodium carbonate and *t*-BuOK as a base in THF to promote the enantioselective esterification through DKR of *trans*-cinnamaldehyde.[82] The corresponding (S)-allylic esters were generated in good yields of 75 to 87% and enantioselectivities ranging from 67 to 97% ee. These products were further converted into pharmaceutically important α-methyl-substituted carboxylic acids. This ruthenium catalyst was also combined with *Burkholderia cepacia* lipase (PS-IM) to promote the acetylation of another β-chloroalcohol evolving through DKR.[83] As shown in Scheme 8.49, the corresponding acetate was obtained in 69% yield and 98% ee under comparable reaction conditions. This process constituted the key step in a total synthesis of the (S)-enantiomer of therapeutically active salbutamol.

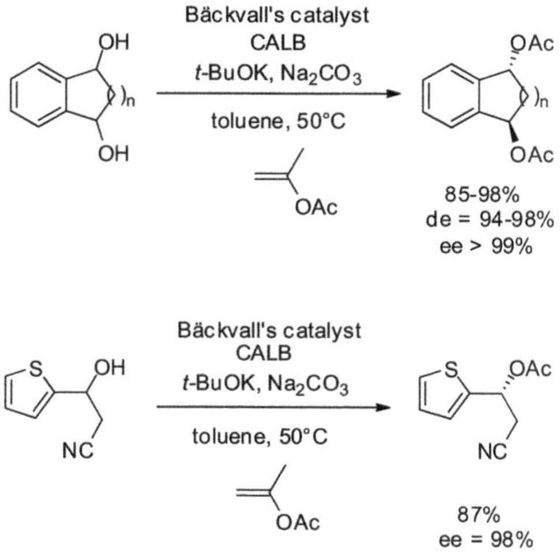

Scheme 8.50   Enzymatic DKRs of diols and a β-hydroxynitrile with Bäckvall's catalyst.

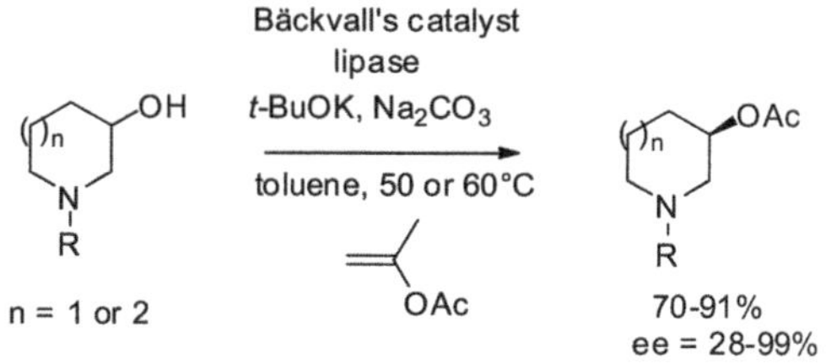

Scheme 8.51   Enzymatic DKR of *N*-heterocyclic 1,2-amino alcohols with Bäckvall's catalyst.

The same authors have also investigated the DKR of bicyclic diols into their diacetates induced by a combination of *Candida antarctica* lipase B (CALB) and the same ruthenium catalyst as above.[84] As shown in Scheme 8.50, these chiral diacetates were produced in excellent yields and almost complete diastereo- and enantioselectivities. The utility of this methodology was demonstrated by its application to the total synthesis of sertraline. In addition, these conditions were also applied to the DKR of a β-hydroxynitrile to provide the corresponding chiral acetate in 87% yield and enantioselectivity of 98% ee, as shown in Scheme 8.50.[85] Again, this protocol was applied to the total synthesis of another biologically active product, such as (*R*)-duloxetine.

In 2011, these authors also developed an efficient DKR of *N*-heterocyclic 1,2-amino alcohols using combinations of Bäckvall's catalyst with various lipases, such as CALB, PS-IM or PS-L1.[86] As shown in Scheme 8.51, various 3-acetoxypyrrolidines and -piperidines were reached in high yields and moderate to excellent enantioselectivities of 28 to 99% ee.

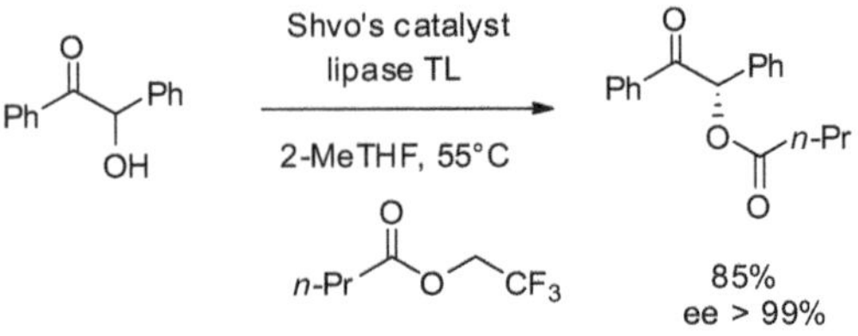

Scheme 8.52   Enzymatic DKR of benzoin with Shvo's catalyst.

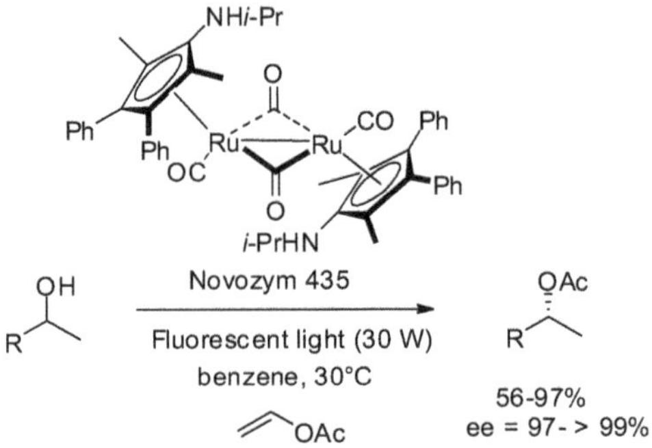

Scheme 8.53   Photoactivated DKR of alcohols through enzymatic and ruthenium catalysis.

In 2011, Shvo's catalyst was used in combination with triacylglycerol lipase (TL) by Alcantara *et al.* in the DKR of benzoin.[87] The reaction was carried out in 2-methyltetrahydrofuran as the solvent in the presence of 5 mol% of Shvo's catalyst, lipase TL, and trifluoroethyl butyrate as acyl donor at 55 °C. Under these conditions, the corresponding (*S*)-butyrate was obtained in 85% yield combined with >99% ee, as shown in Scheme 8.52.

In 2010, Park and co-workers demonstrated that household fluorescent light activated a diruthenium complex to generate a catalytic species highly active for the racemisation of secondary alcohols. This catalytic system was applicable in combination with lipase Novozym 435 to the DKR of various alcohols, providing the corresponding (*R*)-acetates in high yields and excellent enantioselectivities ranging from 97 to >99% ee (Scheme 8.53).[88]

This enzyme was also combined by the same authors with a polymer-bound ruthenium catalyst, and used in the DKR of 1-phenylethanol.[89] The process was performed in the presence of isopropenyl acetate as acyl donor, K_2CO_3 as base in toluene at room temperature, providing the corresponding (*R*)-acetate in 96% yield and 98% ee, as shown in Scheme 8.54. It must be noted that the ruthenium catalyst was prepared by heating a mixture of polystyrene-attached benzoyl chloride and the corresponding ruthenium complex [Ph$_4$(η^4-C$_4$CO)]Ru(CO)$_3$. The utility of this methodology was applied to a total synthesis of rivastigmine, which is an acetylcholinesterase inhibitor.

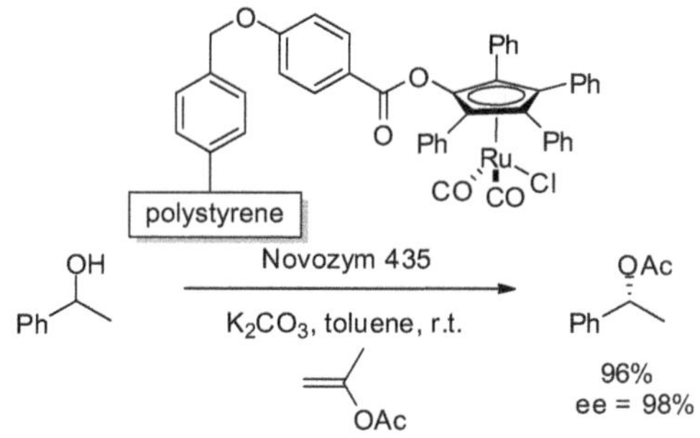

Scheme 8.54 Enzymatic DKR of 1-phenylethanol with a polymer-bound ruthenium catalyst.

8.2.1.2 DKRs of Amines

Enantiomerically pure chiral amines are particularly important for the pharmaceutical and agrochemical industries. Their production *via* DKR is more challenging than that of alcohols, since only a few practical procedures have been developed. Generally, the occurrence of this type of DKR requires harsh conditions, such as a high temperature combined with a long reaction time because their racemisation is more difficult than that of alcohols and they can act as strong ligands for active metal intermediates.[90] Moreover, in these harsh reaction conditions, most enzymes would be denatured, making them unsuitable for DKR. Efficient catalysts for the racemisation of amines are thus much rarer than those for the racemisation of alcohols. Since Reetz and Schimossek reported the use of palladium on charcoal for DKR of amines,[11c] milder conditions for amine racemisation have been developed through the use of ruthenium-, palladium-, and iridium-based catalysts. In the context of ruthenium catalysts, Paetzold and Bäckvall have developed a highly efficient process for the DKR of a variety of unfunctionalised primary amines, which used a combination of an analogue of Shvo's catalyst and CALB, leading to the corresponding amides in high yields and enantioselectivities, as shown in Scheme 8.55.[91]

In 2008, Bäckvall *et al.* employed a closely related methodology for the production of benzyl carbamates, allowing a further release of the free amines under very mild conditions.[92] Indeed, a drawback with other DKR procedures of amines is that the product is a chiral amide, from which the free amine can only be liberated under harsh reaction conditions. As shown in Scheme 8.56, a series of amines could be converted through DKR, in the presence of dibenzyl carbonate as the acyl donor, into the corresponding carbamates in both excellent yields and enantioselectivities. In this case, CALB was associated with Shvo's catalyst.

In addition, another ruthenium catalyst (Scheme 8.57) of the Shvo's type was employed by these authors in combination with CALB to perform the DKR of a number of functionalised primary amines to give the corresponding acetates when using isopropyl acetate as the acyl donor.[93] As shown

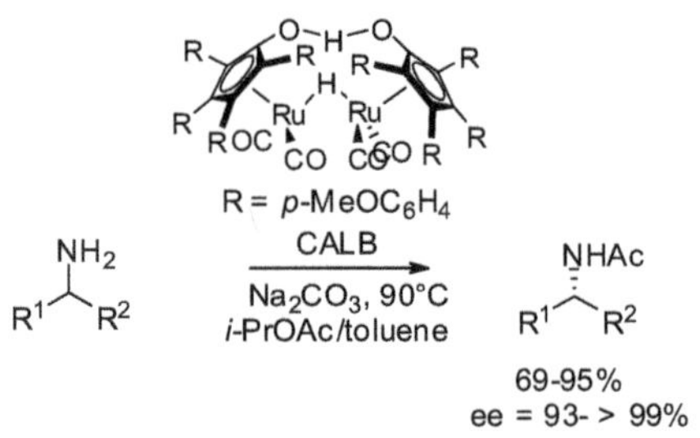

R = *p*-MeOC$_6$H$_4$

69-95%
ee = 93- > 99%

Scheme 8.55   Enzymatic DKR of unfunctionalised primary amines with a Shvo's catalyst analogue.

60-95%
ee = 90-99%

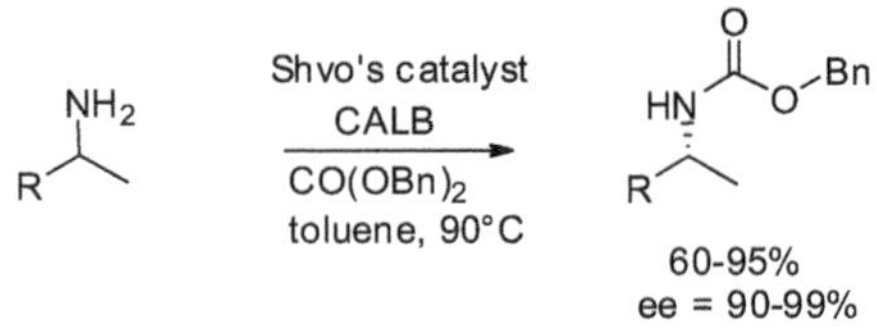

Scheme 8.56   Synthesis of benzyl carbamates.

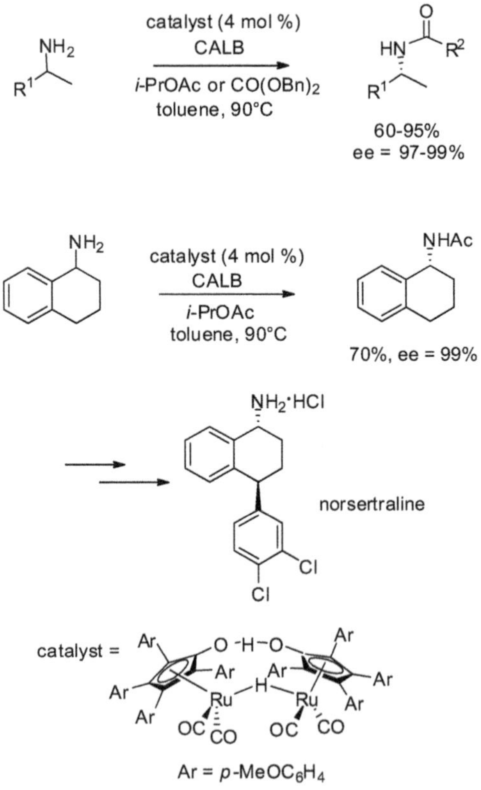

60-95%
ee = 97-99%

70%, ee = 99%

norsertraline

catalyst =

Ar = *p*-MeOC$_6$H$_4$

Scheme 8.57   Enzymatic DKRs of functionalised primary amines with a Shvo's catalyst analogue.

in Scheme 8.57, these products were isolated as almost single enantiomers in all cases and in high yields. This protocol could be extended to the use of dibenzyl carbonate as the acyl donor, allowing the easy release of the free amine from the corresponding carbamates under mild conditions. As shown in Scheme 8.57, these carbamates were produced in comparable yields and enantioselectivities to the acetates. This highly efficient methodology was applied to the synthesis of the selective serotonin reuptake inhibitor, nor-sertraline, employed for the treatment of depression. The first step of this synthesis was the DKR of the readily available 1,2,3,4-tetrahydro-1-naph-thylamine, which yielded the corresponding enantiopure acetate in good yield by treatment with CALB in combination with the catalyst and isopropyl acetate (Scheme 8.57).

In 2010, the same authors reported the DKR of the β-amino ester, ethyl 3-amino-3-phenylpropanoate, by using *Candida antarctica* lipase A (CALA) immobilised in mesocellular foam (GamP-MCF) in combination with the methoxy analogue of Shvo's ruthenium catalyst at 90 °C.[94] It was shown that the use of 2,4-dimethyl-3-pentanol as a hydrogen donor allowed side product formation to be avoided. Thus, the reaction performed in the presence of trifluoroethyl butyrate as the acyl donor provided the corresponding (*S*)-amide in 85% yield and 89% ee. More recently, Gotor *et al.* reported the synthesis of biologically active 1-aryl- and 1-heteroarylpropan-2-amines on the basis of their DKR performed in the presence of a catalytic system consisting of lipase CALB and Shvo's ruthenium catalyst.[95] The process employed ethyl methoxyacetate as acyl donor and, in toluene at 100 °C, provided the corresponding amides in moderate to high yields and high enantioselectivities of up to 97% ee, as shown in Scheme 8.58.

The utility of the DKR is not limited to a selective synthesis of an enan-tiomer; when the reaction occurs along with the creation of a new stereo-genic centre, an enantioselective synthesis of a diastereoisomer is also possible. As an example, Bäckvall *et al.* have studied the DKR of enantiopure β-amino ketones induced by a combination of Shvo's catalyst with *Candida antarctica* lipase B (CALB) in the presence of *p*-chlorophenyl acetate as the acyl donor and hydrogen gas required for the reduction of the ketone.[96] As shown in Scheme 8.59, the corresponding almost diastereo- and enantiopure 1,3-aminoacetates were remarkably afforded in high yields.

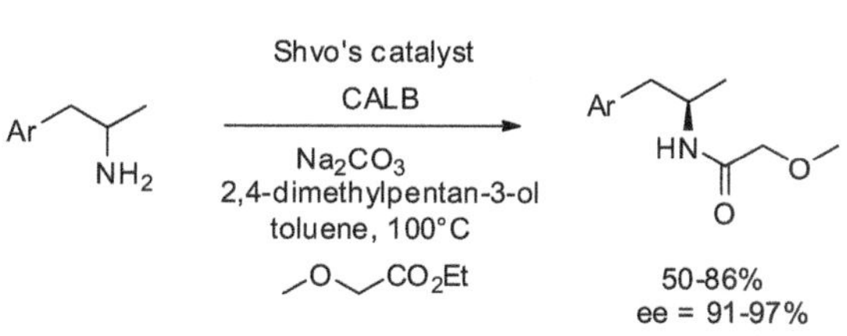

Scheme 8.58 Enzymatic DKR of 1-aryl- and 1-heteroarylpropan-2-amines with Shvo's catalyst.

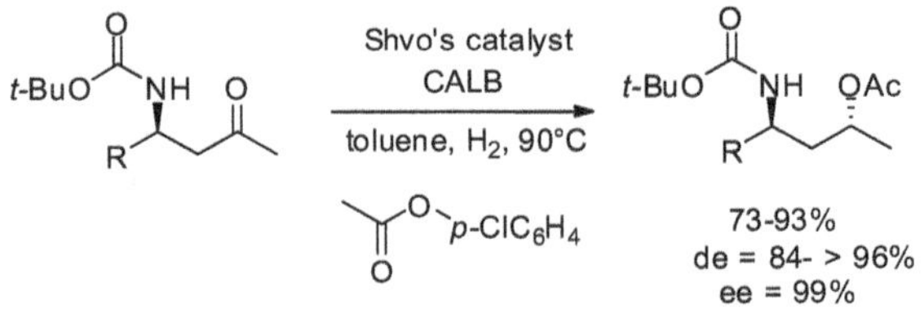

Scheme 8.59   Enzymatic DKR of β-amino esters with Shvo's catalyst.

8.2.2   Enzymatic DKRs using Metals other than Ruthenium

8.2.2.1   DKRs of Alcohols

In addition to ruthenium, other metals have the potential to produce diverse DKR methods. However, although some rhodium, iridium, and aluminium complexes are known to catalyse the racemisation of alcohols, only a few have proved to be compatible with enzymatic reactions. In 1996, Williams *et al.* demonstrated the compatibility of enzymes and rhodium complexes by reporting the first example of racemisation of a secondary alcohol through rhodium-catalysed Oppenhauer oxidation–Meerwein–Ponndorf–Verley; reduction with concomitant acylation of one enantiomer with a lipase from *Pseudomonas fluorescens* (PFL) (Scheme 8.60).[11a]

More recently, Akai *et al.* developed a novel DKR process of allylic alcohols promoted by the combined use of lipase CALB with $[VO(OSiPh_3)_3]$.[97] This complex catalysed the 1,3-transposition of the starting allylic alcohol, resulting in a thermodynamic equilibrium of two regioisomers, which underwent highly enantio- and chemoselective esterification under the action of the lipases (Scheme 8.61). Because the $[VO(OSiPh_3)_3]$-catalysed 1,3-transposition reactions were not sensitive to oxygen and moisture, this DKR method offered the advantage of a facile experimental procedure without the need for special apparatus (anaerobic conditions as for ruthenium-catalysed reactions). Furthermore, it featured a unique preparation of chiral esters of secondary alcohols from the corresponding ketones *via* the readily available tertiary alcohols. Later, Jacobs *et al.* found that vanadium salts, such as $VOSO_4$ and V_2O_4, had activities for the racemisation of benzylic alcohols, but were not active for aliphatic ones, because their activities were based on their acidic properties for the generation of carbenium intermediates.[98] These catalysts were active in the absence of additives, such as bases and hydrogen mediators, in *n*-octane at 80 °C. For example, the DKR of 1-phenylethanol, performed in the presence of $VOSO_4$ combined with CALB and vinyl octanoate, yielded the corresponding (*R*)-ester in 93% yield and 99% ee. The formation of carbenium ion intermediates was suggested because of the results of ^{18}O-labeling experiments.

In 2006, Berkessel *et al.* demonstrated that aluminium could also be highly efficiently combined with lipases to afford the DKR of various secondary alcohols.[99] The best results were obtained when the inexpensive

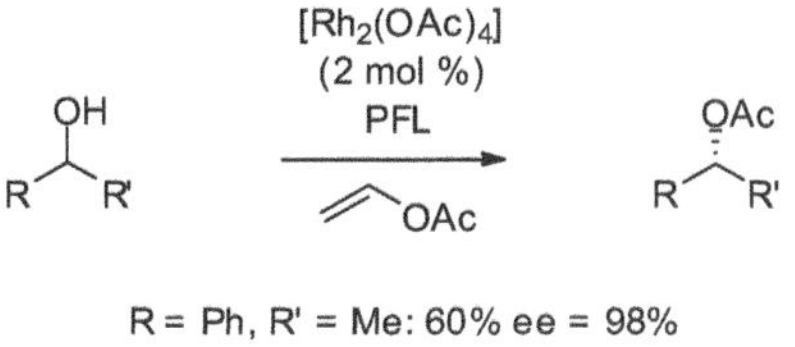

R = Ph, R' = Me: 60% ee = 98%

Scheme 8.60 First rhodium-catalysed enzymatic DKR of alcohols.

80-94%
ee = 97-99%

81-96%
ee = 91-99%

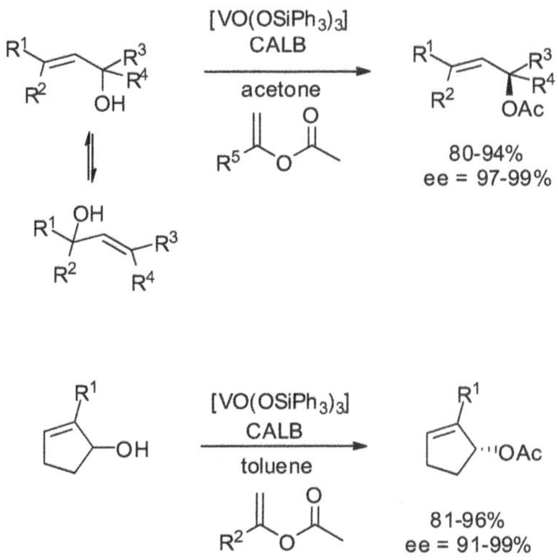

Scheme 8.61 Enzymatic DKRs of allylic alcohols with [VO(OSiPh$_3$)$_3$].

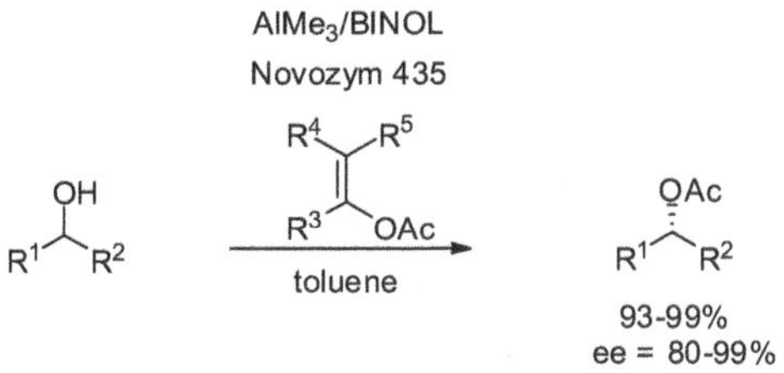

93-99%
ee = 80-99%

Scheme 8.62 Enzymatic DKR of alcohols with AlMe$_3$–BINOL.

aluminium species was readily prepared by the reaction of AlMe$_3$ with a bidentate ligand, such as BINOL, as shown in Scheme 8.62.

In 2010, Bäckvall *et al.* reported, for the first time, the chemoenzymatic DKR of axially chiral allenes.[100] In this work, the DKR of allenic alcohols could be achieved by using a combination of a palladium catalyst, such as [{(IPr)PdBr$_2$}$_2$], with porcine pancreatic lipase PPL in the presence of vinyl butyrate as the acyl donor. The corresponding (*S*)-butyrates were produced in good yields and enantioselectivities of up to 89% ee, as shown in Scheme 8.63.

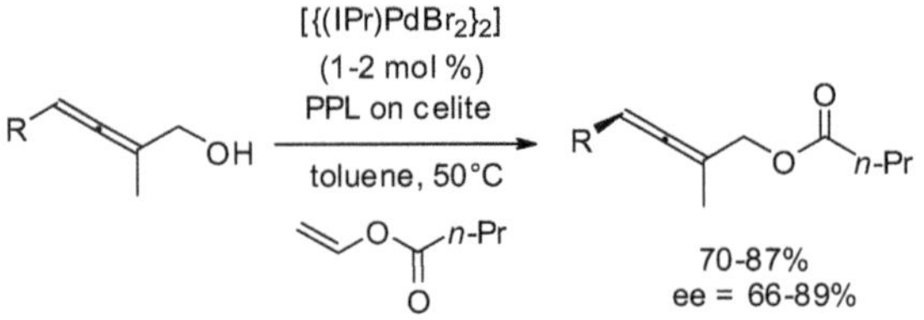

Scheme 8.63   Enzymatic DKR of allenic alcohols with [{(IPr)PdBr$_2$}$_2$].

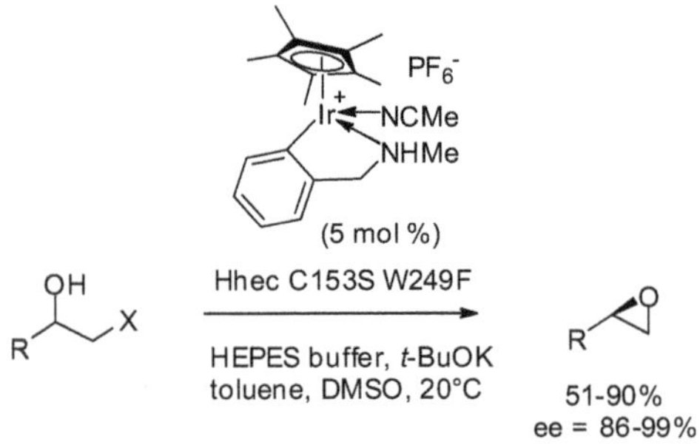

Scheme 8.64   Enzymatic DKR of β-haloalcohols with a cationic half-sandwich iridacycle complex.

On the other hand, several iridium catalysts have been successfully applied to the DKR of secondary alcohols. As an example, Feringa *et al.* have reported the synthesis of a cationic half-sandwich iridacycle complex, which was found to be the most active racemisation catalyst for β-haloalcohols upon activation with a base such as *t*-BuOK.[101] Furthermore, the combination of this water-compatible catalyst with haloalcohol dehalogenase Hhec C153S W249F, incorporating the double mutations Cys 153Ser which increased the enzyme's stability towards oxidation, and Trp249Phe which increased its enantioselectivity especially for aromatic substrates, allowed the first direct DKR of a range of β-haloalcohols to provide the corresponding enantioenriched epoxides to be achieved. As shown in Scheme 8.64, these epoxides were produced in a single step in good to high yields combined with excellent enantioselectivities.

In 2012, Ikariya *et al.* found that a combined catalyst system of bifunctional amidoiridium complexes derived from benzylic amines with lipase CALB could provide a range of chiral acetates from racemic secondary alcohols through DKR with nearly perfect enantioselectivities, as shown in Scheme 8.65.[102]

Cp*Ir^{III}(NHC) complexes are known to be efficient catalysts in the transfer hydrogenation of carbonyl compounds. One of these catalysts has been used by Corberan and Peris in the one-pot enzymatic DKR of a β-branched aldehyde.[103] Therefore, the treatment of this aldehyde by Amano lipase PS-D I and this catalyst at 80 °C in the presence of *p*-chlorophenyl acetate as the acyl

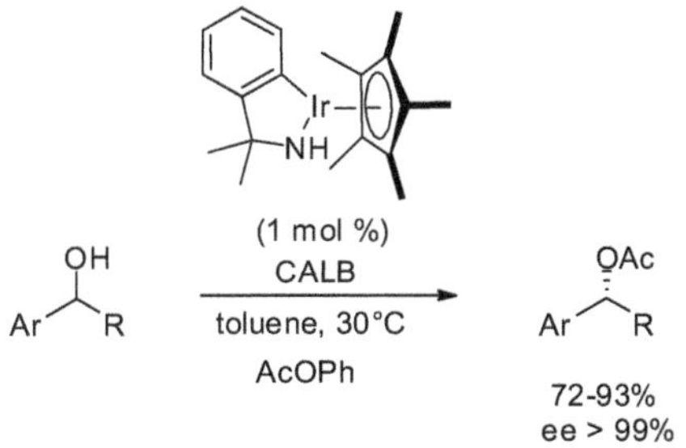

Scheme 8.65 Enzymatic DKR of alcohols with a bifunctional amidoiridium complex.

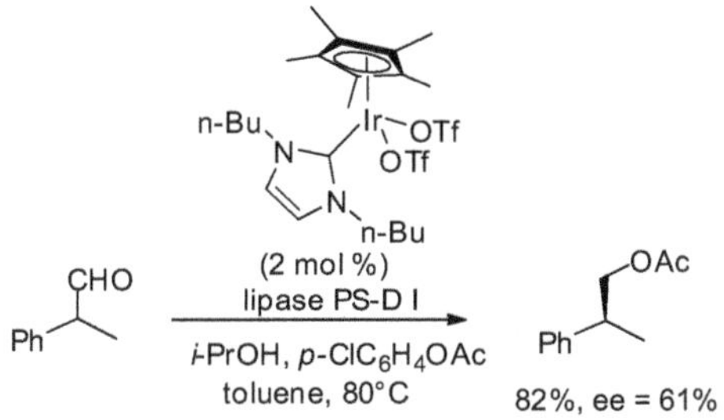

Scheme 8.66 Enzymatic DKR of a β-branched aldehyde with a Cp*Ir^{III}(NHC) complex.

donor provided the corresponding acetate in good yield albeit with moderate enantioselectivity of 61% ee, as shown in Scheme 8.66.

Moreover, Marr *et al.* reported the DKR of 1-phenylethanol using a combination of Novozym 435 with another iridium catalyst used at a remarkably low catalyst loading of 0.1 mol%, which allowed (*R*)-phenylethanol acetate to be achieved in almost quantitative yield and 97% ee, as shown in Scheme 8.67.[104]

Acid zeolites have also been tested for the racemisation of alcohols under biphasic conditions.[105] Their scope was found, however, to be limited to benzylic alcohols, since electron-rich benzylic alcohols were not suitable substrates because of the formation of dimers. Under optimised conditions, based on the use of H-Beta zeolite, CALB lipase, and an excess of vinyl octanoate at 60 °C, enantiopure (*R*)-1-phenylethyl octanoate (>99% ee) was obtained in 90% yield from 1-phenylethanol. In addition, Lozano *et al.* have recently performed the DKR of this alcohol in the presence of acidic zeolite catalysts (CBV400) in an ionic liquid–supercritical carbon dioxide system with a continuous reaction system.[106] Therefore, when Novozym 435 was employed at 50 °C and 100 bars in the presence of vinylpropanoate as the acyl donor, the expected (*R*)-phenylethylpropionate was produced in excellent yield of 98% with enantioselectivity of 97% ee and without any activity loss during 14 days of operation.

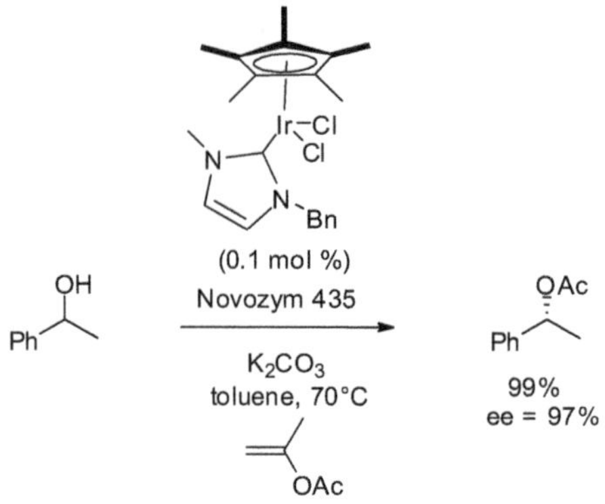

Scheme 8.67   Enzymatic DKR of 1-phenylethanol catalysed with another Cp*Ir^{III}-(NHC) complex.

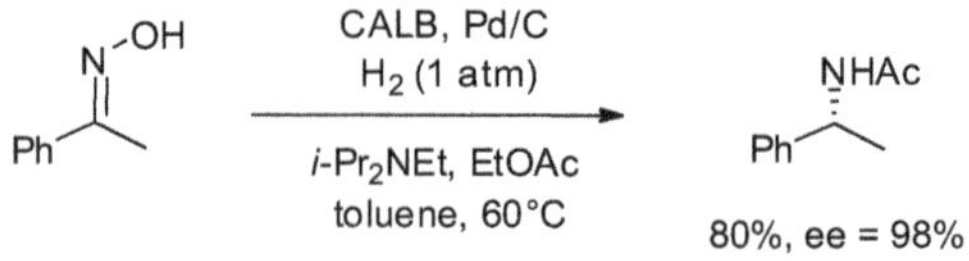

Scheme 8.68   Enzymatic DYKAT of ketoximes with Pd/C.

8.2.2.2   DKRs of Amines

In 1996, Reetz and Schimossek demonstrated the first example of chemoenzymatic DKR for the preparation of enantiopure amines.[11c] Thus, the combination of immobilised CALB as the biocatalyst and palladium on carbon as the racemisation catalyst was used for the synthesis of (R)-N-(1-phenylethyl)acetamide from 1-phenylethylamine in moderate yield (64%) and enantiomerically pure form (99% ee). However, this DKR required a very long reaction time (8 days) at 50–55 °C. Ever since, milder conditions for amine racemisation have been developed through the use of palladium-, and iridium-based catalysts among other ones. As an example, Kim *et al.* have used palladium on charcoal under modified conditions based on those of Reetz and Schimossek to transform ketoximes into chiral amides through the DKR of the intermediate amines (DYKAT), as shown in Scheme 8.68.[107] The authors envisioned that the use of low concentrations of the substrate amine should suppress the side reactions due to the palladium species. Indeed, the yield of (R)-N-(1-phenylethyl)acetamide was significantly increased up to 80% in the indirect amine DKR in which 1-phenylethylamine was generated slowly from the corresponding ketoxime by catalytic hydrogenation.

   Later, Kim *et al.* reported the direct DKR of primary amines using a recyclable Pd nanocatalyst combined with a lipase in the presence of ethyl acetate or ethyl methoxyacetate as the acyl donor.[108] As shown in Scheme 8.69, a series of primary amines and one amino acid amide have

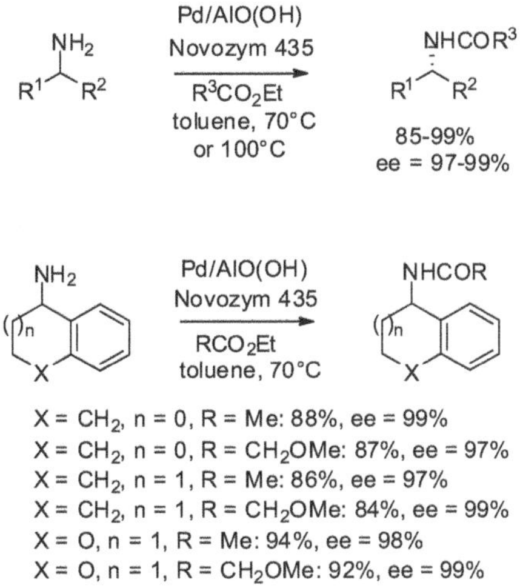

$$R^1 \overset{NH_2}{\underset{}{\bigwedge}} R^2 \xrightarrow[\substack{R^3CO_2Et \\ toluene, 70°C \\ or 100°C}]{\substack{Pd/AlO(OH) \\ Novozym\ 435}} R^1 \overset{NHCOR^3}{\underset{}{\bigwedge}} R^2$$

85-99%
ee = 97-99%

X = CH$_2$, n = 0, R = Me: 88%, ee = 99%
X = CH$_2$, n = 0, R = CH$_2$OMe: 87%, ee = 97%
X = CH$_2$, n = 1, R = Me: 86%, ee = 97%
X = CH$_2$, n = 1, R = CH$_2$OMe: 84%, ee = 99%
X = O, n = 1, R = Me: 94%, ee = 98%
X = O, n = 1, R = CH$_2$OMe: 92%, ee = 99%

Scheme 8.69 Enzymatic DKRs of amines with a palladium nanocatalyst.

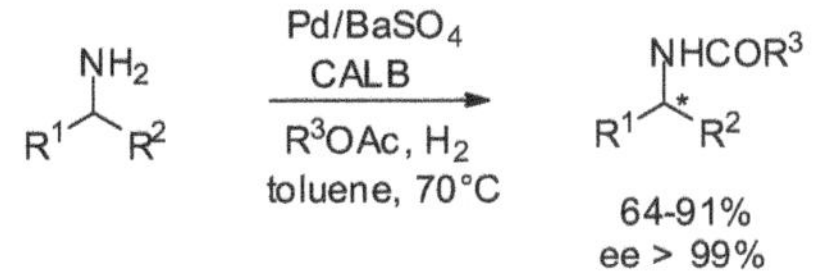

$$R^1 \overset{NH_2}{\underset{}{\bigwedge}} R^2 \xrightarrow[\substack{R^3OAc, H_2 \\ toluene, 70°C}]{\substack{Pd/BaSO_4 \\ CALB}} R^1 \overset{NHCOR^3}{\underset{*}{\bigwedge}} R^2$$

64-91%
ee > 99%

Scheme 8.70 Enzymatic DKR of benzylic amines with Pd/BaSO$_4$.

been efficiently resolved with good yields and high enantioselectivities. The catalyst, Pd/AlO(OH), was prepared as palladium nanoparticules entrapped in aluminium hydroxide. Because this catalyst was highly thermostable, the DKR reactions could be operated at 100 °C with multiple recycling of the catalyst.

In addition, Jacobs *et al.* have developed the DKR of benzylic amines in the presence of a combination of palladium supported on an alkaline earth-type support such as BaSO$_4$ with a lipase.[109] Hence, this heterogeneous catalytic system has allowed various benzylic amines to be transformed into their corresponding enantiomerically pure amides with excellent yields and enantioselectivities of >99% ee, as shown in Scheme 8.70.

Although this type of catalyst was highly active and selective for the racemisation of benzylic amines, the reaction times for DKR were still longer than 24 hours in some cases, and small amounts of side products were formed. In order to improve the performance of these Pd catalysts, these authors employed microwave irradiation as a heating source.[110] Therefore, it was demonstrated that racemisation reactions of benzylic amines under microwave irradiation catalysed by Pd/CaCO$_3$ were faster and more selective. Furthermore, it was checked that the microwave irradiation had no influence on the activity and enantioselectivity of the immobilised *Candida*

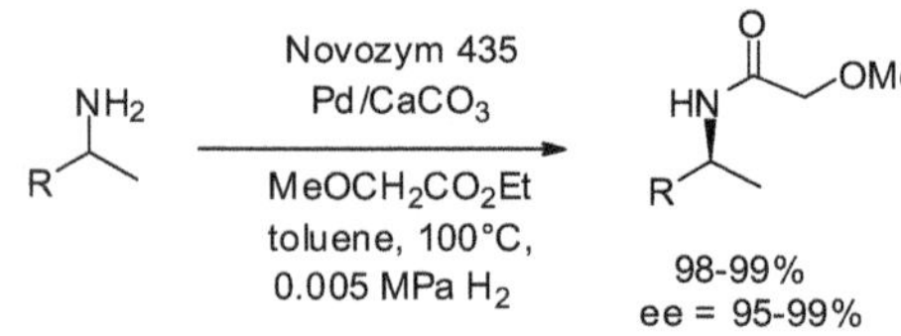

Scheme 8.71   Microwave-promoted enzymatic DKR of benzylic amines with Pd/CaCO$_3$.

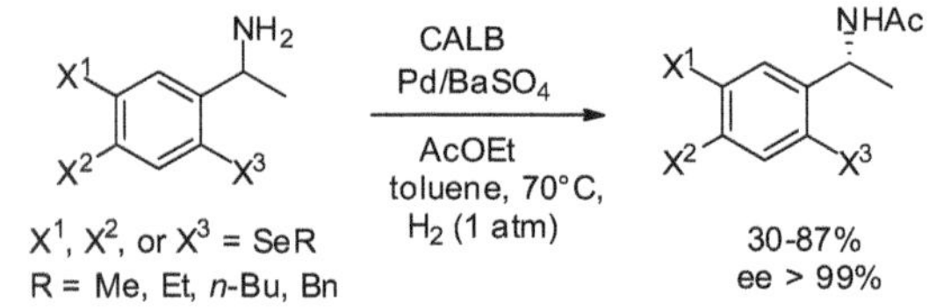

Scheme 8.72   Enzymatic DKR of organoselenium-1-phenylethanamines with Pd/BaSO$_4$.

antarctica lipase B (Novozym 435). Consequently, the microwave-promoted DKR of a range of these amines was achieved by using a combination of this enzyme with Pd/CaCO$_3$ in the presence of ethyl methoxyacetate as the acyl agent at 100 °C under hydrogen atmosphere, furnishing the corresponding amides in high yields and excellent enantioselectivities of up to 99% ee, as shown in Scheme 8.71. In the same context, these authors have demonstrated that Pd on amine-functionalised silica proved to be more selective for the racemisation of 1-phenylethylamine than Pd on alkaline earth supports.[111] The difference in selectivity between various Pd catalysts was determined by the rates of formation of the side products. Therefore, the combination of a Pd catalyst with Novozym 435 for the DKR of 1-phenylethylamine gave the best results when Pd on 3-aminopropyl-functionalised silica or Pd on 3-(1-piperazino)propyl-functionalised silica were used as the racemisation catalysts, namely 93% yield of the corresponding (*R*)-amide with enantioselectivity of 99% ee.

In 2009, Andrade *et al.* demonstrated that the DKR mediated by palladium and lipase could be efficiently applied to selenium-containing amines.[112] As shown in Scheme 8.72, a series of organoselenium-1-phenylethanamines was submitted to DKR catalysed by a combination of Pd/BaSO$_4$ and *Candida antarctica* lipase B in the presence of ethyl acetate as the acyl donor in toluene at 70 °C under hydrogen atmosphere (1 atm), providing the corresponding selenium-containing amides in generally high yields and enantioselectivities of >99% ee in all cases of substrates.

In addition, Kim *et al.* have chosen palladium nanoparticles entrapped in an AlO(OH) matrix as the racemisation catalyst combined with Novozym 435 to achieve the DKR of phenylglycine amide.[113] With this combination of catalysts combined with various acyl donors, such as ethyl methoxyacetate, ethyl acetate, ethyl phenyl acetate, or methyl *p*-hydroxyphenyl

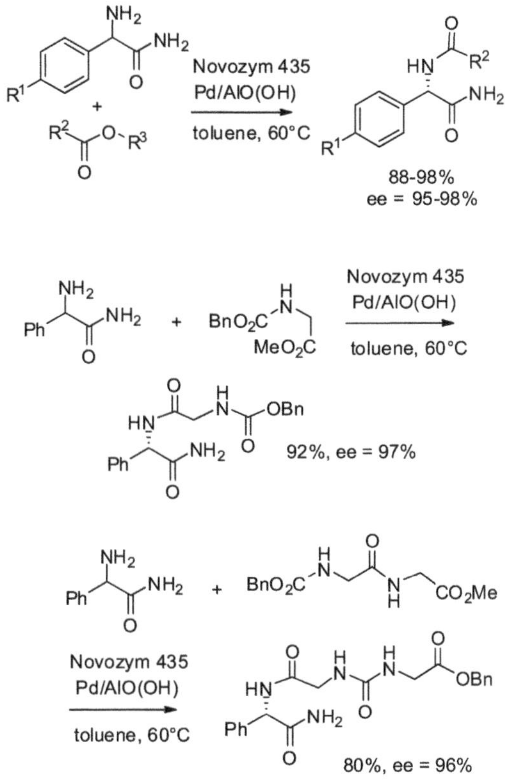

Scheme 8.73 Enzymatic DKRs of phenylglycine amide derivatives with Pd/AlO(OH).

acetate in toluene at 60 °C, the corresponding acylated amides were produced in excellent yields and enantioselectivities, as shown in Scheme 8.73. The most efficient acyl donor, ethyl methoxyacetate, was selected to extend this methodology to a range of phenylglycine amide derivatives to yield the corresponding additional products in comparable excellent yields and enantioselectivities, as shown in Scheme 8.73. Interestingly, the DKR reactions of phenylglycine amide, performed in the presence of *N*-benzyloxycarbonylglycine methyl ester or *N*-benzyloxycarbonylglycylglycine methyl ester as the acyl donor, provided the corresponding di- and tripeptides, respectively, in high yields and excellent enantioselectivities of >96% ee, as shown in Scheme 8.73. Therefore, this highly efficient methodology constituted a new route to enantiopure amino acid derivatives.

Palladium nanocatalyst [Pd⁰/AlO(OH)] was shown by Kim *et al.* to be able to catalyse, in combination with lipase Novozym 435, the DKR of primary benzyl amines.[114] As shown in Scheme 8.74, the corresponding chiral amides were produced in remarkable yields and enantioselectivities ranging from 90 to >99% ee.

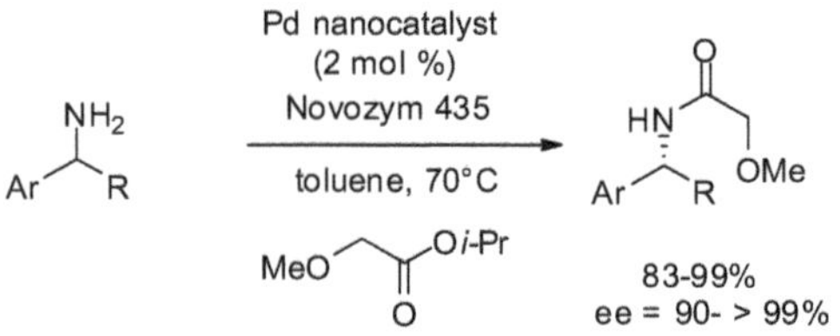

Scheme 8.74 Enzymatic DKR of benzyl amines with a palladium nanocatalyst.

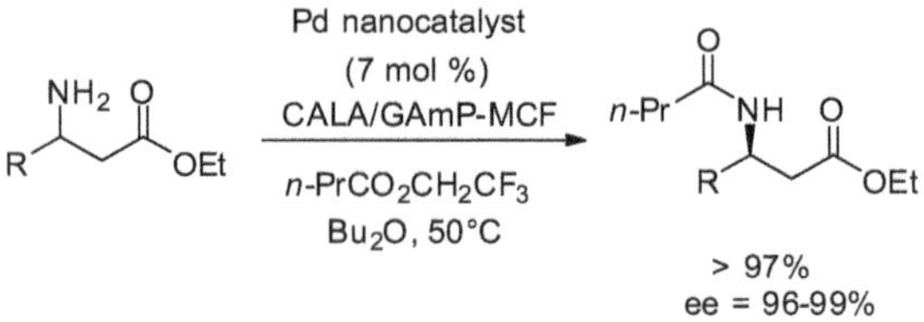

Scheme 8.75 Enzymatic DKR of β-amino esters with a palladium nanocatalyst.

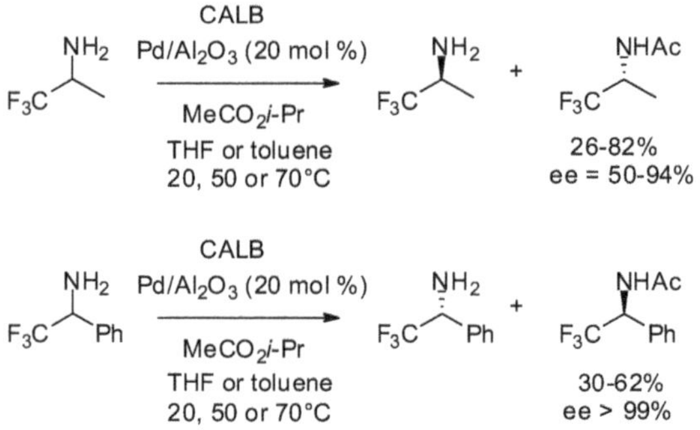

Scheme 8.76 Enzymatic DKRs of α-trifluoromethylated amines with Pd/Al$_2$O$_3$.

In the same area, the DKR of β-amino esters, including aliphatic, aromatic as well as heteroaromatic ones, was performed by Bäckvall *et al.*, using a palladium nanocatalyst [Pd^0/AlO(OH)] in combination with immobilised *Candida antarctica* lipase A (CALA/GAmP-MCF), providing the corresponding chiral amides in both excellent yields and enantioselectivities, as shown in Scheme 8.75.[115]

In 2013, Lin and co-workers reported enzymatic resolution of α-trifluoromethylated amines *via* DKR employing a combination of CALB with palladium on a support of Al$_2$O$_3$.[116] As shown in Scheme 8.76, aromatic a-trifluoromethylated amines as well as aliphatic amines were achieved in up to 62 to 82% yields, respectively, and enantioselectivities of up to >99% ee. Moreover, the catalyst system could be recycled for at least five times retaining 70% conversion and 99% ee value in the case of 2,2,2-trifluoro-1-phenylethanamine as substrate, for example. Interestingly, stereoselectivity

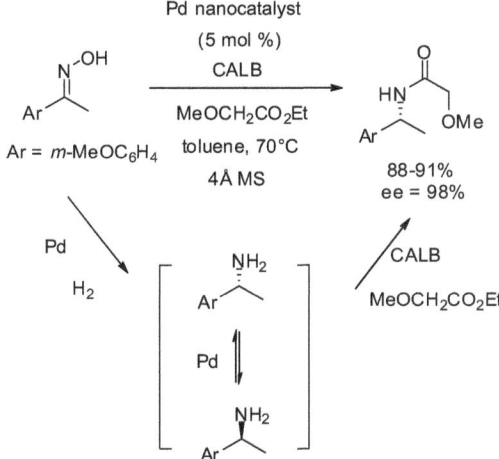

Scheme 8.77   Enzymatic reductive acylation of a ketoxime with a palladium nanocatalyst.

inversion was observed when varying the size of the substitution from methyl to ethyl, namely resolution of 1,1,1-trifluoro-2-propylamine gave optically pure amide with *R*-configuration, while 1,1,1-trifluoro-2-butylamine and other larger amines including 3,3,3-trifluoro-1-phenyl-2-propylamine provided amides with *S*-configuration.

In another area, the asymmetric reductive acylation of ketoxime of *m*-methoxyacetophenone was developed by Kim and co-workers, in 2010.[117] This process was catalysed by a combination of lipase CALB with a palladium nanocatalyst in the presence of ethyl methoxyacetate as an acyl donor, molecular sieves in toluene at 70 °C under 0.1 bar of hydrogen pressure. It allowed the formation of the corresponding almost enantiopure amide in high yields of up to 91% and with enantioselectivity of 98% ee, as shown in Scheme 8.77. The utility of this novel methodology was applied to the total synthesis of the calcimimetic (+)-NPS R-568.

The use of palladium-based catalysts has, however, significant limitations that restrict its industrial applicability including high catalyst loading, limited substrate scope, and high substrate dilution. In this context, Page *et al.* have reported an efficient process for the DKR of a secondary amine using a novel iridium-based amine racemisation catalyst under significantly milder conditions than those described previously.[118] Hence, the combination of [IrCpI$_2$]$_2$ with *Candida rugosa* lipase at 40 °C in toluene allowed the DKR of the racemic secondary cyclic amine, depicted in Scheme 8.78, to be achieved in high yield and enantioselectivity. The reaction was performed in 23 hours on a 3 g scale in the presence of 3-methoxyphenylpropyl carbonate, providing the corresponding chiral product carbonate.

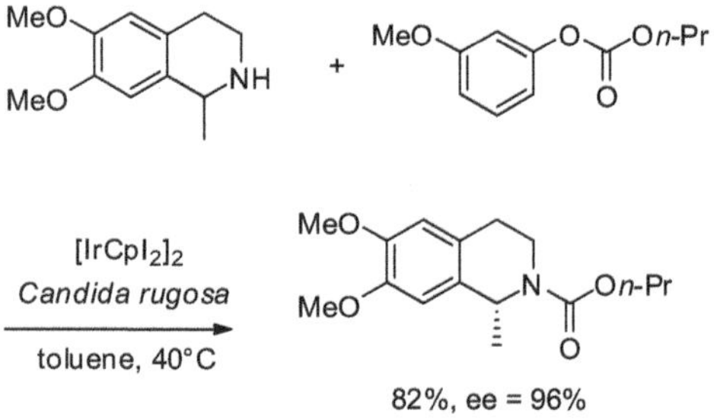

Scheme 8.78   Enzymatic DKR of a cyclic amine with $[IrCpI_2]_2$.

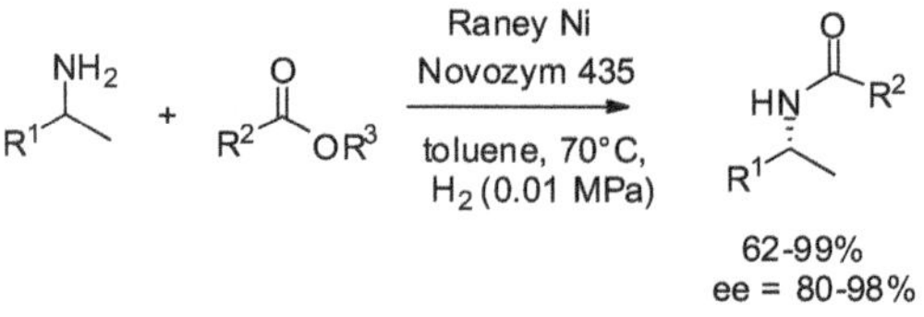

Scheme 8.79   Enzymatic DKR of amines with Raney nickel.

Since palladium on carbon or on alkaline earth supports is generally not effective for the DKR of aliphatic amines, and in the search for less expensive heterogeneous racemisation catalysts, Jacobs *et al.* have shown that heterogeneous Raney nickel could be applied to the racemisation of aliphatic amines in addition to the more usually employed benzylic amines.[119] As an extension, when combined with Novozym 435, Raney nickel allowed the DKR of a range of amines to be efficiently achieved, as shown in Scheme 8.79. For aliphatic amines, racemisation and enzymatic resolution could be combined in one pot, resulting in an efficient DKR process. When ethyl methoxyacetate was used as the acyl donor, the reaction allowed the corresponding amides to be obtained in good yields and excellent enantioselectivities, as shown in Scheme 8.79. For benzylic amines, which reacted less fast with the enzyme, it could be demonstrated that the slow enzymatic conversion of the amine in the presence of the nickel catalyst was the main effect impeding efficient one-pot DKR. Consequently, a two-pot process was proposed in which the liquid was alternatingly shuttled between two vessels containing the solid racemisation catalyst and the biocatalyst. After four such cycles, the corresponding amides were isolated in good yields and high enantioselectivities (Scheme 8.79).

On the other hand, Aron *et al.* have very recently identified zinc complexes of picolinaldehyde as low-cost and environmentally benign catalysts for the racemisation of amino acids.[120] When this type of catalyst was combined with alcalase, it allowed the DKR of a series of amino acid esters to be achieved with enantioselectivities of up to >98% ee, as shown in Scheme 8.80. Aromatic as well as straight- and γ-branched-chain amino

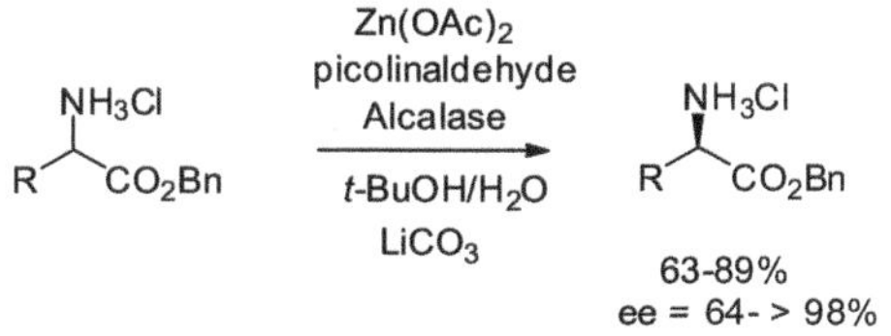

Scheme 8.80   Enzymatic DKR of amines with a zinc complex.

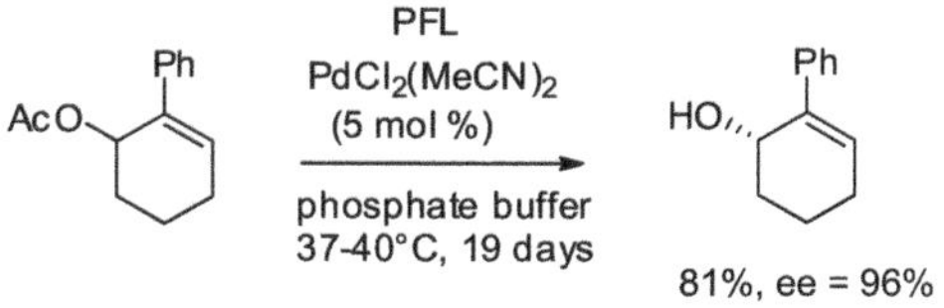

Scheme 8.81   Enzymatic   DKR   of   2-phenyl-2-cyclohexenyl   acetate   with PdCl$_2$(MeCN)$_2$.

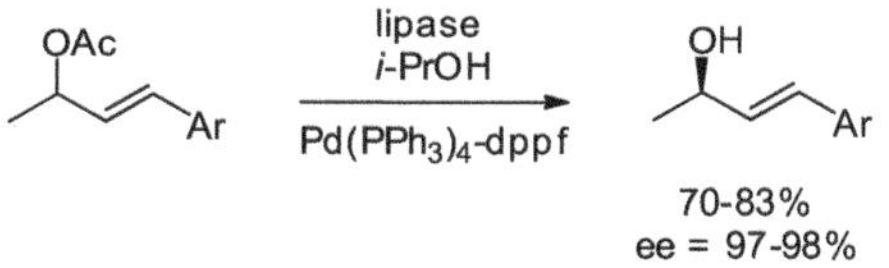

Scheme 8.82   Enzymatic DKR of acyclic allylic acetates with Pd(PPh$_3$)$_4$–dppf.

acids were resolved in good yields with high enantiopurity, whereas β-branched amino acids were poorly resolved.

8.2.2.3   DKRs of Allylic Acetates

In 1996, Allen and Williams demonstrated that the DKR of allylic acetates could be accomplished through coupling Pd-catalysed racemisation and enzymatic hydrolysis of allylic acetates in buffer solution.[11b] However, the DKR under these conditions was limited to cyclohexenyl acetates to yield symmetrical palladium–allyl intermediates. Among them, 2-phenyl-2-cyclohexenyl acetate was the only substrate to have been resolved with good results (81% yield, 96% ee), as shown in Scheme 8.81.

In 1999, Kim *et al.* improved the DKR of allylic acetates significantly by replacing the enzymatic hydrolytic reaction with the enzymatic transesterification reaction and employing Pd(PPh$_3$)$_4$ and 1,1′-bis(diphenylphosphino)ferrocene (dppf) as the racemising catalyst system in THF.[121] In this process, 2-propanol was used as the acyl acceptor. In this case, a palladium(0) complex racemised allylic acetates *via* π-allylpalladium intermediates. The use of the chelating ligand (dppf) decreased the formation of by-products, such as 1,3-dienes, during the DKR. As shown in Scheme 8.82, various acyclic allylic acetates were transformed into their corresponding allylic alcohols at room temperature in good yields and excellent enantioselectivities.

References

1. (a) S. F. Mayer, W. Kroutil and K. Faber, *Chem. Soc. Rev.*, 2001, **30**, 332–339; (b) S. F. Mayer, W. Kroutil and K. Faber, *Pure Appl. Chem.*, 2002, **74**, 2253–2257; (c) A. C. Marr and S. Liu, *Trends Biotechnol.*, 2011, **29**, 199–204.

2. (a) G. H. Müller and H. Waldmann, *Tetrahedron Lett.*, 1996, **37**, 3833–3836; (b) G. H. Müller, A. Lang, D. R. Seithel and H. Waldmann, *Chem. - Eur. J.*, 1998, **4**, 2513–2522.

3. E. Wingstrand, A. Laurell, L. Fransson, K. Hult and C. Moberg, *Chem. - Eur. J.*, 2009, **15**, 12107–12113.

4. V. Gauchot, W. Kroutil and A. R. Schmitzer, *Chem. - Eur. J.*, 2010, **16**, 6748–6751.

5. (a) S. Borchert, E. Burda, J. Schatz, W. Hummel and H. Gröger, *J. Mol. Catal. B: Enzym.*, 2012, **84**, 89–93; (b) E. Burda, W. Hummel and H. Gröger, *Angew. Chem., Int. Ed.*, 2008, **47**, 9551–9554.

6. W. Szymanski, C. P. Postema, C. Tarabiono, F. Berthiol, L. Campbell-Verduyn, S. de Wildeman, J. G. de Vries, B. L. Feringa and D. B. Janssen, *Adv. Synth. Catal.*, 2010, **352**, 2111–2115.

7. C. De Souza de Oliveira, K. T. de Andrade and A. T. Omori, *J. Mol. Catal. B: Enzym.*, 2013, **91**, 93–97.

8. M. J. Finf, M. Schön, F. Rudroff, M. Schnürch and M. D. Mihovilovic, *ChemCatChem*, 2013, **5**, 724–727.

9. (a) H. Pellissier, *Adv. Synth. Catal.*, 2011, **353**, 1613–1666; (b) E. Fogassy, M. Nogradi, D. Kozma, G. Egri, E. Palovics and V. Kiss, *Org. Biomol. Chem.*, 2006, **4**, 3011–3030; (c) E. Vedejs and M. Jure, *Angew. Chem., Int. Ed.*, 2005, **44**, 3974–4001; (d) M. Breuer, K. Ditrich, T. Habicher, B. Hauer, M. Keβeler, R. Stürmer and T. Zelinski, *Angew. Chem., Int. Ed.*, 2004, **43**, 788–824; (e) D. E. J. E. Robinson and S. D. Bull, *Tetrahedron: Asymmetry*, 2003, **14**, 1407–1446.

10. (a) R. Noyori, M. Tokunaga and M. Kitamura, *Bull. Chem. Soc. Jpn*, 1995, **68**, 36–56; (b) R. S. Ward, *Tetrahedron: Asymmetry*, 1995, **6**, 1475–1490; (c) S. Caddick and K. Jenkins, *Chem. Soc. Rev.*, 1996, 447–456; (d) H. Stecher and K. Faber, *Synthesis*, 1997, 1–16; (e) M. T. El Gihani and J. M. J. Williams, *Curr. Opin. Chem. Biol.*, 1999, **3**, 11–15; (f) R. Azerad and D. Buisson, *Curr. Opin. Chem. Biol.*, 2000, **11**, 565–571; (g) F. F. Huerta, A. B. E. Minidis and J.-E. Bäckvall, *Chem. Soc. Rev.*, 2001, 321–331; (h) M. J. Kim, Y. Ahn and J. Park, *Curr. Opin. Biotechnol.*, 2002, **13**, 578–587; (i) H. Pellissier, *Tetrahedron*, 2003, **59**, 8291–8327; (j) O. Pamies and J.-E. Bäckvall, *Chem. Rev.*, 2003, **103**, 3247–3261; (k) M.-J. Kim, A. Yangsoo and J. Park, *Bull. Korean Chem. Soc.*, 2005, **26**, 515–522; (l) J.-E. Bäckvall, in *Asymmetric Synthesis - The Essentials*, ed. M. Christmann and S. Bräse, Wiley-VCH, Weinheim, 2006;

(m) B. Martin-Matute and J.-E. Bäckvall, *Curr. Opin. Chem. Biol.*, 2007, **11**, 226–232; (n) A. Kamal, M. A. Azhar, T. Krisnaji, M. S. Malik and S. Azeeza, *Coord. Chem. Rev.*, 2008, **252**, 569–592; (o) Y. Ahn, S.-B. Ko, M.-J. Kim and J. Park, *Coord. Chem. Rev.*, 2008, **252**, 647–658; (p) B. Martin-Matute and J.-E. Bäckvall, in *Asymmetric Organic Synthesis with Enzymes*, ed. V. Gotor, I. Alfonso and E. Garcia-Urdiales, Wiley-VCH, Weinheim, 2008, pp. 89–113; (q) H. Pellissier, *Tetrahedron*, 2008, **64**, 1563–1601; (r) A. A. Kamaruddin, M. H. Uzir, H. Y. Aboul-Enein and H. N. A. Halim, *Chirality*, 2009, **21**, 449–467; (s) R. Karvembu, R. Prabhakaran, M. M. Tamizh and K. Natarajan, *C. R. Chimie*, 2009, **12**, 951–962; (t) J. H. Lee, K. Han, M.-J. Kim and J. Park, *Eur. J. Org. Chem.*, 2010, 999–1015; (u) H. Pellissier, *Tetrahedron*, 2011, **67**, 3769–3802; (v) *Chirality from Dynamic Kinetic Resolution*, ed. H. Pellissier, Royal Society of Chemistry, Cambridge, 2011; (w) H. Pellissier, *Adv. Synth. Catal.*, 2011, **353**, 659–676.

11. (a) P. M. Dinh, J. A. Howarth, A. R. Hudnott and J. M. J. Williams, *Tetrahedron Lett.*, 1996, **37**, 7623–7626; (b) J. V. Allen and J. M. J. Williams, *Tetrahedron Lett.*, 1996, **37**, 1859–1862; (c) M. T. Reetz and K. Schimossek, *Chimia*, 1996, **50**, 668–669.

12. (a) F. F. Huerta, A. B. E. Minidis and J.-E. Bäckvall, *Chem. Soc. Rev.*, 2001, **30**, 321–331; (b) O. Pamies and J.-E. Bäckvall, *Chem. Rev.*, 2003, **103**, 3247–3261; (c) O. Pamies and J.-E. Bäckvall, *Curr. Opin. Biotechnol.*, 2003, **14**, 407–413; (d) O. Pamies and J.-E. Bäckvall, *Trends Biotechnol.*, 2004, **22**, 130–135.

13. R. Stürmer, *Angew. Chem., Int. Ed.*, 1997, **36**, 1173–1174.

14. A. L. E. Larsson, B. A. Persson and J.-E. Bäckvall, *Angew. Chem., Int. Ed.*, 1997, **36**, 1211–1212.

15. J.-E. Bäckvall, *Asymmetric Synth.*, 2007, 171–175.

16. R. Karvembu, R. Prabhakaran and K. Natarajan, *Coord. Chem. Rev.*, 2005, **249**, 911–918.

17. (a) A. L. E. Larsson, B. A. Persson and J.-E. Bäckvall, *Angew. Chem., Int. Ed.*, 1997, **109**, 1256–1258; (b) B. A. Persson, A. L. E. Larsson, M. Le Ray and J.-E. Bäckvall, *J. Am. Chem. Soc.*, 1999, **121**, 1645–1650.

18. (a) B. A. Persson, F. F. Huerta and J.-E. Bäckvall, *J. Org. Chem.*, 1999, **64**, 5237–5240; (b) R. J. Kazlauskas, A. N. E. Weissfloch, A. T. Rappaport and L. A. Cuccia, *J. Org. Chem.*, 1991, **56**, 2656–2665.

19. F. F. Huerta, Y. R. S. Laxmi and J.-E. Bäckvall, *Org. Lett.*, 2000, 2, 1037–1040.

20. O. Pamies and J.-E. Bäckvall, *J. Org. Chem.*, 2001, **66**, 4022–4025.

21. O. Pamies and J.-E. Bäckvall, *Adv. Synth. Catal.*, 2002, **344**, 947–952.

22. F. F. Huerta and J.-E. Bäckvall, *Org. Lett.*, 2001, **3**, 1209–1212.

23. O. Pamies and J.-E. Bäckvall, *J. Org. Chem.*, 2002, **67**, 9006–9010.

24. O. Pamies and J.-E. Bäckvall, *J. Org. Chem.*, 2002, **67**, 1261–1265.

25. A.-B. L. Runmo, O. Pamies, K. Faber and J.-E. Bäckvall, *Tetrahedron Lett.*, 2002, **43**, 2983–2986.

26. J. H. Koh, H. M. Jung, M.-J. Kim and J. Park, *Tetrahedron Lett.*, 1999, **40**, 6281–6284.

27. D. Lee, E. A. Huh, M.-J. Kim, J. Y. Jung and M.-J. Kim, *Org. Lett.*, 2000, **2**, 2377–2379.

28. M.-J. Kim, Y. K. Choi, M. Y. Choi, M. J. Kim and J. Park, *J. Org. Chem.*, 2001, **66**, 4736–4738.

29. (a) H. M. Jung, J. K. Koh, M.-J. Kim and J. Park, *Org. Lett.*, 2000, **2**, 409–411; (b) H. M. Jung, J. K. Koh, M.-J. Kim and J. Park, *Org. Lett.*, 2000, **2**, 2487–2490; (c) M.-J. Kim, M. Y. Choi, M. Y. Han, Y. K. Choi, J. K. Lee and J. Park, *J. Org. Chem.*, 2002, **67**, 9481–9483.

30. J. H. Choi, Y. H. Kim, S. H. Nam, S. T. Shin, M.-J. Kim and J. Park, *Angew. Chem., Int. Ed.*, 2002, **41**, 2373–2376.

31. J. H. Choi, Y. K. Choi, Y. H. Kim, E. S. Park, E. J. Kim, M.-J. Kim and J. Park, *J. Org. Chem.*, 2004, **69**, 1972–1977.

32. A. Dijksman, J. M. Elzinga, Y.-X. Li, I. W. C. E. Arends and R. A. Sheldon, *Tetrahedron: Asymmetry*, 2002, **13**, 879–884.

33. (a) B. Martin-Matute, M. Edin, K. Bogar, F. B. Kaynak and J.-E. Bäckvall, *J. Am. Chem. Soc.*, 2005, **127**, 8817–8825; (b) B. Martin-Matute, M. Edin, K. Bogar and J.-E. Bäckvall, *Angew. Chem., Int. Ed.*, 2004, **43**, 6535–6539; (c) A. H. Ell, J. B. Johnson and J.-E. Bäckvall, *Chem. Commun.*, 2003, 1652–1653; (d) K. Bogar, B. Martin-Matute and J.-E. Bäckvall, *Beilstein J. Org. Chem.*, 2007, **3**(No. 50); (e) G. Csjernyik, K. Bogar and J.-E. Bäckvall, *Tetrahedron Lett.*, 2004, **45**, 6799–6802; (f) J. Norinder, K. Bogar, L. Kanupp and J.-E. Bäckvall, *Org. Lett.*, 2007, **9**, 5095–5098.

34. *Modern Organofluorine Chemistry*, ed. P. Kirsch, Wiley, Weinheim, 2004.

35. K. Bogar and J.-E. Bäckvall, *Tetrahedron Lett.*, 2007, **48**, 5471–5474.

36. S.-B. Ko, B. Baburaj, M.-J. Kim and J. Park, *J. Org. Chem.*, 2007, **72** 6860–6864.

37. N. Kim, S.-B. Ko, M. S. Kwon, M.-J. Kim and J. Park, *Org. Lett.*, 2005, **7**, 4523–4526.

38. T. H. Riermeier, P. Gross, A. Monsees, M. Hoff and H. Trauthwein, *Tetrahedron Lett.*, 2005, **46**, 3403–3406.

39. C. Roengpithya, D. A. Patterson, E. J. Gibbins, P. C. Taylor and A. G. Livingston, *Ind. Eng. Chem. Res.*, 2006, **45**, 7101–7109.

40. A. Wolfson, C. Yehuda, O. Shokin and D. Tavor, *Lett. Org. Chem.*, 2006, **3**, 107–110.

41. G. K. M. Verzijl, J. G. de Vries and Q. B. Broxterman, *Tetrahedron: Asymmetry*, 2005, **16**, 1603–1610.

42. M.-J. Kim, H. M. Kim, D. Kim, Y. Ahn and J. Park, *Green Chem.*, 2004, **6**, 471–474.

43. (a) S. F. G. M. Van Nispen, J. van Buijtenen, J. A. J. M. Vekemans, J. Meuldijk and L. A. Hulshof, *Tetrahedron: Asymmetry*, 2006, **17**, 2299–2305; (b) J. van Buijtenen, J. Meuldijk, J. A. J. M. Vekemans, L. A. Hulshof, H. Kooijman and A. L. Spek, *Organometallics*, 2006, **25**, 873–881.

44. O. Pamies and J.-E. Bäckvall, *J. Org. Chem.*, 2003, **68**, 4815–4818.

45. O. Pamies and J.-E. Bäckvall, *Adv. Synth. Catal.*, 2001, **343**, 726–731.

46. A.-B. L. Fransson, L. Boren, O. Pamies and J.-E. Bäckvall, *J. Org. Chem.*, 2005, **70**, 2582–2587.

47. P. Kielbasinski, M. Rachwalski, M. Mikolajczyk, M. A. H. Moelands, B. Zwanenburg and F. P. J. T. Rutjes, *Tetrahedron: Asymmetry*, 2005, **16**, 2157–2160.

48. P. Hoyos, M. Fernandez, J. V. Sinisterra and A. R. Alcantara, *J. Org. Chem.*, 2006, **71**, 7632–7637.

49. S. Akai, K. Tanimoto and Y. Kita, *Angew. Chem., Int. Ed.*, 2004, **43**, 1407–1410.

50. B. Martin-Matute, M. Edin and J.-E. Bäckvall, *Chem. – Eur. J.*, 2006, **12**, 6053–6061.

51. A.-B. L. Fransson, Y. Xu, K. Leijondahl and J.-E. Bäckvall, *J. Org. Chem.*, 2006, **71**, 6309–6316.

52. M. Edin, B. Martin-Matute and J.-E. Bäckvall, *Tetrahedron: Asymmetry*, 2006, **17**, 708–715.

53. B. Olofsson, K. Bogar, A.-B. L. Fransson and J.-E. Bäckvall, *J. Org. Chem.*, 2006, **71**, 8256–8260.

54. (a) M. Edin and J.-E. Bäckvall, *J. Org. Chem.*, 2003, **68**, 2216–2222; (b) M. Edin, J. Steinreiber and J.-E. Bäckvall, *Proc. Natl. Acad. Sci. U. S. A.*, 2004, **101**, 5761–5766.

55. B. Martin-Matute and J.-E. Bäckvall, *J. Org. Chem.*, 2004, **69**, 9191–9195.

56. I. Hilker, G. Rabani, G. K. M. Verzijl, A. R. A. Palmans and A. Heise, *Angew. Chem., Int. Ed.*, 2006, **45**, 2130–2132.

57. B. A. C. van As, J. van Buijtenen, T. Mes, A. R. A. Palmans and E. W. Meijer, *Chem. – Eur. J.*, 2007, **13**, 8325–8332.

58. M. Eckert, A. Brethon, Y.-X. Li, R. A. Sheldon and I. W. C. E. Arends, *Adv. Synth. Catal.*, 2007, **349**, 2603–2609.

59. M.-J. Kim, Y. I. Chung, Y. K. Choi, H. K. Lee, D. Kim and J. Park, *J. Am. Chem. Soc.*, 2003, **125**, 11494–11495.

60. L. Boren, B. Martin-Matute, Y. Xu, A. Cordova and J.-E. Bäckvall, *Chem. – Eur. J.*, 2006, **12**, 225–232.

61. (a) B. A. C. van As, J. van Buijtenen, A. Heise, Q. B. Broxterman, G. K. M. Verzijl, A. R. A. Palmans and E. W. Meijer, *J. Am. Chem. Soc.*, 2005, **127**, 9964–9965; (b) J. van Buijtenen, B. A. C. van As, J. Meuldijk, A. R. A. Palmans, J. A. J. M. Vekemans, L. A. Hulshof and E. W. Meijer, *Chem. Commun.*, 2006, 3169–3171.

62. A. Träff, K. Bogar, M. Warner and J.-E. Bäckvall, *Org. Lett.*, 2008, **10**, 4807–4810.

63. P. Krumlinde, K. Bogar and J.-E. Bäckvall, *J. Org. Chem.*, 2009, **74** 7407–7410.

64. J. Mangas-Sanchez, M. Rodriguez-Mata, E. Busto, V. Gotor-Fernandez and V. Gotor, *J. Org. Chem.*, 2009, **74**, 5304–5310.

65. K. Leijondahl, L. Boren, R. Braun and J.-E. Bäckvall, *Org. Lett.*, 2008, **10**, 2027–2030.

66. K. Leijondahl, L. Boren, R. Braun and J.-E. Bäckvall, *J. Org. Chem.*, 2009, **74**, 1988–1993.

67. L. Boren, K. Leijondahl and J.-E. Bäckvall, *Tetrahedron Lett.*, 2009, **50**, 3237–3240.

68. M.-J. Kim, Y. K. Choi, S. Kim, D. Kim, K. Han, S.-B. Ko and J. Park, *Org. Lett.*, 2008, **10**, 1295–1298.

69. D. Mavrynsky, M. Päivio, K. Lundell, R. Sillanpää, L. T. Kanerva and R. Leino, *Eur. J. Org. Chem.*, 2009, 1317–1320.

70. N. Jana, T. Mahapatra and S. Nanda, *Tetrahedron: Asymmetry*, 2009, **20**, 2622–2628.

71. (a) Q. Chen and C. Yuan, *Chem. Commun.*, 2008, 5333–5335; (b) Q. Chen and C. Yuan, *Tetrahedron*, 2010, **66**, 3707–3716.

72. K. S. A. Vallin, D. Wensbo Posaric, Z. Hamersak, M. A. Svensson and A. B. E. Minidis, *J. Org. Chem.*, 2009, **74**, 9328–9336.

73. F. Felluga, W. Baratta, L. Fanfoni, G. Pitacco, P. Rigo and F. Benedetti, *J. Org. Chem.*, 2009, **74**, 3547–3550.

74. P. Hoyos, A. Buthe, M. B. Ansorge-Schumacher, J. V. Sinisterra and A. R. Alcantara, *J. Mol. Catal. B: Enzym.*, 2008, **52–53**, 133–139.

75. P. Marcé, Y. Diaz, M. I. Matheu and S. Castillon, *Org. Lett.*, 2008, **10**, 4735–4738.

76. D. Strübing, P. Krumlinde, J. Piera and J.-E. Bäckvall, *Adv. Synth. Catal.*, 2007, **349**, 1577–1581.

77. M. R. Atuu and M. M. Hossain, *Tetrahedron Lett.*, 2007, **48**, 3875–3878.

78. H. K. Lee and Y. Ahn, *Bull. Korean Chem. Soc.*, 2004, **25**, 1471–1473.

79. K. Benaissi, M. Poliakoff and N. R. Thomas, *Green Chem.*, 2009, **11**, 617–621.

80. M. Päiviö, D. Mavrynsky, R. Leino and L. T. Kanerva, *Eur. J. Org. Chem.*, 2011, 1452–1457.

81. E. V. Johnston, K. Bogar and J.-E. Bäckvall, *J. Org. Chem.*, 2010, **75**, 4596–4599.

82. L. K. Thalen, A. Sumic, K. Bogar, J. Norinder, A. K. A. Persson and J.-E. Bäckvall, *J. Org. Chem.*, 2010, **75**, 6842–6847.

83. A. Träff, C. E. Solarte and J.-E. Bäckvall, *Collect. Czech. Chem. Commun.*, 2011, **76**, 919–927.

84. P. Krumlinde, K. Bogar and J.-E. Bäckvall, *Chem. Eur. J.*, 2010, **16**, 4031–4036.

85. A. Träff, R. Lihammar and J.-E. Bäckvall, *J. Org. Chem.*, 2011, **76**, 3917–3921.

86. R. Lihammar, R. Millet and J.-E. Bäckvall, *Adv. Synth. Catal.*, 2011, **353**, 2321–2327.

87. P. Hoyos, M. A. Quezada, J. V. Sinisterra and A. R. Alcantara, *J. Mol. Catal. B: Enzym.*, 2011, **72**, 20–24.

88. Y. Do, I.-C. Hwang, M.-J. Kim and J. Park, *J. Org. Chem.*, 2010, **75**, 5740–5742.

89. K. Han, C. Kim, J. Park and M.-J. Kim, *J. Org. Chem.*, 2010, **75**, 3105–3108.

90. W.-H. Kim, R. Karvembu and J. Park, *Bull. Korean Chem. Soc.*, 2004, **25**, 931–933.

91. J. Paetzold and J.-E. Bäckvall, *J. Am. Chem. Soc.*, 2005, **127**, 17620–17621.
92. C. E. Hoben, L. Kanupp and J.-E. Bäckvall, *Tetrahedron Lett.*, 2008, **49**, 977–979.
93. L. K. Thalen, D. Zhao, J.-B. Sortais, J. Paetzold, C. Hoben and J.-E. Bäckvall, *Chem. - Eur. J*, 2009, **15**, 3403–3410.
94. M. Shakeri, K. Engström, A. G. Sandström and J.-E. Bäckvall, *Chem-CatChem*, 2010, **2**, 534–538.
95. M. Rodriguez-Mata, V. Gotor-Fernandez, J. Gonzalez-Sabin, F. Reboleddo and V. Gotor, *Org. Biomol. Chem.*, 2011, **9**, 2274–2278.
96. R. Millet, A. M. Träff, M. L. Petrus and J.-E. Bäckvall, *J. Am. Chem. Soc.*, 2010, **132**, 15182–15184.
97. S. Akai, K. Tanimoto, Y. Kanao, M. Egi, T. Yamamoto and Y. Kita, *Angew. Chem., Int. Ed.*, 2006, **45**, 2592–2595.
98. S. Wuyts, J. Wahlen, P. A. Jacobs and D. E. De Vos, *Green Chem.*, 2007, **9**, 1104–1108.
99. A. Berkessel, M. L. Sebastian-Ibarz and T. N. Müller, *Angew. Chem., Int. Ed.*, 2006, **45**, 6567–6570.
100. J. Deska, C. del Pozo Ochoa and J.-E. Bäckvall, *Chem. - Eur. J.*, 2010, **16**, 4447–4451.
101. (a) R. M. Haak, F. Berthiol, T. Jerphagnon, A. J. A. Gayet, C. Tarabiono, C. P. Postema, V. Ritleng, M. Pfeffer, D. B. Janssen, A. J. Minnaard, B. L. Feringa and J. G. de Vries, *J. Am. Chem. Soc.*, 2008, **130**, 13508–13509; (b) T. Jerphagnon, A. J. A. Gayet, F. Berthiol, V. Ritleng, N. Mrsic, A. Meetsma, M. Pfeffer, A. J. Minnaard, B. L. Feringa and J. G. de Vries, *Chem. - Eur. J.*, 2009, **15**, 12780–12790.
102. Y. Sato, Y. Kayaki and T. Ikariya, *Chem. Commun.*, 2012, **48**, 3635–3637.
103. R. Corberan and E. Peris, *Organometallics*, 2008, **27**, 1954–1958.
104. (a) C. L. Pollock, K. J. Fox, S. D. Lacroix, J. McDonagh, P. C. Marr, A. M. Nethercott, A. Pennycook, S. Qian, L. Robinson, G. C. Saunders and A. C. Marr, *Dalton Trans.*, 2012, **41**, 13423–13428; (b) A. C. Marr, C. L. Pollock and G. C. Saunders, *Organometallics*, 2007, **26**, 3283–3285.
105. (a) S. Wuyts, K. De Temmerman, D. E. De Vos and P. Jacobs, *Chem. Commun.*, 2003, 1928–1929; (b) S. Wuyts, K. De Temmerman, D. E. De Vos and P. A. Jacobs, *Chem. - Eur. J.*, 2005, **11**, 386–397; (c) Y. Zhu, K.-L. Fow, G.-K. Chuah and S. Jaenicke, *Chem. - Eur. J.*, 2007, **13**, 541–547.
106. P. Lozano, T. De Diego, C. Mira, K. Montague, M. Vaultier and J. L. Iborra, *Green Chem.*, 2009, **11**, 538–542.
107. Y. K. Choi, M. J. Kim, Y. Ahn and M.-J. Kim, *Org. Lett.*, 2001, **3**, 4099–4101.
108. M.-J. Kim, W.-H. Kim, K. Han, Y. K. Choi and J. Park, *Org. Lett.*, 2007, **9**, 1157–1159.
109. (a) A. D. E. Parvulescu, D. E. De Vos and P. A. Jacobs, *Chem. Commun.*, 2005, 5307–5309; (b) A. N. Parvulescu, P. A. Jacobs and D. E. De Vos, *Chem. - Eur. J.*, 2007, **13**, 2034–2043.
110. A. N. Parvulescu, E. Van der Eycken, P. A. Jacobs and D. E. De Vos, *J. Catal.*, 2008, **255**, 206–212.

111. A. N. Parvulescu, P. A. Jacobs and D. E. De Vos, *Appl. Catal., A*, 2009, **368**, 9–16.

112. L. H. Andrade, A. V. Silva and E. C. Pedrozo, *Tetrahedron Lett.*, 2009, **50**, 4331–4334.

113. Y. K. Choi, Y. Kim, K. Han, J. Park and M.-J. Kim, *J. Org. Chem.*, 2009, **74**, 9543–9545.

114. Y. Kim, J. Park and M.-J. Kim, *Tetrahedron Lett.*, 2010, **51**, 5581–5584.

115. K. Engström, M. Shakeri and J.-E. Bäckvall, *Eur. J. Org. Chem.*, 2011, 1827–1830.

116. G. Chen, B. Xia, Q. Wu and X. Lin, *RSC Adv.*, 2013, **3**, 9820–9828.

117. K. Han, Y. Kim, J. Park and M.-J. Kim, *Tetrahedron Lett.*, 2010, **51**, 3536–3537.

118. M. Stirling, J. Blacker and M. I. Page, *Tetrahedron Lett.*, 2007, **48**, 1247–1250.

119. A. N. Parvulescu, P. A. Jacobs and D. E. De Vos, *Adv. Synth. Catal.*, 2008, **350**, 113–121.

120. A. E. Felten, G. Zhu and Z. D. Aron, *Org. Lett.*, 2010, **12**, 1916–1919.

121. Y. K. Choi, J. H. Suh, D. Lee, I. T. Lim, J. Y. Jung and M.-J. Kim, *J. Org. Chem.*, 1999, **64**, 8423–8424.

Reactions Catalysed by a Combination of Organocatalysts and Enzymes

The combination of organocatalysts and enzymes remains rare, and the first examples of asymmetric tandem reactions catalysed by this type of catalyst combination have been described only recently. For example in 2004, Cordova *et al.* worked out a one-pot procedure involving L-proline as catalyst in the first step of the reaction and the enzyme Amano I (lipase extracted from *Pseudomonas cepacia*) in the second step.[1] As shown in Scheme 9.1, the aldol reaction between an aldehyde and acetone occurred to give the corresponding intermediate β-hydroxy ketone, which was subsequently submitted to kinetic resolution by treatment with the enzyme, affording the corresponding almost enantiopure acetate in moderate yields.

In 2012, Kudo *et al.* exploited another organocatalyst–enzyme catalytic system to promote an asymmetric sequential tandem Friedel–Crafts-type alkylation–α-oxyamination reaction performed in aqueous solvent, allowing the compatibility of the two catalysts.[2] As shown in Scheme 9.2, the first step of the tandem reaction consisted of an enantioselective Friedel–Crafts-type alkylation of an oxygen-functionalised indole or pyrrole with 3-nitro- or 4-nitrocinnamaldehyde catalysed by a chiral resin-supported peptide catalyst, providing the corresponding aldehyde intermediate, which was subsequently submitted to stereoselective α-oxyamination by adding 2,2,6,6-tetramethylpiperidin-1-oxyl (TEMPO) and laccase directly to the reaction mixture. This afforded the corresponding chiral highly functionalised products as mixtures of *syn-* and *anti*-diastereomers in good yields and with moderate diastereoselectivities of 46 to 58% de in favour of the *syn-*diastereomers. On the other hand, the *syn-*diastereoisomers were obtained

RSC Catalysis Series No. 20
Enantioselective Multicatalysed Tandem Reactions
By Hélène Pellissier
© Hélène Pellissier 2014
Published by the Royal Society of Chemistry, www.rsc.org

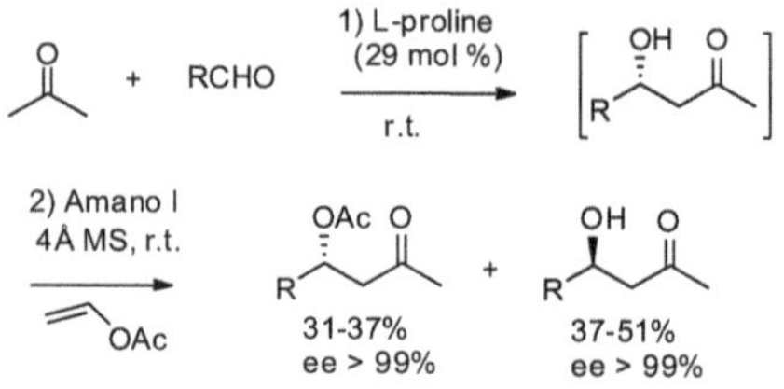

Scheme 9.1 Tandem aldol–kinetic resolution reaction catalysed by chiral amine catalysis and enzyme catalysis.

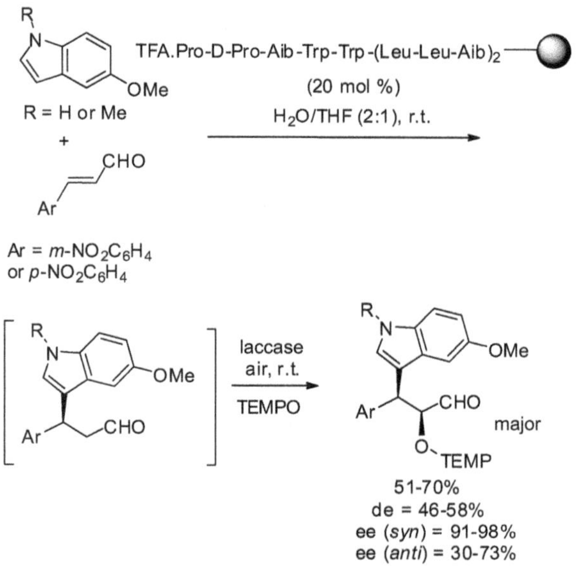

Scheme 9.2 Tandem Friedel–Crafts-type alkylation–α-oxyamination reaction catalysed by chiral resin-supported peptide catalysis and enzyme catalysis.

in high to excellent enantioselectivities ranging from 91 to 98% ee, while the enantioselectivities of the *anti*-diastereomers were found to be lower (30–73% ee).

In addition, several asymmetric tandem reactions catalysed by combinations of enzymes and achiral organocatalysts based on the DKR methodology have been successfully developed in recent years. For example, Yang *et al.* have described highly efficient DKR of secondary aromatic alcohols by using a combination of lipase Novozym 435 and easily available acid resins, such as resin CD8604, as compatible racemisation catalysts.[3] When employing complex acyl donors, such as 4-chlorophenyl acetate, instead of simple acyl donors to inhibit the acid resin-catalysed transesterification, the corresponding chiral acetates were obtained in both good to excellent yields and enantioselectivities of up to >99% ee, as shown in Scheme 9.3. It was noteworthy that the catalytic system could be reused more than 10 times with little loss of yield and ee value.

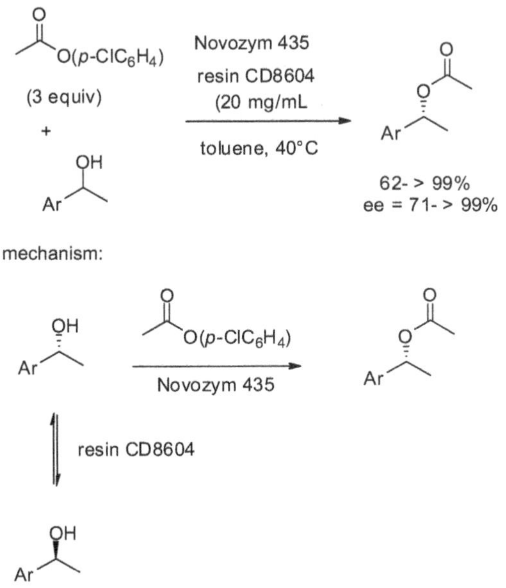

Scheme 9.3   DKR of alcohols catalysed by enzyme catalysis and acid resin catalysis.

Baeyer–Villiger monooxygenases are highly promiscuous enzymes which enable stereoselective conversion of ketones into esters or lactones, and have been identified and characterised for their potential as biocatalysts capable of performing DKR. As an example, Gotor *et al.* have reported the use of 4-hydroxyacetophenone monooxygenase from *Pseudomonas fluorescens* ACB in combination with a weak anion exchange resin, such as Dowex MWA-1, to perform the DKR of benzyl ketones.[4] Indeed, this DKR could be achieved by combining an isolated Baeyer–Villiger monooxygenase-catalysed Baeyer–Villiger oxidation of benzyl ketones with a racemisation induced by resin Dowex MWA-1, providing the corresponding esters in good yields and enantioselectivities of up to 83% ee, as shown in Scheme 9.4. It must be noted that the oxidations were coupled to a second enzymatic reaction catalysed by glucose-6-phosphate dehydrogenase in order to regenerate NADPH.

In 2011, Caddick and co-workers reported a nice asymmetric synthesis of chiral *trans*-4,5-dioxygenated cyclopentenone derivatives based on a sequential organocatalysed rearrangement of pyranones followed by an enzymatic DKR.[5] As shown in Scheme 9.5, the use of DABCO as organocatalyst in combination with lipase AK Amano 20 allowed a series of chiral acetylated cyclopentenones to be achieved from the corresponding pyranones in moderate to high yields and enantioselectivities of up to 96% ee.

In 2012, Pietruszka *et al.* reported a remarkable DKR of alkyl 2,3-dihydrobenzo[b]furan-3-carboxylates achieved by using a combination of lipase Novozym 435 with the Schwesinger base, *tert*-butylimino-2-diethylamino-1,3-dimethylperhydro-1,2,3-diazaphosphorine (BEMP), immobilised on polystyrene to induce racemisation of the ester.[6] The application of a basic

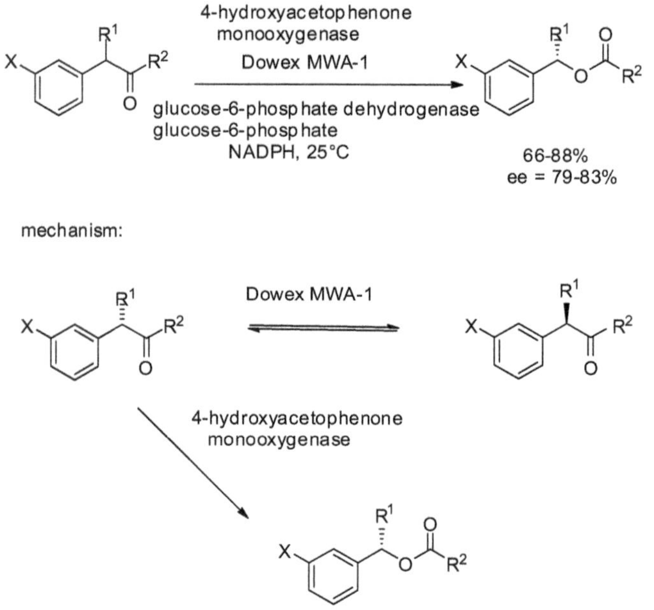

Scheme 9.4   Tandem Baeyer–Villiger reaction–DKR reaction of benzyl ketones catalysed by enzyme catalysis and anion exchange resin catalysis.

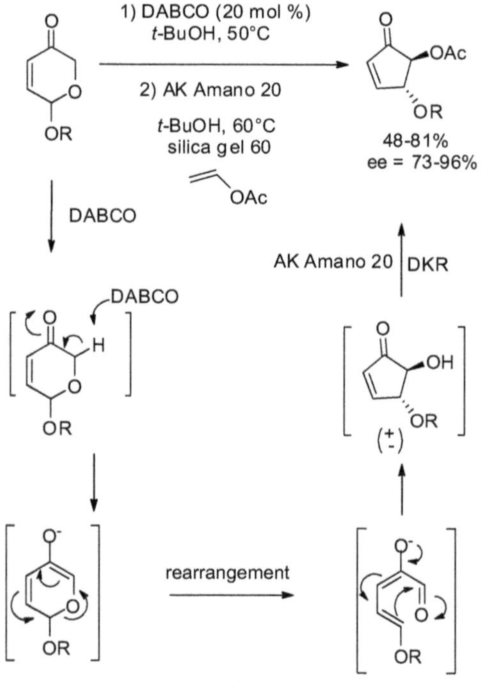

Scheme 9.5   Tandem rearrangement–DKR reaction of pyranones catalysed by amine catalysis and enzyme catalysis.

two-phase system (heptane–buffer) allowed the enzyme and the chiral acid produced to be kept in the aqueous layer, while the ester in the organic layer was continuously pumped through the racemisation column containing an immobilised form of BEMP. Consequently, a constant racemisation of the ester could be obtained as the enantiomerically pure acid accumulated in the aqueous phase, which was finally obtained in high yields of up to 95% and excellent enantioselectivities of up to >99% ee, as shown in Scheme 9.6.

Finally, Bertrand *et al.* have developed a highly efficient chemoenzymatic DKR of primary amines using a combination of lipase CALB and octanethiol

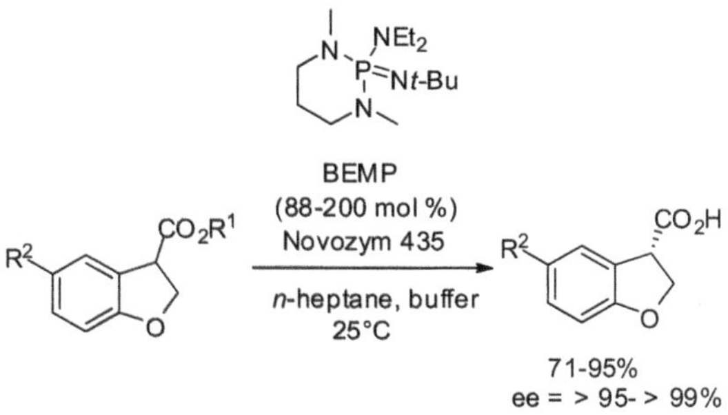

Scheme 9.6 DKR of alkyl 2,3-dihydrobenzo[b]furan-3-carboxylates catalysed by amine catalysis and enzyme catalysis.

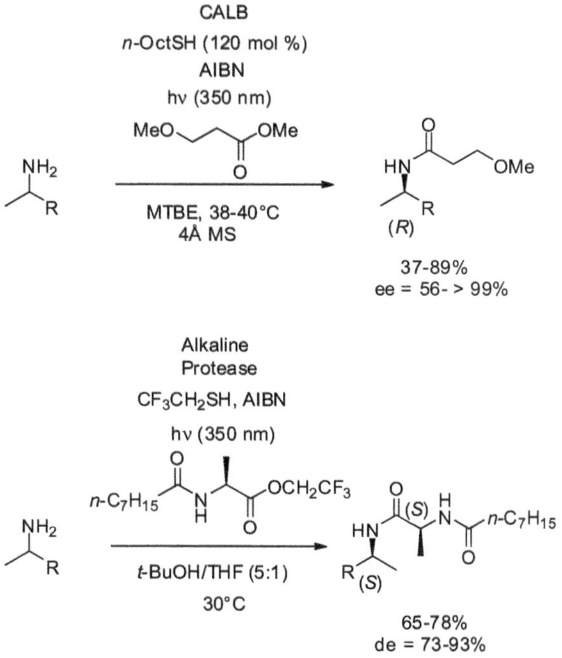

Scheme 9.7 DKRs of amines catalysed by enzyme catalysis and radical racemising agents.

as a radical racemising agent and methyl β-methoxy propanoate as the acyl donor.[7] The process was carried out under photochemical irradiation at 350 nm in glassware in the presence of AIBN. The corresponding (*R*)-amides were achieved in moderate to high yields and enantioselectivities ranging from 56 to >99% ee, as shown in Scheme 9.7. Moreover, the same authors have demonstrated that the stereoselectivity of the DKR process could be switched by using alkaline protease as the enzyme instead of lipase CALB.[8] As shown in Scheme 9.7, when the reaction was carried out with this enzyme in the presence of *N*-octanoyl L-alanine trifluoroethyl ester as a chiral acyl donor and trifluoro ethanethiol as a radical racemising agent under photochemical irradiation at 350 nm, it led to the corresponding (*S,S*)-diamides in good yields and diastereoselectivities of 73 to 93% de. It must be noted that it was decided to include these beautiful processes in this review although they employed substoichiometric quantities of the radical racemising agents.

References

1. M. Edin, J.-E. Bäckvall and A. Cordova, *Tetrahedron Lett.*, 2004, **45**, 7697–7701.
2. K. Akagawa, R. Umezawa and K. Kudo, *Beilstein, J. Org. Chem.*, 2012, **8**, 1333–1337.
3. Y. Cheng, G. Xu, J. Wu, C. Zhang and L. Yang, *Tetrahedron Lett.*, 2010, **51**, 2366–2369.
4. C. Rodriguez, G. de Gonzalo, A. Rioz-Martinez, D. E. Torres Pazmino, M. W. Fraaije and V. Gotor, *Org. Biomol. Chem.*, 2010, **8**, 1121–1125.
5. J. P. M. Nunes, L. F. Veiros, P. D. Vaz, C. A. M. Afonso and S. Caddick, *Tetrahedron*, 2011, **67**, 2779–2787.
6. P. Bongen, J. Pietruszka and R. C. Simon, *Chem. – Eur. J.*, 2012, **18**, 11063–11070.
7. F. Poulhès, N. Vanthuyne, M. P. Bertrand, S. Gastaldi and G. Gil, *J. Org. Chem.*, 2011, **76**, 7281–7286.
8. L. El Blidi, N. Vanthuyne, D. Siri, S. Gastaldi, M. P. Bertrand and G. Gil, *Org. Biomol. Chem.*, 2010, **8**, 4165–4168.

CHAPTER 10
Conclusions

Organocatalysis, metal catalysis as well as biocatalysis represent three growing and firmly established fields in organic chemistry. Due to the explosion of fundamental knowledge guiding reaction design in all these three key fields, merging all forms of catalysis constitutes a challenge to single catalytic systems today. As far as metal catalysis, organocatalysis, and biocatalysis are concerned, each discipline has its own advantages, limitations, and range of applications. On the other hand, the combination of catalysts from different disciplines enables unprecedented transformations not currently possible by use of any catalysis alone, or makes current synthetic methods more economical and practical. As demonstrated in this second section, the combination of different types of catalysts is particularly fruitful for the development of novel enantioselective tandem reactions. In particular, the combination of organocatalysis and transition metal catalysis has successfully grown from its infancy to its adolescence in the last few years. The plethora of combinations highlighted in this section demonstrates the power of exploring the complementary advantages of these two types of catalysis to access structures or activation modes which are hitherto unprecedented along with new types of reactivities. Although metal and organocatalysis individually will always have their own place in synthetic organic chemistry, increasingly there is a need to search for novel dual catalyst systems, allowing the formation of products not accessible by using one of the catalysts alone. In another context, the versatility of the combination of enzymes with metal catalysts is also well demonstrated in this second section of the book, particularly by chemoenzymatic dynamic kinetic resolutions of compounds, such as alcohols, amines, and allylic acetates. On the other hand, the combination of organocatalysts with enzymes remains rare, and the first examples reported in the literature about asymmetric tandem reactions catalysed by this type of catalyst combination are collected in the last chapter of this section.

RSC Catalysis Series No. 20
Enantioselective Multicatalysed Tandem Reactions
By Hélène Pellissier
© Hélène Pellissier 2014
Published by the Royal Society of Chemistry, www.rsc.org

General Conclusion

Nature created multienzymatic systems to accomplish extremely efficient one-pot tandem catalysis. As in an assembly line, tens of enzymes are well organised to transform simple materials to complex molecules with perfect control of selectivity by a series of coupled reactions in the cell. It has long been chemists' endeavor to extend such coordinated catalytic action to artificial processes to make synthetic chemistry more sustainable. Nowadays, owing to the resource-intensive nature of the current synthetic industry, the development of tandem one-pot reactions, avoiding the use of costly and time-consuming protection–deprotection processes as well as purification procedures of intermediates, has become especially important and valuable because society is confronted with bottle-neck problems such as energy and time shortage and environmental pollution.

Moreover, despite impressive advances, researchers are still eagerly looking for new asymmetric catalysed reactions, since only a fraction of the known chemical transformations have an asymmetric version having a wide substrate scope. The identification of the best-performing catalyst is one of the most time- and resource-consuming tasks to be overcome. In addition to the classic "one catalyst one reaction" approach, more recent strategies can be pursued. A single molecule with two or more functional groups, each one having a different catalytic activity, is a multifunctional catalyst. If at least one of the subunits bearing the functional groups is chiral, then the molecule can be defined as a chiral multifunctional catalyst. Examples of such structures can be found both in organocatalysis and transition-metal catalysis. Moreover, a complementary and emerging strategy is multiple catalysis, which is the use of distinct noncovalently bound catalysts. The two-catalyst system, where the products obtained are not accessible by using one of the catalysts alone, clearly reflects the new type of reactivities exhibited in

RSC Catalysis Series No. 20
Enantioselective Multicatalysed Tandem Reactions
By Hélène Pellissier
© Hélène Pellissier 2014
Published by the Royal Society of Chemistry, www.rsc.org

multicatalysed tandem reactions. This undoubtedly indicates the importance of such novel processes in synthetic organic chemistry.

This book clearly demonstrates that the concept of asymmetric multicatalysis, involving organocatalysts, transition metals as well as enzymes, has emerged as a powerful and novel tool in organic synthesis. It shows that a tremendous increase in the combination of various types of catalysts applied to induce asymmetric tandem reactions has occurred in recent years. In addition to the discovery of novel combinations of catalysts, the number of various unprecedented enantioselective tandem reactions is increasing day after day, rendering this concept a novel, revolutionary, emerging and powerful field. Indeed, this combining concept is fascinating since it enables unprecedented transformations not currently possible by using each of the catalysts alone, particularly in the case of cooperative and relay catalysis. Because there are so many different catalysts, the combination of different catalysts seems to be illimited and provides tremendous opportunities to develop novel chemistry. Studies on the mechanisms of these multicatalysed tandem reactions along with considering the molecular structure of the catalytically active species are, however, still highly desirable since they will strongly help accelerate the search to find new catalyst systems.

This book collects the major progress in the general and novel field of asymmetric multicatalysed tandem reactions. The first section of the book deals with asymmetric tandem reactions catalysed by multiple catalysts belonging to the same discipline, while asymmetric tandem reactions catalysed by multiple catalysts from different disciplines are collected in the second section. The first section of the book is divided into five chapters, dealing successively after an introduction (Chapter 1) with reactions catalysed by multiple organocatalysts (Chapter 2), reactions catalysed by two metals (Chapter 3), and multienzyme-catalysed reactions (Chapter 4) followed by conclusions (Chapter 5). The second section of the book, which concerns the enantioselective tandem reactions catalysed by multiple catalysts from different disciplines, is also divided into five chapters, dealing successively after an introduction (Chapter 6) with reactions catalysed by a combination of metals and organocatalysts (Chapter 7) which have been the most developed, reactions catalysed by a combination of metals and enzymes (Chapter 8), and finally reactions catalysed by a combination of organocatalysts and enzymes (Chapter 9), followed by conclusions (Chapter 10). The two catalysts can interact in a cooperative, relay or sequential manner; these three types of catalysis are treated successively in most of the sections of the book.

The future direction of the field of asymmetric multicatalysed tandem reactions is to continue expanding their scope through the employment of novel combinations of (novel) catalysts which will allow novel enantioselective transformations to be discovered. These novel unprecedented one-pot reactions are expected to be rapidly applied to the synthesis of biologically interesting molecules including natural products.

Subject Index

Page numbers in *italics* refer to tables and figures.